T0304546

STRUCTURAL BIOINFORMATICS

An Algorithmic Approach

CHAPMAN & HALL/CRC
Mathematical and Computational Biology Series

Aims and scope:
This series aims to capture new developments and summarize what is known over the whole spectrum of mathematical and computational biology and medicine. It seeks to encourage the integration of mathematical, statistical and computational methods into biology by publishing a broad range of textbooks, reference works and handbooks. The titles included in the series are meant to appeal to students, researchers and professionals in the mathematical, statistical and computational sciences, fundamental biology and bioengineering, as well as interdisciplinary researchers involved in the field. The inclusion of concrete examples and applications, and programming techniques and examples, is highly encouraged.

Series Editors

Alison M. Etheridge
Department of Statistics
University of Oxford

Louis J. Gross
Department of Ecology and Evolutionary Biology
University of Tennessee

Suzanne Lenhart
Department of Mathematics
University of Tennessee

Philip K. Maini
Mathematical Institute
University of Oxford

Shoba Ranganathan
Research Institute of Biotechnology
Macquarie University

Hershel M. Safer
Weizmann Institute of Science
Bioinformatics & Bio Computing

Eberhard O. Voit
The Wallace H. Couter Department of Biomedical Engineering
Georgia Tech and Emory University

Proposals for the series should be submitted to one of the series editors above or directly to:
CRC Press, Taylor & Francis Group
4th, Floor, Albert House
1-4 Singer Street
London EC2A 4BQ
UK

Published Titles

Bioinformatics: A Practical Approach
Shui Qing Ye

Cancer Modelling and Simulation
Luigi Preziosi

Computational Biology: A Statistical Mechanics Perspective
Ralf Blossey

Computational Neuroscience: A Comprehensive Approach
Jianfeng Feng

Data Analysis Tools for DNA Microarrays
Sorin Draghici

Differential Equations and Mathematical Biology
D.S. Jones and B.D. Sleeman

Exactly Solvable Models of Biological Invasion
Sergei V. Petrovskii and Bai-Lian Li

Handbook of Hidden Markov Models in Bioinformatics
Martin Gollery

Introduction to Bioinformatics
Anna Tramontano

An Introduction to Systems Biology: Design Principles of Biological Circuits
Uri Alon

Kinetic Modelling in Systems Biology
Oleg Demin and Igor Goryanin

Knowledge Discovery in Proteomics
Igor Jurisica and Dennis Wigle

Modeling and Simulation of Capsules and Biological Cells
C. Pozrikidis

Niche Modeling: Predictions from Statistical Distributions
David Stockwell

Normal Mode Analysis: Theory and Applications to Biological and Chemical Systems
Qiang Cui and Ivet Bahar

Pattern Discovery in Bioinformatics: Theory & Algorithms
Laxmi Parida

Spatiotemporal Patterns in Ecology and Epidemiology: Theory, Models, and Simulation
Horst Malchow, Sergei V. Petrovskii, and Ezio Venturino

Stochastic Modelling for Systems Biology
Darren J. Wilkinson

Structural Bioinformatics: An Algorithmic Approach
Forbes J. Burkowski

The Ten Most Wanted Solutions in Protein Bioinformatics
Anna Tramontano

Chapman & Hall/CRC Mathematical and Computational Biology Series

STRUCTURAL BIOINFORMATICS
An Algorithmic Approach

Forbes J. Burkowski

CRC Press
Taylor & Francis Group
Boca Raton   London   New York

CRC Press is an imprint of the
Taylor & Francis Group an **informa** business
A CHAPMAN & HALL BOOK

Chapman & Hall/CRC
Taylor & Francis Group
6000 Broken Sound Parkway NW, Suite 300
Boca Raton, FL 33487-2742

© 2009 by Taylor & Francis Group, LLC
Chapman & Hall/CRC is an imprint of Taylor & Francis Group, an Informa business

No claim to original U.S. Government works
Printed in the United States of America on acid-free paper
10 9 8 7 6 5 4 3 2 1

International Standard Book Number-13: 978-1-58488-683-9 (Hardcover)

Visit the Taylor & Francis Web site at
http://www.taylorandfrancis.com

and the CRC Press Web site at
http://www.crcpress.com

To my mother,
Maria Burkowski

Table of Contents

Preface

GOALS OF THIS BOOK

Structural bioinformatics thrives at the crossroads of biology, mathematics, and computer science. Consequently, I have written this book with the main intention of building bridges between these disciplines. A running theme through most of the book is to set various topics and problems within a scientific methodology that emphasizes NATURE (the source of our empirical observations), SCIENCE (the mathematical modeling of the natural process), and COMPUTATION (the science of calculating predictions and mathematical objects based on these mathematical models). Consequently, there is a stress on mathematics because it provides the tools that we need to define the models that act as links between structural biology and computational algorithms. The emphasis is on algorithms, most of which could form the basis of a class assignment. There are approximately 60 problems and exercises across ten chapters. The book can be used as a text that provides the framework for a one-semester undergraduate course.

In a less practical sense, but perhaps just as important, I hope to convince the reader that structural bioinformatics has what might be called "aspects of delight." Many protein structures have an undeniable beauty. Often, this comes from symmetry in the overall structure. The symmetry arises from the reuse of building blocks to form dimers, trimers, etc. Even though it may be argued that this replication is the mere byproduct of efficiency, one DNA sequence leading to the production of several structural parts that combine to make a whole unit, the symmetry of the final result has a pleasing appearance much like that observed in a flower or snowflake. There is also a beauty in mechanism: The transient interplay of molecular surfaces that result in a chemical reaction, show the wonderful results of evolution at the atomic level (see, for example, the RNA–RNA interaction for pseudouridylation briefly described in Section 5.1). The magic that is life takes place in the cavities of macromolecular surfaces. I would also hope that the reader appreciates more abstract notions of beauty, as occasionally seen in the mathematics of structural bioinformatics. For example, in the derivation of the optimal rotation to maximize the overlap of two proteins in 3-D space, the analysis is long but eventually reduces to the requirement that we solve a very simple and elegant equation with only three variables: $C = R\lambda$.

INTENDED AUDIENCE

The book is mainly intended for students, typically third- or fourth-year undergraduates, who have completed introductory courses in bioinformatics and wish to go farther in the study of structural biology. Prerequisites include linear algebra, calculus, bioinformatics ("Biology 101," sequence analysis, etc.), computer programming, and introductory algorithms. Beyond this, the book should provide practitioners of bioinformatics with the basic mathematical techniques underlying some of the core algorithms in structural bioinformatics.

BOOK CONTENT

This book is not meant to be a technical handbook or research monograph. Nor is it an overview of the entire area. Instead, I have provided some introductory material in the first few chapters, followed by some algorithms and topics that are representative of structural bioinformatics. Selected topics are relatively few in number but usually covered to a depth that goes beyond that of introductory texts that give a broader overview of bioinformatics. If your favorite algorithm is missing, then I apologize. Topics dealing with protein structure prediction have been omitted because it is difficult to deal with very complicated algorithms that would never be implemented in class assignments.

From my perspective, a second edition of the book should contain material on drug design, more material on structure comparison, geometric hashing, and protein flexibility (perhaps normal mode analysis). However, I welcome readers to share thoughts on what they consider important.

Chapter 1 introduces the notions of NATURE, SCIENCE, and COMPUTATION as realms of discourse. In practicing structural bioinformatics, we deal with empirical data derived from nature, we build scientific models, and, finally, we do computations. All these activities are subject to error. The cautionary plea of this chapter is to be careful about the overall process. Data are subject to empirical error and calculations are compromised by computational error. More importantly, the models themselves are often designed to embrace approximations. This is fine, but one should be aware of the tradeoffs.

Chapter 2 is a relief from the extended sermon of Chapter 1 and introduces protein and RNA structure. This is essentially a review of material that may have been lightly touched upon in earlier courses. The concepts of primary, secondary, tertiary, and quaternary structure are introduced and explained. There are lots of figures and motivational topics.

Chapter 3 is a quick overview of data related topics: Where does one find data related to structural bioinformatics? What tools are available for visualization of molecular structure? What are some of the primary applications available to researchers and students? The emphasis is on data and applications that can be obtained by free download.* This chapter is simply an overview of a few topics. An entire book could be written on these topics and, of course, it would rapidly be out of date.

Chapter 4 is a review of dynamic programming. It covers the methodology usually found in an undergraduate course in algorithms and then goes on to the longest common subsequence problem as the first bioinformatics problem that can be solved using dynamic programming.

Chapter 5 covers RNA secondary structure prediction. It describes the Nussinov algorithm and the MFOLD algorithm. The science model for each algorithm is carefully explained. Both algorithms are wonderful examples of dynamic programming applied in a way that is not immediately obvious.

Chapter 6 deals with protein sequence alignment. Although the book concentrates on structural issues, the notion of sequence alignment is indispensable to the study of structure, and so this chapter provides a review of the topic. It is also another example of dynamic programming at work.

Chapter 7 covers the basic mathematical calculations needed for measuring the geometric properties of macromolecules. Derivations of formulae for bond length, bond angle, and dihedral angles are presented. Calculation of dihedral angles can be somewhat tricky, and so special care is taken for this topic.

---

* My conversation with a salesperson selling software for drug development:

  FJB: Oh!... Uh.... You may have misunderstood my request. I was asking for the *academic* price.

  Salesperson: Sir, $16,000 is the academic price. The full retail price is about $80,000!

Chapter 8 deals with transformation of coordinates. These transformations facilitate translation and rotation of molecules in 3-D space. The chapter assumes knowledge of linear algebra and provides a detailed derivation of the rotation matrix.

Chapter 9 introduces structure comparison, alignment, and superposition. This is a continuation of material begun in the previous chapter. This topic area is somewhat unsettled, and so the coverage is a bit limited. Topics were chosen simply by personal preference, and these choices do not imply any relative merit of the algorithms being covered.

Chapter 10 provides an introduction to machine learning. In recent years it has become clear that algorithms dealing with regression and classification are required tools of the trade in bioinformatics. We start with the notions that are important for regression, and this is done in a way that leads us naturally into the topic of support vector machines for classification. Both regression and classification are becoming more and more useful for protein analysis and drug design.

The Appendices offer some review of the mathematics that is needed. This is primarily linear algebra.*

Readers are encouraged to visit the website: http://structuralbioinformatics.com. It contains text errata, some extra material on structural bioinformatics, and general information related to the text.

---

* The mathematical review in the appendices is probably excessive. This is undoubtedly my overreaction to student comments found on the course evaluation for my Structural Bioinformatics course at the University of Waterloo. Indignant student: "He actually expects you to remember first-year calculus and algebra—and that was 3 years ago!"

Acknowledgments

The book contains about 123 figures. Almost all the molecular graphics were generated using UCSF Chimera. The primary reference for this molecular modeling system is "UCSF Chimera—A visualization system for exploratory research and analysis" by E. F. Pettersen, T. D. Goddard, C. C. Huang, G. S. Couch, D. M. Greenblatt, E. C. Meng, and T. E. Ferrin (*J. Comput. Chem.*, 25 [2004], 1605–1612).

I also wish to express thanks to Anna Tramontano for reading the manuscript, making suggestions, and correcting some poorly formed judgments. She did this when the material was about two thirds complete. So, if you come across erroneous material you should assume that this was not in the version that Anna read.

I take full responsibility for errors and will gladly hear from readers who may have helpful criticisms.

Finally, I would like to thank my loving wife Karen for putting up with all the extra demands that this project has required.

F. J. B.

CHAPTER 1

The Study of Structural Bioinformatics

What a piece of work is man!
How noble in reason!
How infinite in faculties!
In form and moving, how express and admirable!
In action how like an angel!
In apprehension, how like a god!
The beauty of the world! The paragon of animals!
And yet, to me, what is this quintessence of dust?

WILLIAM SHAKESPEARE, *HAMLET* II, II

1.1 MOTIVATION

It can be argued that the structures within biological organizations represent the most beautiful and complex entities known to man. As you read this sentence, photons hit receptors in the retinas of your eyes and initiate a rapid cascade of events involving innumerable firings of synapses within your brain. At some lowest level of description, the formation of a thought in your brain is facilitated by changes of individual molecular conformations and by the reactions between the structural interfaces of large macromolecules in the brain.

Although we understand that this must be so, the exact nature of the events leading to the formation of a thought is still not fully understood. Science has made wondrous advances since the time of Shakespeare, but

1

there is still a lot to be discovered in understanding "this quintessence of dust" that encloses the human mind.

The next time you see a pianist playing a piece of music at a rapid pace, stop to consider that what you are witnessing is a complex biological system in action. There is the fantastically rapid transmission of signals along nerve bundles and the complicated logic of protein-based neurotransmitters in the synapses linking the nerves. In addition to this, imagine the intricate behavior of muscle fibers with their various proteins such as the myosins and actins interacting with precision and timing to generate each and every musical note.

The ultimate goal of structural biology is to fully understand the intricate reactions and interactions of all molecules in a living cell. We want to describe the "inner symphony" that transpires when a cellular system reacts to external influences, be they sensory, chemical, thermal, or otherwise. We also want to understand the discordant events that arise from disease states that exhibit reaction pathways that may be deleterious to the biological system as a whole.

Although the wonder and awe of biological systems are appropriately expressed in metaphor and poetry, the precise description and analysis of these systems will require science: the careful execution of experimental methods that depend on biochemistry, computer science, and mathematics.

Molecular structure is at the heart of life processes. The structure of macromolecules such as proteins and RNA-protein assemblages endow them with functional capabilities that carry on these processes. In particular, proteins are *active* molecules. As enzymes they facilitate other reactions at the minute local level, and as participants within muscle tissue they provide more global effects such as the movement of a limb. They are marvelous little machines. Viewed as such, we can attempt to describe their behavior with mathematical models that attempt to capture the dynamic characteristic of their changing geometry. This is much more difficult than simply writing down the one-dimensional sequence of amino acids that make up the protein. Structural analysis will demand some statistical analysis and judicious approximations. Furthermore, a simple mechanical view of the protein's dynamics will not take into account the intricacy of electron movement and charge distributions that a chemist might want to consider.

Even so, working with the mechanical/geometric aspects of structure will still present a challenge and will give us valuable insights into the workings of these molecules. Continuing on, going deeper into the complexity of the cell, we can consider the organization of entire sets of proteins, their differences, their similarities, and how they interact with each other. As we try to formulate models for systems of this magnitude, our mathematical tools must move from simple geometrical description to strategies involving machine learning and statistical analysis. We have to endure adverse tradeoffs: modeling larger complex systems means that we may abandon guarantees related to the precise behavior of smaller molecules. For example, our algorithm might discover a drug lead, but it is only a lead. The screening algorithm is suggesting that, from a statistical perspective, this virtual molecule shows some promise of binding to a large and complex protein, but only the chemist will find out for sure, and even then there are other issues such as unanticipated side effects.

So, the study of structural bioinformatics demands a lot: linear algebra and vector analysis, multivariate calculus, probability and statistics, biochemistry, cell biology, algorithms, programming, and software design. It is not a recommended path for the technically timid.

Despite all these demands, there are great rewards in this area of science. You are studying the results of evolutionary creation at its most compelling level. There is within this subject beauty and elegance. The radial symmetry of a beta barrel, the secretive workings of the GroEL chaperonin, and the motor-like action of adenosine triphosphate (ATP) synthetase are only a few examples. Even human immunodeficiency virus (HIV) protease has a certain beauty despite its deadly services.

Think about snow. As tiny water droplets accumulate in the atmosphere and undergo a freezing drift toward the earth, each flake takes a somewhat random path through slightly different environments, and the result is that each flake has a unique appearance. Despite the randomness, hydrogen bonding and the tetrahedral character of water molecules bestow a beautiful hexagonal symmetry on the snowflake. This is a fantastic property for so simple a molecule: H_2O—not much to it when you compare it with huge macromolecules. Even so, we can see this small, almost insignificant event as a metaphor for molecular evolution in the large. As the snowflake falls, random events constrained within a framework of physical laws lead to a symmetric, beautiful structure at a visible macroscopic level. Now step back and consider something more than water droplets. Consider a primordial soup of amino acids and several hundred other types of small molecules. Wait for a few billion years, and again the system drifts along subject to random moves such as DNA mutations but still constrained by the rules of chemistry (including those ever-present hydrogen bonds), and again we get pattern, symmetry, and intricate mechanisms spinning up into tissue, bone, and even those little receptors that trap the photons coming off this page.

This book cannot explain all aspects of the functionality of biological structures; we can only devise some mathematical models that are mere approximations of various mechanistic molecular behaviors. Along the way, we hope to discover more about molecular interactions, and eventually we might learn how to build drugs that alleviate suffering from cancer or wipe out AIDS. There are no guarantees for success, but we can at least make some initial steps toward this quest.

1.2 SMALL BEGINNINGS

"I make friends."

J. F. SEBASTIAN, *BLADE RUNNER*, 1982

Our current technology is far from that depicted in the 1982 science-fiction movie *Blade Runner*. At a poignant moment in the movie, J. F. Sebastian, a "genetic designer," comes up with his memorable quote, and we realize that he can design and create little creatures that can perform with human-like abilities.

When it comes to building new life forms, or simply modifying the behavior of existing cellular components, we are far from possessing the technology used by J. F. Sebastian. Our situation is more like that of a junior programmer who is hired by a corporation that has developed large libraries of software applications to serve its needs. Even though the programmer may want to design and implement his or her own mission-critical applications, he or she is initially restricted to the narrow responsibilities of program maintenance and the debugging of code written earlier by other people.

In a comparable way, we are limited when it comes to modifying functionality in living systems. We can sometimes accelerate some cell reaction, but more often, the medicinal chemist is concerned with the inhibitory capability of a drug candidate. Some drugs, for example, those associated with chemotherapies prescribed for the treatment of cancer, are extremely aggressive with very harsh side effects. We might dream of designing the equivalent of sophisticated molecular scalpels to correct some wayward cell process, but with current limited technologies, we sometimes have to settle for building the equivalent of molecular sledgehammers.

However, technologies can improve at a rapid pace with dizzying accomplishments. When I wrote my first computer programs in 1963, it was for an IBM 1620 with an IBM 1311 disk unit. The disk drive unit was approximately the same size as a top-loading washing machine and held about 2 MB of storage. People were thrilled to have such an advanced storage unit. After all, a single unit could hold the equivalent of 25,000 punch cards! Trade articles at the time speculated on the extreme challenge of designing a disk drive that could store more than this. They claimed that the drive already was a marvel of engineering because the read/write heads were only a few microinches off the surface of the platter. If some futurist was around to tell these people that disk drives 40 years later would have a storage capacity that was the equivalent of 100,000 of these units, but with a platter size of 3.5 inches, that person would have been dismissed as an overoptimistic fool. Alas, we now take so much for granted! Your laptop in 2007 has the same storage capacity as a lineup of IBM 1311 disk units set side by side, stretching on for 50 miles. Incidentally, future projections for the year 2020 estimate 86 terabytes (TB; 86,000 GB) on your desktop and 3 TB on your portable music player. So, if you are a young undergraduate student now, you will eventually have a poignant nostalgia for today's technologies. This will happen when your grandchildren fail to understand how your ancient iPod could do anything reasonable with a measly 160 GB of storage.

Who then can really predict what proteomics, nanotechnologies, and medicinal chemistry can bring over the next 50 years? One goal of this book is to convince you that structural bioinformatics will play a role in these endeavors. Ultimately, it is all about function. We have to understand how macromolecules interact with both small molecules and other macromolecules. However, function is determined by structure, so it is our necessary starting point on this journey.

1.3 STRUCTURAL BIOINFORMATICS AND THE SCIENTIFIC METHOD

"Follow the Force"

YODA, *STAR WARS*

$F = Gm_1m_2/r^2$

ISAAC NEWTON, *MATHEMATICAL PRINCIPLES OF NATURAL PHILOSOPHY*, 1687

So let it not look strange if I claim that it is much easier to explain the movement of the giant celestial bodies than to interpret in mechanical terms the origination of just a single caterpillar or a tiny grass.

IMMANUEL KANT, *NATURAL HISTORY AND THE THEORY OF HEAVEN*, 1755

When we develop algorithms to do computations in structural bioinformatics, we are hopefully using techniques that are based on scientific methodology. For example, to solve a problem, we might write a program that implements an algorithm based on a scientific description of a natural process and then run tests to verify that we are getting accurate results from that program. The algorithm is usually derived from a mathematical model that gives us a computational representation of some physical process. For example, we could predict the position for some ligand in a protein-binding site by developing an algorithm that uses chemical properties and geometric analysis to derive an optimal low-energy position for the ligand. As an example of a more complicated level of a mathematical model, we could consider a scientific theory such as quantum mechanics if we needed to calculate the potential energy surface of a molecule.

In the subsections to follow, we formalize the ideas that are only loosely described in the previous paragraph. We will assume that our goal is the solution of some particular problem related to a natural process in the real world. Solving the problem involves three levels of activity:

1. We can observe this natural process at work and collect meaningful data.

2. We can develop scientific explanations supporting mathematical models that describe the inner workings of the natural process.

3. Using these empirical and theoretical sources we can design a computational strategy (an algorithm) to solve the problem.

1.3.1 Three Realms: NATURE, SCIENCE, and COMPUTATION

In the Figure 1.1 we see three interrelated boxes that represent the three realms that are of concern to us. The idea is that these boxes represent three different areas of discourse. In later chapters we will use them to organize our conceptualization, modeling, and solving of a problem. We describe each of these in turn.

The NATURE box represents some particular real-world natural process that we wish to study in order to solve some problem related to this process. In some cases, we know very little about the workings of this natural process, and so it is often regarded as a "black box." Our ultimate strategy is to evaluate or at least approximate the output of this box, which we designate using the variable f. Accepting the risk of oversimplification, we will assume that the natural process produces various quantities that can be measured, and these results are retained in some way, for example, as a set of components in a vector denoted in the figure using the variable f. We further assume that there is a dependency between f and various other quantities that are part of the natural process. For the sake of argument we will assume that this dependency is represented by the notation $f(x, x')$ where both x and x' represent vectors containing components that act as independent variables for f. The vector x holds quantities that can be both observed and measured. Some of the x quantities will be related to properties that are under the control of the observer (inputs to the box), or they are measurable quantities related to existing properties that are simply part of the natural process. These quantities will be used in the scientific analysis and will act as input to the computational procedure. The vector x' holds quantities that do not appear in the scientific analysis and are thus unavailable to the computational procedure. There are two possible reasons for components of x' to be ignored by the scientific analysis: our level of understanding is such that we simply do not know of their existence or we do know about these variables but have chosen to ignore them in order to get a simpler scientific description of the natural process. In the latter case, it must be recognized that the scientific description covered in the next paragraph is approximate and consequently limited in terms of application.

The SCIENCE box represents a collection of statements that scientifically characterize the workings of the natural process with the intention of aiding in the development of the algorithm that is used to solve the problem related to the natural process under consideration. In Figure 1.1, this is represented by the formulation of a function $h(x)$ that defines a reasonable approximation of the function $f(x, x')$. The contribution of this scientific content toward the development of the needed algorithm can vary significantly from one problem to the next. In fact, the problem may be so difficult that we cannot provide any analysis, and the box is almost empty except for a simple assumption about the natural process. In this case the algorithm must work directly from observations that are taken from the natural process. At the other extreme, we may have a very complete analysis of the process, but the model is so computationally complicated (consider quantum mechanics) that we cannot formulate a tractable solution to the problem. Ideally, the box contains a useful scientific analysis characterized by a hypothesis that has a firm theoretical foundation supporting a well-defined mathematical model that provides insight for the development of an algorithm to solve the given problem. Variations on this hypothesis, model, and theory type of analysis is discussed in the next subsection. The formulation of $h(x)$ may be

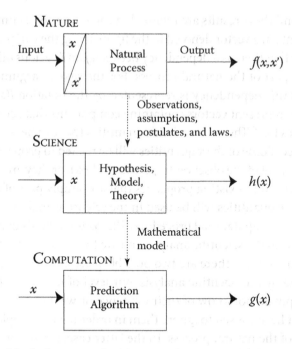

FIGURE 1.1 Block diagrams for NATURE, SCIENCE, and COMPUTATION.

very complicated. The scientific characterization may be done using differential equations, integral equations, or any other mathematical description. In some situations, especially when nonlinear equations are involved, an explicit derivation of $h(x)$ may be impossible, and we can only define $h(x)$ implicitly or not at all.

Finally, the COMPUTATION box holds the specification of the algorithm used to solve the problem. It will define the inputs of the algorithm and the sequence of computations necessary to produce the required output that provides a solution to the given problem. These specifications will typically describe any data structures that are used by the algorithm.

We regard a computer program as an *implementation* of the algorithm. The program will be written in some particular computer language in contrast to the algorithm that is specified by pseudocode or some higher level formalism that is independent of any particular computer language.

1.3.2 Hypothesis, Model, and Theory

By a model is meant a mathematical construct which ... is expected to work—that is, correctly describe phenomena from a reasonably wide area.

Furthermore ... it must be rather simple.

JOHN VON NEUMANN, 1955

In Figure 1.1, the SCIENCE box is the most important because it provides a rational explanation about the workings of the natural process. Using scientific insights as a basis for our studies, we can have a much higher level of confidence in the realistic significance of our computational efforts. When discussing scientific statements, we will use terms such as *hypothesis*, *model*, and *theory*. In an effort to provide more clarity in these discussions, it is worthwhile to establish definitions of these terms so that our scientific methods are more transparent and subject to further improvement.*

As this book is about algorithms, our treatment of terms such as *hypothesis*, *model*, and *theory* will be adapted to provide a mathematical setting that is suitable for computational goals. Although these terms all relate to scientific statements, they tend to differ with respect to the level of certainty in the pronouncement. To make the distinctions more memorable and to put them in a scientific setting that is easily understood, we can consider these terms as they apply to the scientific description of the solar system and how that description has changed over the last two millennia or so. Although it may seem strange to discuss planetary orbits in a structural bioinformatics book, it is useful to do so because, as noted in the quote by Kant, it is a comparatively simple system when compared to a complex biological system.

A **hypothesis** is a well-informed explanation about a scientific phenomenon. Ideally, it is a clear statement about cause and effect in some physical realm. However, it may be subject to revision, and typically has little credibility without the support of a mathematical model that incorporates a scientific theory.

For example, a seemingly reasonable hypothesis is that the earth is at the center of the universe and the sun goes around the earth. This hypothesis would explain why the sun appears to rise in the morning and set in the evening. Historically, the geocentric hypothesis of Aristotle and Ptolemy prevailed for almost 2000 years. It was aided by religious notions that our world

---

* The reader should be warned that terms such as *hypothesis*, *model*, and *theory* vary somewhat in their use and may have slightly different meanings in scientific literature. For example, Humphreys [Hu95] describes how computational science has had an impact on the scientific method to the extent that the concept of a model has changed with the advent of computers.

certainly *must* be at the center of the universe.* These notions lasted until the 16th century when Copernicus formulated a different hypothesis. Unfortunately, the geocentric hypothesis could not explain the "retrograde motions" observed for planets such as Mars that seemed to move back and forth against the star field background. To support the shaky geocentric hypothesis, extra explanations were proposed describing a complicated system of "epicycles" that attempted to justify the observed retrograde phenomena.

A **model** provides us with a stronger explanation and is often associated with a hypothesis that has more predictive power.

For example, Copernicus came up with the hypothesis that the planets, including the earth, orbited the sun. This almost eliminated the desperate and erroneous epicycle explanations of the retrograde motions. With a bit of generosity we will say that Copernicus had a *model* of the solar system because his explanation did at least provide a useful mathematical description of solar system events: planets follow paths around the sun, and these paths are perfect circles. At this time, it made perfect sense to have circular orbits because religious views of nature essentially demanded that God would certainly have heavenly bodies move in perfect circles.†

Because Copernicus was stuck in the perfect circular orbits model, there was still a need for epicycle adjustments although not to the same extent as with the geocentric hypothesis. Science had to wait for Johannes Kepler to get a more accurate model of the solar system. Based on large amounts of data collected by Tycho Brahe, Kepler was able to determine that orbits were elliptical, and he published this result in 1609. It clearly provided a better mathematical model with a much greater predictive capability.‡

---

* Actually, a sun-centered solar system was proposed in 200 B.C. by Aristarchus of Samos, but he was ignored because his ideas came into conflict with those of the more famous and hence more credible Aristotle.

† As a digression, it may be noted that, when the heliocentric explanation was first published, just after the death of Copernicus in 1543, the publisher felt it necessary to extend something of an apology for the deviation from a geocentric explanation. The views of Copernicus were a "mere hypothesis" and only had a value in that they "seemed to simplify" astronomical calculations.

‡ Another digression: Before working with Brahe's data, Kepler made many attempts to provide a model that demonstrated principles that he considered to be in harmony with certain ideals of perfection. At the time of Kepler, only six planets were known. For Kepler it made perfect sense that only six planets existed because there were five Platonic solids (tetrahedron, cube, octahedron, dodecahedron, and icosahedron)! He initially believed that the six planets moved in circular orbits that were separated by five astronomically large spheres, each having contact with the vertices of some particular, huge, invisible Platonic solid. It is to Kepler's credit that he was able to forsake a model that was driven by a desire to second guess the idealized motives of a supernatural power and eventually let astronomical observations support a clear hypothesis and mathematical model.

Finally, a scientific **theory** provides an explanation that can be confirmed by repeated experimental validation. A theory may be altered or modified as more becomes known, but this is fairly uncommon because the credibility of the explanation is typically very solid and in agreement with the facts known at the time.

Getting back to our solar system example, we would say that Newton's explanation of planetary orbits gives us a mathematical model of planetary motion because it is based on a theoretical statement about gravitation that we now consider to be the "Universal Law of Gravitation" formulated as the inverse square law. Bringing in the effects of gravity gives us an explanation that mathematically justifies the elliptical orbit model proposed by Kepler.

Considering the previous discussion about hypothesis, model, and theory, let us summarize the various scientific statements that have been made regarding planetary orbits:

	Hypothesis	**Model and Theory**	**Support**
Ptolemy	Geocentric	Overly complicated epicyclical model	None (dogmatic assertion of geocentricity)
Copernicus	Heliocentric circular orbits	Circular orbits with small epicyclical adjustments	None (dogmatic assertion that orbits must be perfect circles). To be fair: observable data was roughly consistent with circular orbits
Kepler	Elliptical orbits	Equations for elliptical orbits	Model was shown to be consistent with observable data
Newton	Elliptical orbits	Equations for elliptical orbits	Law of gravitation shows that elliptical orbits are to be expected

The reader should appreciate that the law of gravitation provides a fundamental basis for the two-body model and can also serve to support other more complicated models. Thus, if the system deals with more than two bodies (for example, the sun plus several planets), then the inverse square law can still be used, but planetary orbits will not be elliptical. This is true because of the gravitational interactions among the various planets in the system. However, because the other planets have small masses in comparison to the sun, and their gravitational influences diminish in proportion to the inverse square of the separating distances, the ellipse determined by the two-body analysis is still useful as a very accurate approximation.

1.3.3 Laws, Postulates, and Assumptions

If science is the bedrock on which we build algorithms, then it is important to understand just how firm that bedrock is. In the previous discussion it was stated that elliptical orbits depended on the *law* of gravitation. We cannot prove this law. Newton is asking us to accept his concept of "force acting at a distance." Despite the lack of proof, this law has been accepted as a fundamental truth of nature for over 300 years and serves to give us "classical" models describing the behavior of objects in space.

As discussed by McQuarrie and Simon [Ms97, p. 19], classical Newtonian theory could not be applied to an electron in an orbit around a nucleus. Nonclassical assumptions had to be made for this situation, and Bohr came to the rescue by assuming the existence of stationary electron orbits. This was a clear demonstration that the classical model had limitations and did not apply to very small entities.

In general, the foundation of scientific statements ultimately rests on laws, postulates, or assumptions that are stated without proof. A very powerful theoretical explanation of natural processes at the atomic level is given by quantum mechanics. Again, it must be established by declaring some basic assumptions that essentially act as axioms on which we build a foundation of theory to explain the observables seen at the atomic level. The interested reader may consult [Ms97, chapter 4] to read the details. These postulates cannot be proven. However, they produce a set of scientific statements that agree with experimental data to a remarkable degree. However, as the complexity of the natural system increases, the equations proffered by quantum mechanics become extremely difficult to solve.

Most problems in protein bioinformatics have such complexity. We do not have the luxury of developing a grand set of scientific statements using a few fundamental laws, as is done in Newtonian mechanics. In lieu of fundamental laws or postulates, various simple assumptions are usually made about a natural process in order to formulate a mathematical model that can direct a computational strategy.

Unfortunately, there are occasions when assumptions are made with the sole intention of developing a mathematical model that is amenable to computation with only a perfunctory nod given to the realities of the natural process. Statements such as "We will assume that the data was generated by a normal distribution" may be quite valid or may require further scrutiny and investigation. The message here is straightforward: Even though we may not attain the ideal state of providing fundamental laws or postulates as a basis for our mathematical models, we should always endeavor to make assumptions clear and, as much as possible, related to the physical realities of the processes under investigation.

1.3.4 Model Theory and Computational Theory

In order to bring rigor into our discussions, we often employ carefully worded mathematical statements and theorems. However, one must be careful that an abundance of theory is not overextended in the sense of providing more credibility than is warranted.

For example, is there a theory that helps the Copernican model of circular orbits? We can imagine a huffy Copernicus insisting on perfect circular orbits while carrying a copy of Euclid's treatise on geometry. Even though Euclid's *Elements* has several theorems related to circles, these theorems give us facts *only* about the mathematical model, and they do very little to justify the appropriateness of the model for planetary orbits. Copernicus can use the equations of circles to calculate a rough approximation of planet position, but the rigor behind Euclid's theorems should not enhance the veracity of the model. In fact, the equations of circles, when combined with the observational data, should clearly indicate a lack of consistency with reality and an indication that perfect circles are *not* the appropriate model. However, using data to refute the favored geocentric model was not high on the list of priorities for scientists of the time. Even Kepler's first model tried to maintain the circular orbit theory. As noted Feynman in [FE67, p. 16], he considered the possibility of circular orbits with the sun off center. Eventually, the elliptical orbits model prevailed, and he was able to develop computational procedures that utilized the equations of an ellipse to plot planetary motion and thus predict the future positions of planets in space.

In general, a theoretical statement made in a problem-solving effort typically falls into one of two categories:

1. *Model theory* is related to the mathematical model and serves to provide more mathematical insights about the model but does not provide proof justifying the intrinsic truth of the model. Nonetheless, it

is expected that such theory produces extensions of the mathematical model that are consistent with the observations taken from the natural process.

2. *Computational theory* is related to the problem-solving algorithm. It will make claims about the algorithm itself, for example, convergence of an iterative algorithm or the accuracy of the output. As with the model theory, it also does nothing to justify the veracity of the mathematical model.

As a cautionary note, it is worthwhile to stress the importance of these issues because they are especially relevant when we discuss machine-learning strategies. Almost all machine-learning theory deals with computational strategy and says little if anything about the underlying natural process that is being emulated. For example, in algorithms that use statistical learning, it is necessary to train a predictor by extracting samples or observations from the underlying natural process. There are theorems that establish the accuracy of such a predicting function based on the number of samples collected for the training set. Although such a theorem helps establish the accuracy of the prediction function, it says nothing about the inner workings of the natural process itself.

1.3.5 Different Assumptions for Different Models

To provide an easily understood example of a problem that has three easily characterized realms of discourse, the previous subsections considered the problem of predicting planetary motion. We considered elliptical orbits as mathematical models and noted that a more complicated model could be proposed if we went beyond a two-body system.

Other problems could easily be constructed to illustrate the use of various models with different levels of complexity and accuracy. For example, suppose we wanted to calculate the final velocity of a person without a parachute hitting the ground after falling out of a plane flying at an altitude of 1000 meters. We could use the standard equations that work with acceleration due to gravity and assuming freefall in a vacuum [SE90, p. 51]:

$$v^2 = v_0^2 - 2g(y - y_0). \tag{1.1}$$

With $v_0 = 0$, $y = 0$, $y_0 = 1000$ m, and $g = 9.80$ m/s^2, we get $v = 140$ m/s. However, this model does not take into account air drag [SE90, p. 140]. Using this more appropriate model described by the equation

$$v_t = \sqrt{\frac{2mg}{C\rho A}}$$ (1.2)

with m = mass of person, C = drag coefficient, and ρ = density of air, we get a more realistic assessment of the terminal velocity as only 60 m/s. So, it is clear that the assumption of freefall in a vacuum has a significant impact on the choice of the model and the final answer that it provides.

To get a firmer idea about the significance of the three realms described earlier but within a bioinformatics context, we sketch another problem in the next section and describe the contents of each box in Figure 1.1.

1.4 A MORE DETAILED PROBLEM ANALYSIS: FORCE FIELDS

The central problem of molecular dynamics is to determine the relative motion of all atoms in a set of molecules. It is essentially the N-body problem that Newton pondered when dealing with a collection of planetary bodies but is much more formidable in its complexity. This extra complexity is caused by the following issues:

a. Although the classical mechanics model can be used to describe both planetary motion and molecular motion, the latter involves a much more complicated description of the forces between two bodies. Moreover, forces between two atoms can be both attractive and repulsive (to be described later).

b. The system contains not just a few interacting bodies but perhaps hundreds or even thousands of bodies (atoms).

c. As noted earlier, for a planetary system, one can derive a very reasonable approximation by modeling the motion of a planet using a two-body system, namely, the planet and the sun, with its very large gravitational effect. Other planets can be neglected. For a biological system, several atoms may be in close proximity and, consequently, it is not as easy to eliminate their influence in the calculations.

The problem may be stated as follows: How can we best represent the forces that act on an atom within a molecular ensemble? We now describe this problem under the Nature, Science, and Computation headings.

1.4.1 NATURE

There are many observables that can be extracted from a molecular system. This includes the empirical assessments of the various force constants that are used in the force field formulas described in the next section. As noted in Mackerell [MA98], data may be collected from the gas phase using electron diffraction, microwave, infrared, and Raman techniques, to name just a few. We can also utilize the results of x-ray analysis. In the case of a ligand bound to a protein, an x-ray analysis of the cocrystallized molecules can be used to derive the 3-D coordinates of all atoms in the molecules. In particular, we can determine the distance between the various atoms, and these distances can be used in the mathematical model that calculates the potential energy of the atomic ensemble.

1.4.2 SCIENCE

As already noted, our most accurate description of the atomic ensemble is via quantum mechanics, but this does not facilitate a tractable computational strategy. Instead, our molecular mechanical analysis will consider the atoms to be a collection of masses interacting with each other through various forces. In fact, we will assume that the total energy of the system is minimized when the atoms assume a stable position that is characterized by the minimization of

$$E_{total} = E_{bond} + E_{angle} + E_{dihedral} + E_{vdw} + E_{coulomb}. \tag{1.3}$$

Other formulas may be developed employing more or fewer terms, depending on the needs of the application. As we will see later, there are several useful models that can be employed in the modeling of the energy function.

In general, the terms on the right hand side of the equation fall into two categories:

a. Bonded atoms: These are energy terms related to atoms that share a covalent bond.

b. Nonbonded atoms: These are energy terms related to atoms that are not bonded but are close enough to interact.

1.4.2.1 Energy Terms for Bonded Atoms

E_{bond} is comprised of all the various bond-stretch terms. Each is described by a simple quadratic function:

$$K^b(r - \bar{r})^2 \tag{1.4}$$

where r is the distance between two bonded atoms. This formula is derived from Hooke's law and it models the potential energy as if the two bonded atoms were connected by a spring. If the bond is not "strained," it will have an equilibrium length of $\bar{r}$. If the bond is stretched or compressed, the deviation from $\bar{r}$ measured as $r - \bar{r}$ will increase the potential energy in a quadratic fashion. The bond-stretch force constant K^b has been determined by empirical observations, and it will take on various values depending on the atom types specified for the two atoms that are involved. For example, as reported by Cornell et al. [Cc95],* $\bar{r} = 1.507$ Å and $K^b = 317$ kcal/(mol Å²) when the bond is C–C, whereas $\bar{r} = 1.273$ Å and $K^b = 570$ kcal/(mol Å²) when the bond is C=N.

Because E_{bond} will be the sum of all such quadratic contributions, we can write

$$E_{bond} = \sum_{(i,j) \in B} K_{ij}^b (r_{ij} - \bar{r}_{ij})^2 \tag{1.5}$$

where

$\quad K_{ij}^b$ is the bond stretch force constant that would be used when $atom(i)$ is bonded to $atom(j)$,

$\quad r_{ij}$ is the distance between $atom(i)$ and $atom(j)$,

$\quad \bar{r}_{ij}$ is the distance between $atom(i)$ and $atom(j)$ at equilibrium, and

$\quad B$ is the set of pairs: $\{(i,j) \mid atom(i) \text{ is bonded to } atom(j)\}$.

E_{angle} is the sum of all bond-angle bending terms, and each is modeled as a quadratic spring function with an angle-bending force constant K^a:

$$K^a (\theta - \bar{\theta})^2 \tag{1.6}$$

where θ is the bond angle formed by three atoms, and $\bar{\theta}$ is the value of the unstrained or equilibrium angle. As with K^b, the value of the force constant will depend on the type of atoms involved. For example, again reported by Cornell et al. [Cc95], $\bar{\theta} = 112.4° = 1.96$ radian with $K^a = 63$ kcal/(mol radian²) when the angle is formed by C–C–C, and $\bar{\theta} = 121.9° = 2.13$ radian and $K^a = 50$ kcal/(mol radian²) when the angle is formed by C–N–C.

---

* Table 14 of this paper presents well over 200 of the many parameters that can be used in this calculation.

Because E_{angle} is the sum of all such quadratic contributions, we can write

$$E_{angle} = \sum_{(i,j,k) \in A} K_{ijk}^a (\theta_{ijk} - \bar{\theta}_{ijk})^2 \qquad (1.7)$$

where

  K_{ijk}^a is the force constant used for the bond angle formed when *atom(i)* and *atom(k)* are bonded to *atom(j)*,

  θ_{ijk} is the bond angle made when *atom(i)* and *atom(k)* are bonded to *atom(j)*,

  $\bar{\theta}_{ijk}$ is the bond angle made when θ_{ijk} is at equilibrium, and

  A is the set of triplets: $\{(i, j, k) | atom(i)$ and *atom(k)* are bonded to *atom (j)*$\}$.

$E_{dihedral}$ involves the contributions made by the dihedral angle terms. This is a bit more complicated and is modeled as

$$\frac{K^d}{2} [1 + \cos(n\phi - \gamma)] \qquad (1.8)$$

where ϕ is the dihedral angle formed by a chain of four consecutive bonded atoms, n is the periodicity of the dihedral angle (see [Cc95]), and γ is the phase offset. For example, $K^d/2 = 14.50$ kcal/mol, $n = 2$, and $\gamma = 180°$ when the dihedral angle is formed by the four consecutive atoms X–C–CA–X, where X is any atom, C is a carbonyl sp^2 carbon and CA is an aromatic sp^2 carbon.

As before, we sum over all contributions:

$$E_{dihedral} = \sum_{(i,j,k,l) \in D} \frac{K_{ijkl}^d}{2} [1 + \cos(n_{ijkl}\phi_{ijkl} - \gamma_{ijkl})] \qquad (1.9)$$

where

  K_{ijkl}^d is the force constant used for the dihedral angle formed by the four consecutive atoms indexed by i, j, k, and l,

  n_{ijkl} is the periodicity of this dihedral angle,

  ϕ_{ijkl} is the dihedral angle formed by the consecutive atoms indexed by i, j, k, l,

γ_{ijkl} is the phase offset for this dihedral angle, and

D is the set of quadruplets that specify all the dihedral angles, namely: $\{(i, j, k, l)|i, j, k, \text{ and } l \text{ are the indices of four consecutive bonded atoms}\}$

1.4.2.2 Energy Terms for Nonbonded Atoms

E_{vdw} is due to the attractive/repulsive forces that are characterized as van der Waals interactions. The van der Waals interaction between two atoms is given by the "12-6 potential":

$$\frac{A}{r^{12}} - \frac{B}{r^6} \tag{1.10}$$

where r is the distance between the two atoms. We assume that the atoms are in different molecules, or, if they are in the same molecule, then they are separated by at least three bonds. From the formula it is clear that, for very small r, the r^{-12} term contributes a large increase in the potential energy, but it has very little impact for larger values of r where the potential is negative but very close to zero. Values for constants A and B also depend on the atom types of the two interacting atoms. The derivation of their values is somewhat involved, and we will omit these details which can be found in Cornell et al. [Cc95, Table 14 footnote]. Adding up all contributions we get

$$E_{vdw} = \sum_{j \in F} \sum_{[i<j]} \left[\frac{A_{ij}}{r_{ij}^{12}} - \frac{B_{ij}}{r_{ij}^6} \right] \tag{1.11}$$

where

A_{ij} and B_{ij} are the Lennard–Jones 12-6 parameters (see [Cc95]) for $atom(i)$ interacting with $atom(j)$,

r_{ij} is the distance between $atom(i)$ and $atom(j)$,

F is the full set of all indices for the atoms, and

$[i < j]$ means that index i is less than index j and also that $atom(i)$ and $atom(j)$ are either in separate molecules or, if in the same molecule, they are separated by at least three bonds.

$E_{coulomb}$ is due to the electrostatic forces that exist between two charged atoms. As before, these nonbonded interactions are only calculated for atoms that are separated by at least three bonds. For two atoms indexed by i and j, this is characterized by the formula

$$K_c = \frac{q_i q_j}{\varepsilon r_{ij}} \tag{1.12}$$

where ε is the dielectric constant. Variables q_i and q_j are the atomic charges of the interacting atoms that are separated by a distance of r_{ij}. This formula can be derived from Coulomb's law that expresses the electrostatic force as

$$F \propto \frac{q_i q_j}{r_{ij}^2}. \tag{1.13}$$

The force is positive (indicating repulsion) if the two atoms have the same charge; otherwise, it is negative and produces an attractive interaction. The K_c constant is a conversion factor needed to obtain energies in units of kcal/mol. Using the esu (electrostatic charge unit) to define a unit of charge as in [Sc02, p.254], we can derive the value of the conversion factor as $K_c = 332$ (kcal/mol)(Å/esu²). Partial charge values for various atoms in both amino acids and nucleotides are reported in [Cc95] based on [CcB95]. For now, we will assume that $\varepsilon = 1$. Other values of ε will be discussed in the next section. Adding up all contributions, we get

$$E_{coulomb} = K_c \sum_{j \in F} \sum_{[i<j]} \frac{q_i q_j}{\varepsilon_{ij} r_{ij}} \tag{1.14}$$

where

K_c is 332 (kcal/mol)(Å/esu²),

q_i, q_j are the partial charges for *atom(i)* and *atom(j)*, respectively,

r_{ij} is the distance between *atom(i)* and *atom(j)*,

ε_{ij} is the dielectric constant for the *atom(i)* and *atom(j)* "environ-ment" (assumed to be 1 for now),

F is the full set of all indices for the atoms, and

$[i < j]$ means that index i is less than index j, and also that *atom(i)* and *atom(j)* are either in separate molecules or, if in the same molecule, they are separated by at least three bonds.

1.4.2.3 Total Potential Energy

Combining all terms just defined into one grand total, we get

$$E_{total} = E_{bond} + E_{angle} + E_{dihedral} + E_{vdw} + E_{coulomb}$$

$$= \sum_{(i,j) \in B} K^b_{ij}(r_{ij} - \bar{r}_{ij})^2 + \sum_{(i,j,k) \in A} K^a_{ijk}(\theta_{ijk} - \bar{\theta}_{ijk})^2$$

$$+ \sum_{(i,j,k,l) \in D} \frac{K^d_{ijkl}}{2}[1 + \cos(n_{ijkl}\phi_{ijkl} - \gamma_{ijkl})] \qquad (1.15)$$

$$+ \sum_{j \in F} \sum_{[i<j]} \left[\frac{A_{ij}}{r_{ij}^{12}} - \frac{B_{ij}}{r_{ij}^{6}} \right] + K_c \sum_{j \in F} \sum_{[i<j]} \frac{q_i q_j}{\varepsilon_{ij} r_{ij}}.$$

As mentioned earlier, this empirical energy calculation depends on many parameters, and it is typical that the program performing the calculation will have rather extensive tables to provide values for $K^b_{ij}, \bar{r}_{ij}, K^a_{ijk}, \bar{\theta}_{ijk}, K^d_{ijkl}, n_{ijkl}, \phi_{ijkl}, \gamma_{ijkl}, A_{ij}$, and B_{ij}, all of which depend on the atom types assigned to the atoms involved. There are several versions of these tables. The tables included in Cornell et al. [Cc95] can give the reader a general idea about the information that is required. There are several parameter sets that are now available for various applications. They have names such as AMBER, CHARMM, GROMOS, MMFF, and OPLS.

1.4.3 COMPUTATION

Although somewhat daunting, the empirical force field just described would seem to be a rather straightforward calculation. The program would do all the appropriate table lookups, make the necessary substitutions, and do the indicated calculations. Unfortunately, there are some issues that complicate the situation:

a. Repetitive energy calculations: There are various applications, such as molecular dynamics, for which it is necessary to repeatedly do this computation across several, perhaps billions of, time steps. Each small change in the conformation or position of a molecule requires a new evaluation of the potential energy.

b. Many atoms in the molecule ensemble: A large number of atoms in a molecular ensemble will greatly increase computation time, especially for the double sums associated with E_{vdw} and $E_{coulomb}$. Note that if we do the computation as indicated, then the number of computations for these terms increases as the square of the number of atoms in the system.

Because of these issues, computational chemists have proposed several strategies designed to reduce the computational load. Before proceeding with a description of typical strategies, it should be noted that the given formula for E_{total} is itself a simplified version of other more complicated formulas that strive for higher accuracy. For example, some "Class II force fields," such as MMFF, use

$$K^{b_2}(r-\bar{r})^2 + K^{b_3}(r-\bar{r})^3 + K^{b_4}(r-\bar{r})^4 \tag{1.16}$$

instead of $K^b(r-\bar{r})^2$ with a similar enhancement for $K^a(\theta-\bar{\theta})^2$. It is also possible to replace $\frac{K^d}{2}[1+\cos(n\phi-\gamma)]$ with $[K^{d_1}(1-\cos\phi)+K^{d_2}(1-\cos2\phi)+K^{d_3}(1-\cos3\phi)]/2$.

In an effort to simplify E_{total} even further, some research studies will assume that bond lengths and bond angles are constant. In these applications the *change* in potential energy will be due to changes in molecular positions or conformational changes in molecules that produce variations in dihedral angles. Consequently, the modified energy calculations will be due to $E_{dihedral}$, E_{vdw}, and $E_{coulomb}$ only.

However, neglecting E_{bond} and E_{angle} will not significantly reduce computation time. Using the notation of asymptotic analysis, the number of bonds and the number of bond angles are both $\Theta(n)$, where n is the number of atoms. The most important time savings are related to computational strategies that try to reduce the calculation times of E_{vdw} and $E_{coulomb}$ because both are $\Theta(n^2)$. This is usually done by using a *spherical cutoff* strategy for E_{vdw} and $E_{coulomb}$. When the distance r_{ij} between two atoms exceeds a particular threshold, say 10 Å, then their interaction contribution to E_{vdw} or $E_{coulomb}$ is considered to be zero. The approach works quite well for E_{vdw} because the 12-6 Lennard–Jones potential approaches zero very rapidly and can be safely neglected when r_{ij} is over 9 or 10 Å. The handling of $E_{coulomb}$ requires more caution. To understand the computational trouble stemming from the $E_{coulomb}$ calculation, consider a particular atom A near the center of a large globular protein. For the sake of argument, we can consider the other atoms that interact with A to reside in concentric spherical shells of thickness 1 Å. Although the Coulomb potential falls off as the inverse of the distance, the average number of charged atoms per shell increases with the volume of the shell, a quantity that is proportional to r^2. Consequently, even though the potential of each interacting atom decreases as one gets farther from A, the number of these charges goes up.

As noted in Schreiber and Steinhauser [Ss92], the cumulative sum over the shells does not converge to a final value in a monotonic fashion. Instead, as the radius of the shells grows, the electrostatic contributions oscillate in sign. It should be stressed that, even though the number of atoms in a shell increases, their net resultant is usually reduced, on average, due to cancellation effects. Consider, for example, atom A situated between two atoms both of the same charge but on opposite sides within the same shell. With these issues in mind, the electrostatic potential is typically regarded as a slowly decaying potential, and the strategies for reducing computation time are at times somewhat controversial. So, it is not surprising to see the 1992 paper entitled "Cutoff size does strongly influence molecular dynamics results on solvated polypeptides" [Ss92] followed by the 2005 paper entitled "Cutoff size need not strongly influence molecular dynamics results for solvated polypeptides" [Ba05].

From a computational perspective, the value of the dielectric constant ε also requires more discussion. It is a dimensionless factor that has the value 1 when the interacting atoms are in a vacuum. When the interacting atoms are in a polarizable medium such as water, the Coulombic force is reduced because the intervening water molecules have a screening effect that weakens the electrostatic force between the two atoms. This reduction is usually modeled by increasing the value of ε to 80 when the solvent is water.* The value of 80 should be used for the dielectric constant when the space between the two charges is so large that it admits many water molecules. As this space decreases, for example, in the simulation of a ligand binding to a cleft in a protein, it is reasonable to expect that the two charges would eventually have vacuum between them, and so, at that time, $\varepsilon = 1$. However, when and how should this change take place? Researchers typically adopt one of two methodologies for the modeling of electrostatic charge in a solvent:

1. Explicit solvent: The molecular ensemble explicitly includes *all* the water molecules. In this case we simply use $\varepsilon = 1$, but the calculation

---

* Some research papers use 78 instead of 80.

must include all the atoms, and the extra water molecules will lead to much longer computation times.

2. Implicit solvent: This approach is also referred to as *continuum modeling*. The idea is to remove the water molecules, thus reducing the atom count, and this leads to a decrease in computation time. However, the dielectric effect of the water molecules is retained by replacing ε with a dielectric function (also called a screening function).

As mentioned earlier, a coarse version of the dielectric function would be $\varepsilon = 80$, but this is inappropriate for very small r_{ij} values. A more realistic screening function has a sigmoid shape. This approach has been used by Hingerty et al. [HR85], and the sigmoid function is given as

$$\varepsilon(r) = 78 - 77 \left(\frac{r}{2.5}\right)^2 \frac{e^{r/2.5}}{(e^{r/2.5} - 1)^2}. \tag{1.17}$$

A plot of the sigmoid dielectric function is given in Figure 1.2.

This completes our extensive example of a problem that is described within the NATURE, SCIENCE, and COMPUTATION realms.

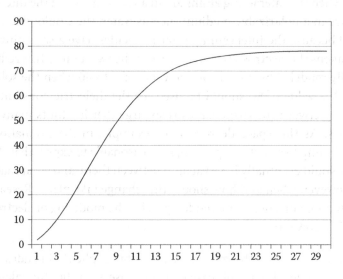

FIGURE 1.2 The dielectric function used by Hingerty et al.

1.5 MODELING ISSUES

Entia non sunt multiplicanda praeter necessitatum.

WILLIAM OF OCKHAM, C1285-1349

Everything should be made as simple as possible, but not simpler.

ALBERT EINSTEIN, 1879-1955

Marge: "You liked *Rashomon*."
Homer: "That's not how *I* remember it."

THE SIMPSONS, EPISODE: "THIRTY MINUTES OVER TOKYO"

The discussion in the last section presented a realistic but complex problem in computational chemistry. The major mathematical issues of the problem were segregated into two categories: The SCIENCE description dealt with the mathematical model of the empirical potential function, and the COMPUTATION description dealt with modifications of the model so that calculations could be accomplished in a timely fashion. For some problems, the separation between the two areas of discourse can be somewhat fuzzy. For example, it has already been noted that the given formulation of E_{total} is already a simplified version that has been selected to reduce the complexity of the calculation. In general, the choice of a model will be a negotiation that requires compromise. The model will be a description that attempts to embody the intrinsic and often formidable complexity of a natural process in the NATURE box while still providing a tractable starting point for the numerical calculations of the COMPUTATION box—calculations that are subject to finite resources and limited execution time. Consequently, we may be forced to accept a trade-off: less time for computation leading to less accuracy in the mathematical model. The primary goal, then, is the construction of good models. As much as possible, we strive for faithful representation while trying to maintain computational efficiency.

In pursuing this goal, it is worthwhile to discuss three aspects of a mathematical model that have been associated with the following three names: Rashomon, Ockham, and Bellman. Breiman [BR01a] discusses these issues in the context of statistical modeling, but the ideas are general enough to be extended to our description of SCIENCE models.

1.5.1 Rashomon

In the Rashomon story,* four people witness a murder and eventually testify in court. The explanations provided by the witnesses are all mutually contradictory, but they are also equally plausible. The philosophical questions lurking beneath the story are related to *truth*. What is truth? How can we know when an explanation is true? More recently, the "Rashomon effect" has been applied to various explanations in scientific understanding, for example, [Br01a] and [He88]. In the empirical potential energy problem, we saw the Rashomon effect in the Science discussion when it was noted that several parameter sets are available.

In general, the Rashomon effect is observed for the Science box of Figure 1.1 when there are several different but reasonable mathematical models. More precisely, we would have several different $h(x)$ functions each with the same or very similar error rate when compared with observables taken from the Nature box. Better examples of the Rashomon Effect would be illustrated by models that employ different equations, not just a change in parameters. Similar considerations would apply for the Computation box. For example, the explicit and implicit solvent models described in the previous section provided reasonable accuracy and good results—the choice of computational model depended mainly on execution time.

1.5.2 Ockham

Ockham's razor is the statement: "Create no more entities than are necessary." It is often considered a very reasonable strategy in the selection of a model when one is presented with various models of varying complexity. Let us go back to our previous example of planetary motion for a moment. Feynman [Fe67, p. 18] states:

> The next question was—what makes planets go around the sun? At the time of Kepler some people answered this problem by saying that there were angels behind them beating their wings and pushing the planets around an orbit.

Eventually, Newton offered another explanation based on the inverse square law of gravitation. In some sense, this "action at a distance" explanation is almost as mystical as a belief in angels; after all, no one can really

---

* *Rashomon* is a 1950 movie directed by Akira Kurosawa. Somewhat later (1964), Paul Newman starred in *The Outrage*, a movie with a similar plot structure.

explain *why* we have this gravitational force; it is just there. But using Ockham's razor, Newton's explanation is much more preferable. If one tries to support the angel hypothesis, one is confronted with a host of complicated questions: How big are these angels? How many are required per planet? What are their wings beating against?

When applied to mathematical modeling, the typical utilization of Ockham's razor leads us to remove as many components (or variables) as possible from the model while still maintaining the power of the model to represent the observables seen in the natural process. Unfortunately, as noted in [Br01a], the accuracy of a model and its simplicity are often in conflict. This is especially true when simplification goes too far and the accuracy of the model is weakened because it lacks necessary information.

1.5.3 Bellman

The "curse of dimensionality" is a phrase coined by Richard Bellman in 1961. If Ockham's razor is a plea for simple mathematical models, then Bellman's curse of dimensionality is a warning about the dangers of using overcomplicated models.

The description of the SCIENCE box in Figure 1.1 considered the function $h(x)$ to be defined over a set of vectors. Let us assume that the dimension of these vectors is n. In many problems, it will be required that an algorithm derive an approximation for $h(x)$, let us call it $g(x)$, based on observed functional values that are dependent on various values of x within the natural process. The curse of dimensionality arises when m, the number of observed samples, is not sufficiently large, and the algorithm cannot derive an accurate approximation function $g(x)$. In many applications, the dependency of $g(.)$ on the components of x is not easily understood. In some cases, inappropriate assumptions are made about the importance of components in x, and so it may have components that matter very little if at all in the determination of $g(.)$. An "overgenerous" selection of components for x will simply increase the dimensionality of x, causing an inflation of the volume of the mathematical space containing the samples. In this case, the density of sample points in the space becomes so rarefied that it is impossible to compute a reasonable $g(x)$ approximation.

To get a visual appreciation of the issue, consider an n-dimensional space containing data samples. Suppose that the data are contained in a hypercube with sides of unit length, and suppose further that we require a density of samples that are spaced at intervals of 0.1 units along all dimensions if we are to derive an accurate $g(x)$. If $n = 1$, we need 10 samples to cover a line of unit length, 100 samples to cover a square with sides of unit length, etc. If $n = 9$, this density requirement is asking for one billion samples, which is typically far more than most data sets provide. Bearing in mind that the estimated number of atoms in the entire known universe is only about 10^{80}, it can be appreciated that there will be serious problems well before $n = 80$. However, many mathematical models deal with x vectors that do have dozens or even hundreds of components. Consequently, there are only a few options that are reasonable:

a. Try to keep n as small as possible.

b. Devise algorithms that will work with sample data that are extremely sparse in a high dimensional space.

c. Use methodologies that combine both (a) and (b).

To address issue (a), researchers have developed various dimensionality reduction strategies, whereas others such as Breiman have devised algorithms such as random forests that optimistically view high dimensionality as a "blessing" [BR01b]. These strategies are beyond the scope of this text. We mention them in case the reader wants to pursue additional material.

1.5.4 Interpretability

Another issue in the formulation of a mathematical model is interpretability—the ability to inspect $g(x)$ with the objective of discovering how $g(.)$ depends on the components of x. For example, if $g(x)$ is predicting the binding affinities of ligands that are drug candidates, then knowing how the components of x are related to high affinities would help us to design new molecules for further testing. However, not all computational procedures allow this. The dependency of $g(.)$, on x may be so complicated that it is impossible to describe. For example, this is typical of neural nets that compute a value for $g(.)$ but the computations are so complicated that it is usually impossible to understand the role that is played by any single component of the input vector x.

The goal of interpretability is controversial, some researchers contending that striving for interpretability is overreaching when that goal leads to methods that weaken the accuracy of $g(x)$. The merits of this argument will depend on the needs of the application, and so we will defer any further discussion except to say that some algorithms do have this trade-off (good accuracy but not interpretable) and the application user should be aware of this limitation.

1.5.5 Refutability

The famous scientist Karl Popper (see [Po65]) asserted that a theory or hypothesis should be refutable or falsifiable if it is to stand as an assertion with scientific merit:

> The criterion of the scientific status of a theory is its falsifiability, or refutability, or testability.

If a proposition is accompanied by a reasonable test, the credibility of the proposition is enhanced if it can successfully meet the challenge of the test. For example, Einstein predicted that the mass of the sun could bend the light coming from distant stars. This gravitational lens effect was later observed, and so his general theory of relativity was not refuted. Refutability enhances the credibility of a scientific theory; it does not prove it beyond a shadow of doubt. As Einstein said:

> No amount of experimentation can ever prove me right; a single experiment can prove me wrong.

If there are no tests of refutability, then a theory is weaker. If it is beyond testability, then it can be accused of being arbitrary or dogmatic.

Another famous example of refutability was given by the British scientist J. B. S. Haldane when asked to pose a test that would refute evolution. His terse answer: "Rabbit fossils in the Precambrian."

1.5.6 Complexity and Approximation

By adopting a scientific methodology, we can produce powerful mathematical models suitable for computational analysis. Nonetheless, the disclaimer uttered by Kant and quoted earlier still remains to haunt us. Although the modern science of biology has discovered a great deal about that caterpillar and that tiny blade of grass, we also appreciate that there is still a lot to learn. Even these seemingly insignificant life forms involve complex systems with formidable complexity and subtle interactions.

This complexity will demand that various simplifications take place. In general, our analysis of a natural system will involve the analysis of a mathematical model that attempts to account for the phenomenon being studied. One must understand the limitations of the mathematical model in order to use it properly. For example, consider a very simple system: an object attached to the end of a spring. A textbook on elementary physics will give us Hooke's law that proposes a mathematical relationship between F, the restoring force exerted by the spring on the object, and x, the elongation of the spring. The simple linear relationship $F = -kx$, where k is the spring constant, is a reasonable first approximation of the system for small values of x, but it fails to accurately model the system when the spring is stretched to the breaking point.

Why is a simplified model necessary? Here are some reasons and associated examples:

- *A starting point for analysis*: A system may be so complicated that we make some simplifying assumptions just to get started with a reasonable analysis. For example, molecular phylogeneticists typically assume the existence of an evolutionary tree in order to build a somewhat simple mathematical model that can be subjected to various tree-based strategies of analysis. To accomplish this, they simply ignore the possibility of *reticulate evolution* that can arise due to lateral gene transfer in evolutionary processes. Introducing this type of gene transfer would require that the tree be replaced by a network, the structure of which is more difficult to infer from the existing evolutionary data. As another example, we can cite the simplified representation of a protein's conformation by using a data structure that keeps the coordinates of the alpha carbon atoms in the backbone chain while ignoring all other atoms. For some applications, this is a reasonable approximation if one only needs a rough estimate of protein shape.

- *Computational efficiency and modeling methods*: Our most accurate modeling of a molecule is provided by quantum physics. Using *ab initio* methods in computational quantum chemistry, we solve Schroëdinger's equation to calculate potential energy surfaces, transition structures, and various other molecular properties. Although we claim to have "solved" these equations, it is necessary to admit that the solution typically involves simplifying assumptions such as the Born–Oppenheimer approximation or the Hartree–Fock approximation. Despite these approximations, the solution strategies are computationally very expensive even for small systems involving tens of atoms. To handle systems with thousands of atoms, we typically resort to molecular mechanics, working with models that employ classical physics, relying on force-field equations with empirically determined parameters. For many applications, this is quite adequate, and the computational time is considerably reduced when compared with approaches that depend on quantum physics.

- *Discretization in the representation of continuous properties*: When we measure some changing phenomenon in a biological system, we are typically sampling the system as it undergoes some continuous change. Consequently, we get a discrete sequence of measurements that are meant to represent the entire continuous process. Consider a protein molecule that undergoes a change in its physical conformation when it binds with a ligand. In drug design, we might want to understand the exact nature of this change. Our modeling of the conformational change could be stored within a computer memory as a sequence of structures, each representing a "snapshot" of the conformation at some particular time. As long as the time intervals between snapshots are small enough, we can capture the movement reasonably well. Discretization is frequently used in science and engineering. The next time you are at 35,000 feet and looking out the window at that jet aircraft wing, you can contemplate that the partial differential equations used in the analysis of the wing surface were subject to a similar type of discretization, which, of course, can only approximate the continuous system. However, the error of the approximation is so small that it does not hamper final objectives, allowing aeronautical engineers to design aircraft with remarkable abilities. For example, the Airbus A330-200 has a maximum take-off weight that is over 500,000 pounds, and the wingspan is only 60.3 meters.

In summary, we see that mathematical models often involve approximations of one type or another. These are acceptable as long as we fully understand the nature of the approximation and its limitations. The limitation for Copernicus was that he would not (perhaps could not) see circular orbits as only an approximation. By insisting that circular orbits were "the Truth," it was necessary to concoct additional "adjustments" for the circle model so that it would more accurately predict data that was to match observations consistent with elliptical orbits.

1.6 SOURCES OF ERROR

Considering the modeling examples described in the previous sections, we can understand that the computational output from a computer program will be subject to various errors that accumulate. It is important to understand the sources of these errors if we are attempting to reduce their effect. In some cases, we simply acknowledge the presence of error as an acceptable shortcoming of the modeling approach or computational strategy that is being employed. Whether we are trying to reduce error or live with it, we should understand the source of error in order to gauge the magnitude of an error.

Referring back to Figure 1.1, we see that our main objective is the efficient approximation of a function $f(x, x')$ considered to be the "true" function that provides the solution to our scientific problem. After collecting data from the NATURE box, we generate a mathematical model that represents our scientific understanding of the natural process that generates f. This model includes the function $h(x)$ that we use as a function to approximate f. Often, the derivation of h is so complicated that it must also be approximated, this time by a function $g(x)$ that is considered as part of the COMPUTATION box.

An abstraction of the situation is depicted in Figure 1.3. In this figure, we see the function $f(x, x')$ as a single point in a high-dimensional infinite space that contains all such functions.

Our scientific analysis generates a function $h(x)$ that we consider to be a single point in a hypothesis space that contains an infinite number of such functions, some of which are reasonable approximations for $f(x, x')$. Considering the Rashomon effect, we might have several such functions. In the figure, we see a particular $h(x)$ that would represent a very good

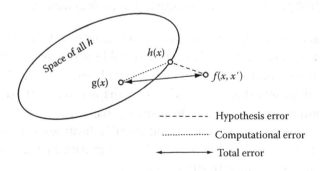

FIGURE 1.3 Sources of error in modeling.

approximation for $f(x, x')$ as it is a member of the space of all h that is closest to f. However, because it is not exactly the same as f, there is an error in the approximation. The magnitude of this *hypothesis error* is represented by the length of the dashed line connecting points $h(x)$ and $f(x, x')$. As mentioned earlier, it may not be possible to compute an exact version of $h(x)$, and so the figure contains a point representing the function $g(x)$, which approximates $h(x)$. The magnitude of this computational error is represented by the length of the dotted line. This computational error may be due to one or more of the following: round-off error, early termination of an iterative algorithm, discretization, or the intrinsic approximation error of an algorithm that uses a heuristic strategy to compute g. The solid line with the arrows represents the total error due to the approximation of f by g.

1.7 SUMMARY

We conclude this chapter by summarizing the main points of this scientific methodology:

- We want to solve a problem that requires the prediction of outputs of a function $f(x, x')$ that is known only through experiments and observations taken from a natural process (the NATURE box).

- In the most general case, the function $f(x, x')$ is part of a "black box" that hides from us the full details of the dependency of f on x and x'. Consequently, it must be approximated by a function that is within an infinite hypothesis space. The selection of this hypothesis space is dependent on the models, information, and knowledge that are assembled to form the SCIENCE box.

- The exact representation of the optimal $h(x)$ is also unknown, but we will use an algorithm that computes $g(x)$, which acts as an approximation of $h(x)$.

- Because $h(x)$ is unknown, our best strategy for the reduction of hypothesis error relies on the ability to choose the best hypothesis space possible. It must be rich enough in content to have a function $h(x)$ that is reasonably "close to" the target function $f(x, x')$. However, the hypothesis space must be simple enough to allow the computation of $h(x)$ in an efficient manner (the computation must run in polynomial time).

- Finally, even though $g(x)$ can be computed in polynomial time, we might still require a modification of the model that sacrifices a small amount of accuracy in exchange for a faster completion of the $g(x)$ computation.

1.8 EXERCISES

1. Suppose you are given the 2-D structure of a small molecule that has various stereoisomeric conformations. The term "2-D structure" means you know how atoms in the molecule are bonded but you do not know the 3-D positions of these atoms for a low-energy conformation of the molecule. To make this discussion less abstract, suppose that this "new" molecule is a simple $C_6H_{12}O_6$ sugar in its cyclical form. Now consider the following computational problems:

   a. Predict the coordinates of all the atoms for a stereoisomer of glucose (say, α-D-glucose).

   b. Predict the minimum potential energy of the various stereoisomers in a vacuum.

   c. Predict the minimum potential energy of the various stereoisomers in a crystal environment that includes water molecules (see [KK83]).

   d. Predict the melting point of a solid that is a pure form of some isomer.

   e. Predict the density of a solid that is a pure form of some isomer.

   Briefly describe the NATURE, SCIENCE, and COMPUTATION issues that arise for each problem. The problems have been ordered in terms of increasing difficulty, and your discussion should describe why the problems get progressively more difficult.

2. Both wet-lab experiments and computational analysis are prone to approximation errors and, in some rare cases, the final results may be completely wrong. Explain why we should place more trust in the laboratory experimental results.

3. Two assistant researchers, Ella Mentarry and Mel Praktiss, have proposed two mathematical models to explain the observed output of some natural process. Mel makes the following declaration: "We will pick the model that best represents the experimental data." Now

consider the following experimental data pairs with output $f(x)$ measured for particular values of input observable x:

i	1	2	3	4
x_i	1.00	1.80	2.00	4.00
$f(x_i)$	6.00	9.18	9.40	16.50

We will assume that $f(x_i)$ does contain some experimental error. The assistants produce the following models:

Mel's Model: $\quad f_M(x) = 1.330x^3 - 9.257x^2 + 21.86x - 7.936$

Ella's Model: $\quad f_E(x) = 3.473x + 2.630$

For Mel's model, the sum of the squares of deviations from the observed output is

$$\sum_{i=1}^{4} [f_M(x_i) - f(x_i)]^2 \approx 0,$$

whereas the Ellas model has

$$\sum_{i=1}^{4} [f_E(x_i) - f(x_i)]^2 \approx 0.131.$$

Note that Mel has constructed a polynomial to go through the data points. If you apply the methods and principles discussed in this chapter, which of the models would you choose as the best model for the natural process? Feel free to challenge Mel's declaration. Plotting the functions $f_M(x)$ and $f_E(x)$ should give some relevant observations about the best choice.

4. Read the following research paper:

J. A. BOARD JR., L. V. KALÉ, K. SCHULTEN, R. D. SKEEL, AND T. SCHLICK. Modeling biomolecules: Larger scales, longer durations. *IEEE Computing Science and Engineering* 1 (1994), 19–30.

Various problems are discussed in this paper. Select a problem and describe the issues that come under the headings of NATURE, SCIENCE, and

COMPUTATION. You may need to extend your discussion with material based on other research papers.

REFERENCES

[Ba05] D. A. C. BECK, R. S. ARMEN, and V. DAGGETT. Cutoff size need not strongly influence molecular dynamics results for solvated polypeptides. *Biochemistry*, **44** (2005), 609–616.

[Br01a] L. BREIMAN. Statistical modeling: the two cultures. *Statistical Science*, **16** (2001), 199–231.

[Br01b] L. BREIMAN. Random forests. *Machine Learning Journal*, **45** (2001), 5–32.

[Cc95] W. D. CORNELL, P. CIEPLAK, C. I. BAYLY, I. R. GOULD, K. M. MERZ, JR., D. M. FERGUSON, D. C. SPELLMEYER, T. F. FOX, J. W. CALDWELL, AND P. A. KOLLMAN. A second generation force field for the simulation of proteins, nucleic acids, and organic molecules. *Journal of the American Chemical Society*, **117** (1995), 5179–5197.

[Fe67] R. FEYNMAN. *The Character of Physical Law*. M.I.T. Press, 1967. (First published by The British Broadcasting Company, 1965).

[He88] K. HEIDER. The Rashomon effect: when ethnographers disagree. *American Anthropologist*, **90** (1988), 73–81.

[Hr85] B. E. HINGERTY, R. H. RITCHEY, T. L. FERRELL, AND J. E. TURNER. Dielectric effects in biopolymers: the theory of ionic saturation revisited. *Biopolymers*, **24** (1985), 427–439.

[Hu95] P. HUMPHREYS. Computational science and scientific method. *Minds and Machines*, **5** (1995), 499–512.

[Kk83] L. M. J. KROON-BATENBURG AND J. A. KANTERS. Influence of hydrogen bonds on molecular conformation: molecular-mechanics calculations on α-D-glucose. *Acta Cryst.*, **B39** (1983), 749–754.

[Ma98] A. D. MACKERELL, JR. Protein Force Fields, In *Encyclopedia of Computational Chemistry*. Eds.: P. V. R. Schleyer, N. L. Allinger, T. Clark, J. Gasteiger, P. A. Kollman, H. F. Schaefer III, P. R. Schreiner. John Wiley, Chichester, U.K., 1998.

[Ms97] D. A. MCQUARRIE AND J. D. SIMON. *Physical Chemistry: A Molecular Approach*. University Science Books, 1997.

[Po65] K. POPPER. *Conjectures and Refutations: The Growth of Scientific Knowledge*. Routledge, London, 1965.

[Sc02] T. SCHLICK. *Molecular Modeling and Simulation: An Interdisciplinary Guide*. Springer-Verlag, New York, 2002.

[Ss92] H. SCHREIBER AND O. STEINHAUSER. Cutoff size does strongly influence molecular dynamics results on solvated polypeptides. *Biochemistry*, **31** (1992), 5856–5860.

[Se90] R. SERWAY. *Physics for Scientists and Engineers with Modern Physics*. 3rd edition, Saunders College Publishing, Philadelphia, PA, 1990.

CHAPTER 2

Introduction to Macromolecular Structure

Since the proteins participate in one way or another in all chemical processes in the living organism, one may expect highly significant information for biological chemistry from the elucidation of their structures and their transformations.

EMIL FISCHER
Nobel Lecture, 1902

2.1 MOTIVATION

Life processes require several molecular families such as carbohydrates, lipids, DNA, RNA, and protein.* This final category of macromolecules is arguably the most important because it has the ability to contribute the most diverse functionality due to the structural complexity of its individual members. In fact, proteins act as enzymes that facilitate chemical reactions, regulators of cell processes, signal transducers enabling signals to be imported and exported into and out of a cell, transporters of small molecules, antibodies, structural elements that bind cells together, and motor elements to facilitate movement.

---

* In this text, we assume the reader is already familiar with the "Molecular Biology 101" topics such as cell structure, the central dogma of molecular biology, the genetic code, and the roles played by macromolecules such as DNA and RNA.

As such, proteins are the versatile building blocks and active molecules that form the components of all living systems. Proteins always act in concert with other molecules using a wide range of chemical reactions and mechanical reconfigurations. With over a billion years of evolution, these complex interactions have become sophisticated enough to accomplish all the coordinated events that constitute the life processes of a living cell. These evolutionary processes have derived structures, such as proteins, which are much more active than simply static macromolecular assemblies. Their composition must ensure proper folding to a stable functioning conformation with an eventual allowance for proteolysis. Proper functioning will mean that the protein interacts with other molecules in a very selective manner, demonstrating a high level of "molecular recognition."

We study protein structure and its dynamic changes so that we can better understand these various aspects of protein function. These studies will involve mathematical modeling of protein structure using geometric analysis, linear algebra formulations, and statistics.

2.2 OVERVIEW OF PROTEIN STRUCTURE

Before discussing protein functionality, let us review the main aspects of protein structure by considering the various levels of description: amino acid primary sequence, secondary structure, tertiary structure, and quaternary structure. To make these topics a bit more interesting for the many readers that already understand these concepts, we include a bit of history.*

2.2.1 Amino Acids and Primary Sequence

During the 1800s, chemists knew about amino acids and realized that each consisted of a side chain (also called an R group) attached to an alpha carbon that is also bonded to amine and carboxyl functional groups as shown in Figure 2.1.

The notion that proteins were linear chains of amino acids was proposed in 1902 by Emil Fischer and Franz Hofmeister. Both understood that amino acid residues were joined by a condensation reaction (see Figure 2.2).

By 1906, approximately 15 amino acids were known to be constituents of proteins, but there was considerable debate about the maximum length that a protein could achieve. It was considerably later that proteins were understood to be very long chains of amino acids. In fact, lengths vary

---

* For more historical information see [D105] for structural molecular biology and [Tr01] for proteins.

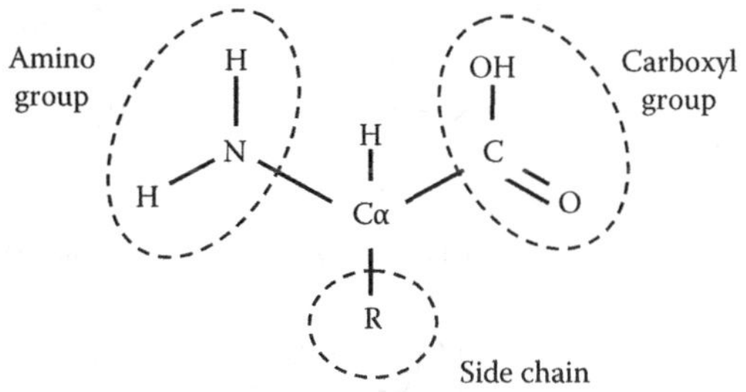

FIGURE 2.1   A typical amino acid.

from fewer than 20 amino acids to more than 5000 amino acids. It is interesting to note that, despite the earlier discovery of the peptide chain, many scientists, Fischer included, were initially quite skeptical that proteins could have a very lengthy macromolecular structure. Even as late as 1952, Fredrick Sanger regarded the "peptide theory" as very likely to be true but nonetheless only a hypothesis that had yet to be proven [FR79]. The identities of all 20 amino acids were not completely known until 1940.

We now understand that proteins are comprised of chains of such amino acids, and this chain can involve several hundred amino acids. The first level of structural description of a protein is called the **primary sequence** and is simply a list of these amino acids mentioned in their order of appearance, as one scans the protein starting with the amino end of the protein and proceeding to the carboxyl end.

In terms of chemical functionality, each of the 20 amino acids has its own "personality." In an effort to simplify descriptions of amino acid behavior, it is convenient to place amino acids into various categories. There are several ways to do this. As an example, Figures 2.3 and 2.4 place residues into four categories: nonpolar hydrophobic, polar uncharged, polar-charged positive, and polar-charged negative.

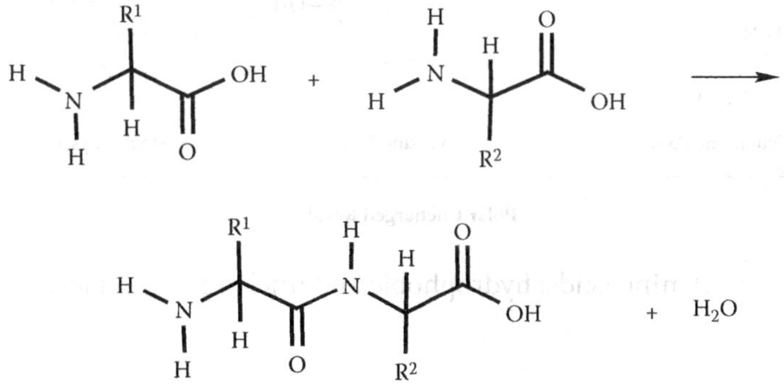

FIGURE 2.2   Condensation reaction.

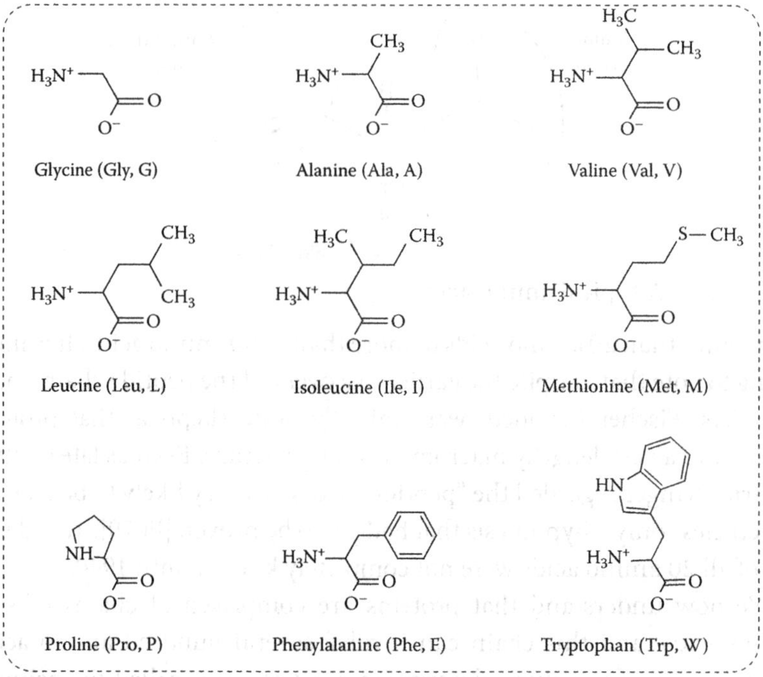

Nonpolar (Hydrophobic) Residues

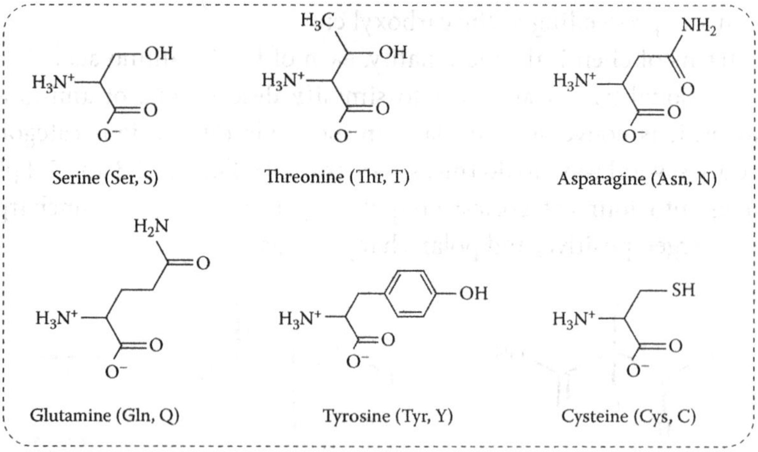

Glycine (Gly, G)

Alanine (Ala, A)

Valine (Val, V)

Leucine (Leu, L)

Isoleucine (Ile, I)

Methionine (Met, M)

Proline (Pro, P)

Phenylalanine (Phe, F)

Tryptophan (Trp, W)

Polar Uncharged Residues

Serine (Ser, S)

Threonine (Thr, T)

Asparagine (Asn, N)

Glutamine (Gln, Q)

Tyrosine (Tyr, Y)

Cysteine (Cys, C)

FIGURE 2.3 Amino acids: hydrophobic and uncharged residues.

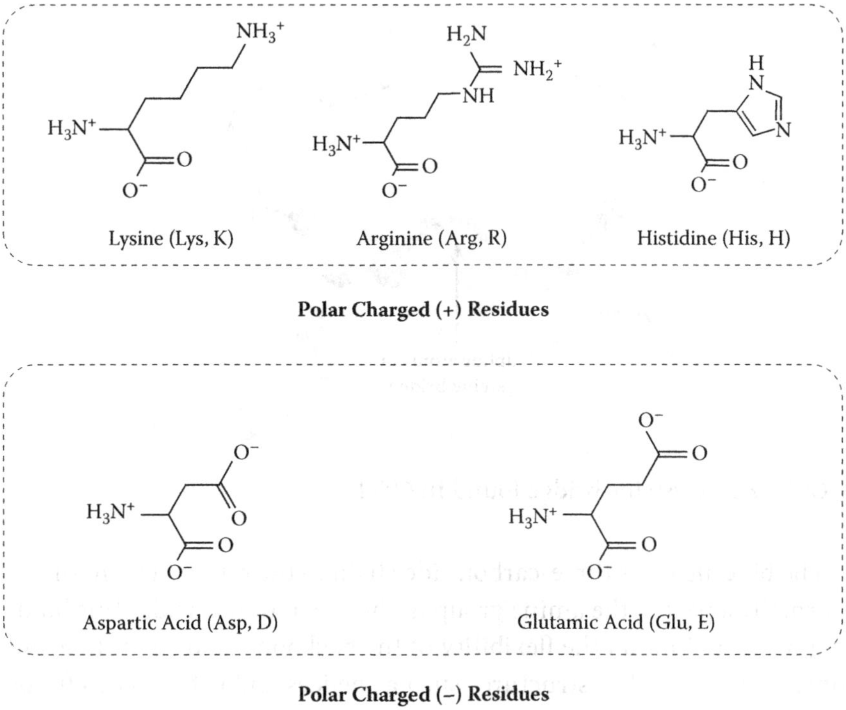

Lysine (Lys, K)   Arginine (Arg, R)   Histidine (His, H)

Polar Charged (+) Residues

Aspartic Acid (Asp, D)   Glutamic Acid (Glu, E)

Polar Charged (−) Residues

FIGURE 2.4 Amino acids: charged residues.

All of the residues are shown in zwitterion form (an ionic molecule with separate positive and negative charges having a net charge of zero).

The given categorization should be regarded as somewhat fuzzy. For example, histidine has a relatively neutral pKa, and so fairly small changes in cellular pH will change its charge. So, it could be placed in the polar uncharged group but, in Figure 2.4, it is put into the polar charged group because it can achieve a positive charge under physiological conditions. Similarly, glycine often appears in the hydrophobic category, but its side chain is essentially nonexistent, and so some biochemists will consider it to be in a category all by itself. It should be noted that hydrophobicity is of particular interest to researchers studying structural bioinformatics because it has been demonstrated that the hydrophobic properties of amino acids contribute to the 3-D conformations of proteins.

Some amino acids have "special features" that also contribute to structural conformation. Proline has a peculiar structure due to the

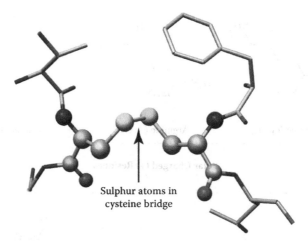

FIGURE 2.5   Cysteine bridge found in 6PTI.

cyclic binding of its three-carbon side chain to the nitrogen atom that is normally a part of the amine group in the other amino acids. This binding places a limit on the flexibility of the backbone, and so proline will often cause a bend in structures such as helices and beta strands (to be considered in the next section). It is more likely to appear at the edges of such secondary structures or in loops that go between them. Another amino acid with important structural contributions is cysteine. It is possible for two cysteine residues to form a disulfide bridge by means of an oxidative reaction:

$$R\text{-}SH + SH\text{-}R' \rightarrow R\text{-}S\text{-}S\text{-}R' + 2H^+ + 2e^-.$$

This allows two cysteine residues in different parts of the protein to form a bride that helps to provide some extra stability for the molecule. Figure 2.5 shows a cysteine bridge that has been extracted from the BPTI (bovine pancreatic trypsin inhibitor) with PDB code 6PTI.

In this figure, the cysteine residues including the two central sulfur atoms have been shown using a "ball and stick" visualization style, whereas the four residues adjacent to the cysteine residues are simply shown in "stick" form.

It has been noted that small proteins tend to have more cys–cys bridges than larger proteins. An interesting example is snake venoms that often have an unusually high preponderance of cys–cys bridges to enhance stability. Venom from the Malaysian krait contains the protein 1VYC

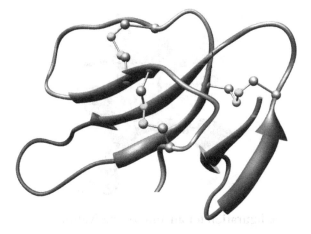

FIGURE 2.6   Cysteine bridges in bucain (a krait venom).

that has three such bridges, even though it is only 65 residues in length (see Figure 2.6).

In this figure, we see that the protein uses cysteine bridges to provide more positional stability to various loop substructures that would tend to be more floppy without the bridges.

The alpha carbon of all amino acids, except for glycine, constitutes a chiral center. In most biological systems, this chirality is described as being in the L-configuration (see Figure 2.7). Amino acids with a D-configuration are not found in the metabolic pathways of eukaryotic cells but have been observed in the structure and metabolism of some bacteria.*

In Figure 2.7, we see that bonds attached to the alpha carbon between the NH_2 and COOH groups are coming up from the page, whereas the other two bonds are going down into the page as they go from the alpha carbon to the hydrogen and beta carbon of valine. The D-configuration of this amino acid would be the mirror image of what is seen in Figure 2.7.

As a final note to end this section, we mention that almost all life-forms are restricted to the 20 amino acids listed earlier. However, recently it has been discovered that two other amino acids can appear, although rarely. They are selenocysteine and pyrrolysine. These unusual amino acids should not be considered as normal cysteine and lysine residues that have been chemically modified due to some posttranslational mechanism. The microbes that utilize these amino acids actually have specific transfer

---

* A list of D-amino acid residues in the Protein Data Bank can be found on the Internet at http://www-mitchell.ch.cam.ac.uk/d-res.html.

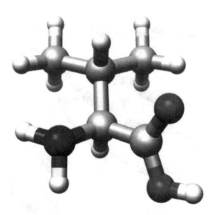

FIGURE 2.7   L-configuration for an amino acid (valine).

RNAs for them, and they are inserted into the growing peptide chain just like any other amino acids [MG00].

2.2.2   Secondary Structure

The next level of structural description is referred to as **secondary structure**. The idea behind secondary structure is to simply characterize residues of the primary sequence, putting them into categories that reflect the most obvious description of the shape of the peptide chain. There is no attempt to establish the physical positions of residues in 3-D space as is done in the specification of tertiary structure. Generally, there are three categories of secondary structure: alpha helices, beta strands, and loops.

The last two types of structure are mainly facilitated by the formation of hydrogen bonding between atoms in the backbone. The nitrogen atom in the backbone acts as a hydrogen donor, and the carbonyl oxygen acts as a hydrogen acceptor.

2.2.2.1 Alpha Helices

Alpha helices are formed when the backbone of the amino acid chain winds into a helical conformation with hydrogen bonds used to stabilize the structure. Figure 2.8 shows the formation of one such hydrogen bond. In this figure, the alpha carbon atoms have been numbered so that one can see how the acceptor and donor are related to one another. Specifically, the carbonyl oxygen of residue #1 acts as the hydrogen acceptor for the nitrogen donor in residue #5. The alpha helix is distinguished by this pattern: residue n interacting with residue $n + 4$ through a hydrogen bond.

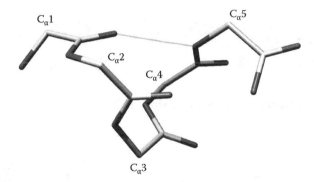

FIGURE 2.8 (See color insert following page 72) Formation of a hydrogen bond in an alpha helix.

This arrangement gives the alpha helix 3.6 residues per turn. So, if one views the helix from a perspective, which is parallel to its axis, one sees an angular increment of approximately 100° between successive alpha carbon atoms.

It is also possible to have helical conformations that are more elongated. The 3_{10} helix has a hydrogen bond between the carbonyl oxygen of residue n and the nitrogen atom in residue $n + 3$. This gives 3 residues per turn and is a type of secondary structure that occurs (rarely) in very short segments or at the ends of an alpha helix. The very rare pi helix has a hydrogen bond between residues n and $n + 5$ with 4.4 residues per turn. In this case, the helix is wide enough to have a hole along the axis of the structure, and this is less energetically favorable when compared to the more compact alpha helix.

Looking at Figure 2.8, we can pretend to grasp the helix with the right hand so that fingers are pointing in a direction that is the same as that specified by an increase in residue number. If the thumb is pointing toward the carboxyl end of the protein, then we have a right-handed helix; otherwise, it is designated to be a left-handed helix. The latter is also a rare event because it forces L-amino acid side chains into an excessively close proximity with carbonyl groups.

Figures 2.9, 2.10, and 2.11 show an alpha helix that has been extracted from the protein 1QYS. The first figure shows only the backbone atoms with longer roughly horizontal lines representing the hydrogen bonds. The side chains have been removed because they would certainly obscure a view of the helical nature of the structure. Even with this simplification, it is still difficult to assess the helical structure, and so Figure 2.10 presents a ribbon model of the structure. Hydrogen bonding as defined

FIGURE 2.9   Hydrogen bonds in an alpha helix.

FIGURE 2.10   Ribbon diagram of a helix with hydrogen bonds.

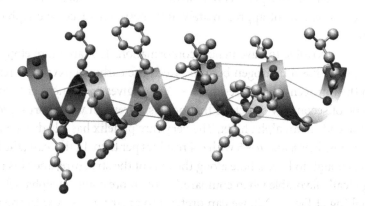

FIGURE 2.11   Ribbon diagram with side chains.

by the Chimera visualization software has been left intact. In many other figures to come, we will see ribbon diagrams that use a plain "linguini" type of ribbon to illustrate a helix. Finally, to provide another visual representation, Figure 2.11 shows a ribbon structure with the side chains sprouting off the helix.

It is interesting to note that Linus Pauling determined the structure of the protein helix in 1948. He is quoted as saying, "I took a sheet of paper and sketched the atoms with the bonds between them and then folded the paper to bend one bond at the right angle, what I thought it should be relative to the other, and kept doing this, making a helix, until I could form hydrogen bonds between one turn of the helix and the next turn of the helix, and it only took a few hours doing that to discover the alpha-helix." Pauling realized that the side chains would point outward from the protein helix.* In 1953, Pauling and Corey [Pc53] published a paper that proposed a somewhat similar structure for DNA based on a triple helix structure. The reader may consult a report written by Stephen Lawson (see http://lpi.oregonstate.edu/ss03/triplehelix.html) to get a balanced perspective on this paper "which would turn out to be one of the most famous mistakes in 20th-century science." In 1954, Pauling was given the Nobel Prize for Chemistry to recognize his work on molecular structure. Pauling was also known for his significant efforts on behalf of world peace and other issues related to humanitarian goals and social justice.

2.2.2.2 Beta Strands

Similar to alpha helices, beta strands are secondary structures formed by hydrogen bonding. In this case, however, the consecutive amino acids interact with other consecutive amino acids that may be quite distant in the primary sequence. The hydrogen bonds provide a sequence of linkages between two chain segments, which may have one of two orientations with respect to one another, as follows:

- Parallel beta strands; the consecutive hydrogen bonds are such that the indices of amino acids on either end of the bonds are increasing in the same direction.

- Antiparallel beta strands; the consecutive hydrogen bonds are such that the indices of amino acids on either end of the bonds are increasing in opposite directions.

The following figures provide more details.

In Figure 2.12, we see the backbone atoms of three beta strands. Side chains have been removed for clarity, but the carbonyl oxygen atoms have been inserted in order to get the hydrogen bonds. The amino acids for each strand have their indices increasing, as we go from left to right. This is indicated by the amide group on the far left and the carboxyl group on the extreme right. Notice how the hydrogen bonds are formed, the

---

* As noted by Anna Tramontano [Tr08]: the reason why Pauling could predict the alpha helix first was because he was not trapped in thinking that the number of residues per turn had to be an integer number. This shows how common assumptions could mislead a lot of people! Also, he was the one who demonstrated the planarity of the peptide bond, which is essential to come up with the model.

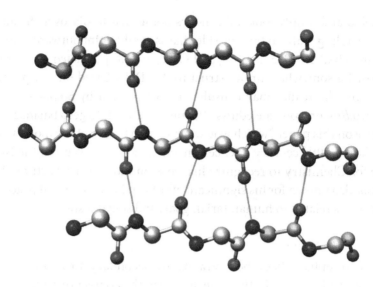

FIGURE 2.12   Hydrogen bonds in parallel beta strands.

carbonyl oxygen atoms being hydrogen bond acceptors, whereas the nitrogen atoms of the adjacent strand act as hydrogen donors.

Figure 2.13 presents a more simplified diagram of the same three strands, each with a rounded-ribbon arrow indicating the direction of increasing indices in the primary sequence.

Figure 2.14 shows the same backbone atoms as those seen in Figure 2.12 but with oxygen atoms and hydrogen bonds removed. The view has been rotated so that the strands are almost behind one another and collectively they define a beta "sheet." The alpha carbon atoms have been colored black so that they are more obvious in the figure. The importance of this view

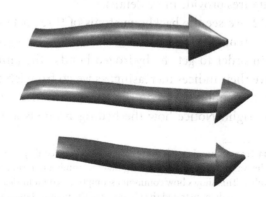

FIGURE 2.13   Ribbon diagram for parallel beta strands of Figure 2.12.

FIGURE 2.14   Side view of the previous ribbon diagram showing pleated conformation.

is that it shows how the strands tend to rise and fall in unison. In this way, the strands form a *pleated* sheet (sometimes described as a corrugated sheet). A significant consequence of pleating is that the side chains cannot originate from an alpha carbon that is in a valley of the pleat if they are to avoid steric clashes. In other words, they will seem to sprout from the alpha carbons in such a way as to avoid the atoms on either side of the alpha carbon. Consequently, the side chains originate from the peaks of the corrugations, and so they appear to line up in rows when the sheet is observed on edge. This lineup effect is shown in Figure 2.15.

Figure 2.15 shows the same three ribbons but with side chains present and drawn as "stick" structures. The scene has been rotated so as to make it more apparent that the three strands form a "sheet." As just described, the side chains appear to sprout in rows that are determined by the pleats of the sheet, and because of the nature of the pleats, these rows of residues alternate above and below the sheet. The ribbons are drawn in such a way as to eliminate the pleated structure of the sheet, but the flatness of the ribbons brings out another significant observation: the sheet is not perfectly flat; it has a slight twist. This twist is not a peculiarity of this protein*; rather, it is typical of many beta sheets.

Let us now consider a set of beta strands that have an antiparallel conformation. All of the figures for this discussion have been prepared by viewing an antiparallel beta sheet within the protein 1QYS. As before, the beta strands make up a sheet, but now they run in directions that alternate as illustrated in Figure 2.16. The hydrogen bonds that stabilize this sheet are shown in Figure 2.17. The reader should compare the appearance of

---

* Figures 2.12 to 2.15 were done with UCSF Chimera, selecting a beta sheet from protein 1AY7.

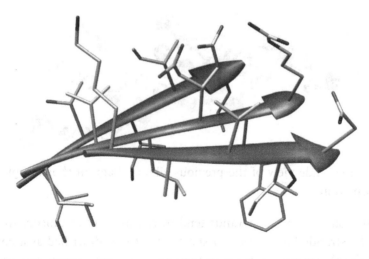

FIGURE 2.15   Ribbon diagram with side chains for parallel beta strands of Figure 2.12.

these hydrogen bonds with that of Figure 2.12. In general, the successive hydrogen bonds that join two beta strands in an antiparallel configuration tend to be more uniform in their direction. They do not exhibit the slightly slanted appearance of hydrogen bonds in a parallel beta-strand sheet as seen in Figure 2.12. The antiparallel construction of the sheet illustrated in Figure 2.17 has a pleated structure that is very similar to that seen in beta strands, which make up parallel sheets. If the structure in Figure 2.17 is viewed from a line of sight, which is almost along the

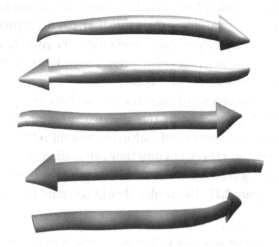

FIGURE 2.16   Ribbon diagram for antiparallel beta strands.

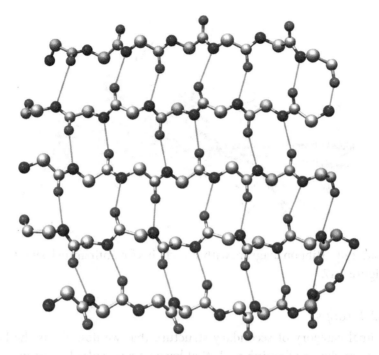

FIGURE 2.17 Hydrogen bonds in antiparallel beta strands.

surface of the sheet, the observer would see an arrangement of strands that is very similar to that of Figure 2.14. Because of this pleated formation, we can observe the same patterns as before when looking at the orientations of the side chains. They seem to sprout off the sheet in rows, and the successive rows alternate in their directions above and below the sheet. This is seen in Figure 2.18.

Sheet twist can also be observed for antiparallel beta strands. In fact, researchers have collected empirical evidence that such deformations are typically larger for antiparallel sheets than for parallel sheets (see [Em04] and [Hc02]). Figure 2.18 shows an illustration of this twist for the antiparallel case.

It should be stated that the ribbon diagram is very suitable when illustrating a twisted sheet. A diagram that shows all the atoms (without ribbons) is very complicated, and the sheet structure becomes hidden in a forest of atoms that hide this pattern.

It is possible for a beta sheet to have a mix of both parallel and antiparallel strand pairs. An example of this can be seen in the salivary nitrophorin protein 1YJH.

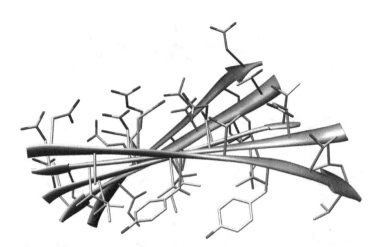

FIGURE 2.18   Ribbon diagram with side chains for antiparallel beta strands of Figure 2.17.

2.2.2.3 Loops

The final category of secondary structure that we describe is the loop. Loops are chains of amino acids that have no particular hydrogen bonding patterns with other parts of the protein. There are proteins that consist of nothing more than a loop stabilized by disulfide bridges. A good example of this is flavoridin (PDB code 1FVL), which is a blood coagulation inhibitor from the habu snake. Another example is (PDB code 1ANS), a neurotoxin from the sea anemone, *Anemonia sulcata*. This protein has three disulfide bridges. Although it does not contain any helices or beta sheets, it does have seven beta turns.

In Figure 2.19, the loop is portrayed as a ribbon diagram, whereas the disulfide bridges have been drawn in ball-and-stick form.

Proteins that are "all loop" are fairly rare. Most proteins will have beta sheets and helices forming a hydrophobic core, and these secondary structures will be interconnected by loop segments. An example of this is given in Figure 2.20 where we see short loops connecting the adjacent strands of an antiparallel beta sheet within the *E. coli ribonuclease* 1V74. This is the simplest way to connect such strands; it is possible to find other more complicated interconnections for adjacent strands involving helices and other beta strands.

For a sheet made of parallel beta strands, the adjacent strands may be connected with simple loops. In many cases, one may find that helices and other beta strands are used to connect adjacent antiparallel strands.

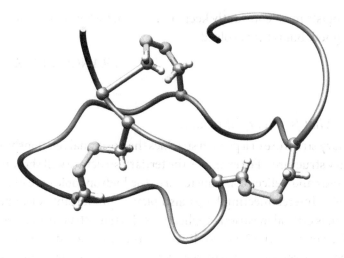

FIGURE 2.19   Loop structure and stabilizing disulfide bridges of the neu-rotoxin 1ANS.

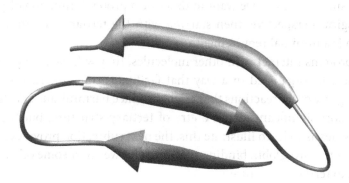

FIGURE 2.20   Loops connecting adjacent beta strands in an antiparallel sheet.

2.2.3 Tertiary Structure

Perhaps the most remarkable features of the molecule are its complexity and its lack of symmetry. The arrangement seems to be almost totally lacking in the kind of regularities which one instinctively anticipates, and it is more complicated than has been predicted by any theory of protein structure.

J. C. KENDREW ET AL., DESCRIBING MYOGLOBIN, 1958

Biologists must constantly keep in mind that what they see was not designed, but rather evolved.

FRANCIS CRICK, 1988

2.2.3.1 What Is Tertiary Structure?

The tertiary structure of a protein specifies the positional relationships of all secondary structures. For example, the tertiary structure will show how the beta strands and helices combine to form the hydrophobic core of a globular protein. Interconnecting loops also play an important role, and their placement is critical in various applications. Tertiary structure is essentially established once all the 3-D coordinates of the atoms are determined.

In the next section, we will see that tertiary structures can combine to form larger quaternary structures. However, many proteins are single peptide chains and can be described by specifying the tertiary structure of that single chain. If we want to describe a protein's functionality from a biological perspective, then starting with the tertiary structure of the protein is a minimal requirement.

As proteins interact with other molecules, they will have binding sites that must be organized in a way that facilitates some interaction while avoiding any other reactions that might produce harmful side effects. This is the most significant raison d'être of tertiary structure, but there are other issues as well. To illustrate this, the next subsection provides a quick sketch of the myoglobin binding site, and later we cover some other topics related to tertiary structure.

2.2.3.2 The Tertiary Structure of Myoglobin

The 3-D structure of myoglobin provides an interesting example of a tertiary structure. It has a biological functionality that has been extensively studied and also has the distinction of being the first protein structure to be determined by x-ray analysis. This was reported by Kendrew et al. in 1958 (see [KB58]). The structure of the myoglobin molecule was not deposited into the Protein Data Bank until later, because the PDB was not established until 1971. Myoglobin appeared there in 1973 with identifying code 1MBN.

Figure 2.21 provides a view of the entire molecule. It has 153 residues that form eight helices interspersed with short loops. These helices are combined into a special arrangement called a *globin fold*. It consists of

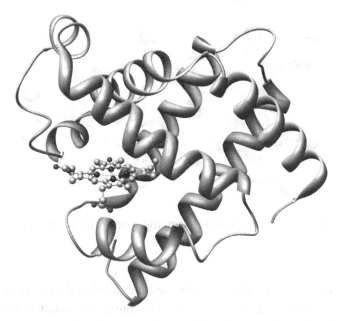

FIGURE 2.21   Tertiary structure of myoglobin (PDB code 1MBN).

a hydrophobic pocket that contains a structure called a *prosthetic heme group* (see Figure 2.22). The heme group for this figure was extracted from the PDB file for oxymyoglobin (PDB code 1MBO), which has a molecule of oxygen hovering over the iron atom at the center of the heme group. The function of myoglobin is to carry oxygen in muscle tissue. More specifically, it stores oxygen and carries it to the mitochondria for oxidative phosphorylation. For x-ray analysis, both proteins (1MBN and 1MBO) were extracted from the muscle of sperm whale (*Physeter catodon*), a mammal that has this protein in great abundance because of the necessity to stay under water for long periods of time.

At the center of the heme structure is an iron atom that is coordinated by the nitrogen atoms of the four pyrrole rings. In normal myoglobin, the iron atom is ionized to the Fe^{2+} (ferrous) state. The heme group is not symmetric and contains two highly polar propionate side chains, which appear on the right side of Figure 2.22. A close inspection of Figure 2.21 will reveal that these carboxylic acid side chains (ionized at physiological pH) avoid the interior of the hydrophobic pocket. This is also shown with more clarity in Figure 2.23. In this presentation, a surface has been put over the protein, and it has been colored so that portions of the surface near hydrophobic residues (as specified in Figure 2.3) are white, whereas

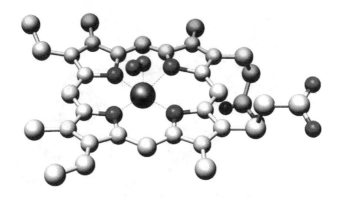

FIGURE 2.22 (See color insert following page 72) Heme group from 1MBO.

any surface over a nonhydrophobic residue is colored blue. If you look carefully at Figure 2.23, you can see a blue bump immediately under the iron atom. This is the NE2 nitrogen atom of a histidine residue (H93). There is another blue bump just above the oxygen molecule due to another histidine residue (H64). These will be discussed further when we look at Figure 2.25.

Figure 2.24 is provided so that you can get a better appreciation for the construction of the heme binding pocket. This diagram shifts the

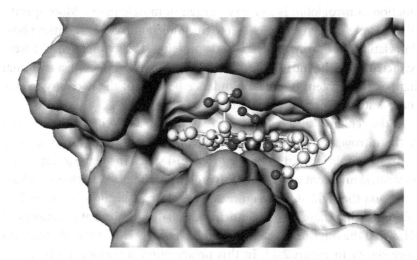

FIGURE 2.23 (See color insert following page 58) Heme group in its binding pocket.

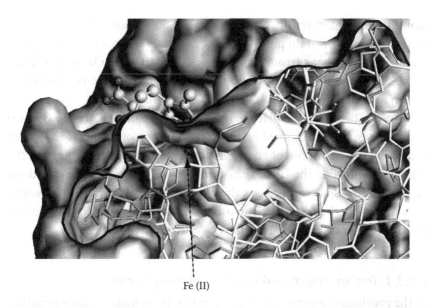

FIGURE 2.24   (See color insert following page 72) Heme group in pocket—cutaway view.

perspective of the observer so that the heme group is being reviewed from a position that is below the plane of the heme. Furthermore, a portion of the surface has been cut away to reveal the residues that create the surface. The heme group (above the surface but in the pocket) is drawn in a ball-and-stick fashion, whereas all structures below the surface are drawn in stick form. Hydrophobic residues are colored white, and all other residues are colored in blue. The dashed arrow indicates the position of the iron

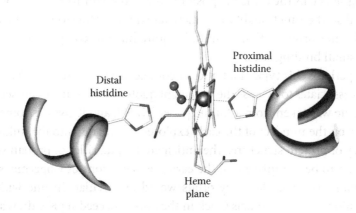

FIGURE 2.25   Heme group and histidine residues.

atom. If you look carefully, you can see H93 just below and to the left of the iron atom. The main point of this diagram is to see the several hydrophobic residues that surround the underside of the heme binding pocket. It should be noted that these nonpolar residues protect the Fe^{2+} from oxidation to the Fe^{3+} (hematin) state that will not bind O_2.

Figure 2.25 shows the placement of the iron atom relative to the two histidine residues mentioned earlier. Coordination of the heme iron atom is done by the NE2 atom of H93 (called the "proximal" histidine) and the NE2 of H64 (called the "distal" histidine). The distal histidine crowds in on the binding space and causes the oxygen molecule to bind at an angle. This makes the bond to the iron atom weaker, and so O_2 is more easily removed when needed. More details can be found in [PH80].

2.2.3.3 Tertiary Structure Beyond the Binding Pocket

In the previous subsection, we have seen that myoglobin tertiary structure provides a binding site that not only holds the heme group with its nonpolar residues but also has a particular geometry that manages the storage of O_2 in a very precise fashion. It is fairly clear that all the residues near the binding site participate in the support of these activities. However, there are many other residues that are considerably farther away from the binding site. It is reasonable to ask about the participation of these residues in the primary functionality just described. Are they really needed? To add more significance to the question, it should be noted that the protein cytochrome B_{562} from *E. coli* is only 106 residues in length, and it also has a binding site with a heme group. To be fair, it should be noted that the B_{562} binding site lacks the critically positioned distal histidine seen in myoglobin, and so the functionality is somewhat simpler. However, it is possible to find other proteins that seem to be quite huge in comparison to their rather small binding sites.

It is perhaps this type of additional complexity that prompted Kendrew and his associates to express the statement quoted at the start of this section. This quote was taken from a *Nature* paper [KB58], and it gives us some perspective on the mindset of the early explorers venturing into the unknown territory of protein structure. The anticipated regularities of protein structure seem to be reminiscent of the earlier expectations of Copernicus and his assumption that planetary orbits would be circular. In line with the admonition made by Francis Crick in the quote succeeding Kendrew's, our current thinking provides a somewhat different perspective on the structure of the myoglobin molecule: It is the result of an evolutionary process driven by functional requirements with no concern for elegant design.

At the macroscopic level, we see many examples of this function-driven evolution with problematic design principles. For example, breathing and food ingestion share a common tract. Reproductive systems and urine elimination share the urethra in male mammals. There are drawbacks to this tissue sharing: breathing will be accompanied by choking if food is taken into the trachea. In the case of shared reproductive and urinary systems, we can consider the unfortunate psychological associations due to such an arrangement. Even though tissue sharing provides a kind of time-sharing efficiency, this type of "tissue parsimony" does not always occur in the evolutionary process: everyone is born with a vestigial appendix that is not needed, and men have nipples.

At the molecular level, we may also see redundant substructures in a protein simply because evolutionary processes have maintained them as legacy contributions inherited from an earlier protein (see [KD99] in which the authors entertain the idea that a particular folding topology may reflect a *vestigial structural remnant* of an ancestral SH2-containing protein).

Even though there is the possibility that various helices or strands might simply constitute vestigial "excess baggage," there are other more significant reasons to have extensive and very complicated structure in a protein.

The functionality of tertiary structure relates to the following issues:

1. *Formation and management of the binding site.* The residues (usually hydrophobic) create a pocket that promotes the docking of the protein with a ligand molecule. As just discussed in the myoglobin example, the positions of particular atoms in the binding site may be critical for the optimal functioning of the reaction. There is also the necessity of forming a binding site that does not combine with other ligands in a way that produces a deleterious effect. Protein evolution can do only so much in this regard and may not protect a cell against toxins that are relatively new to the organism. For example, hemoglobin will avidly take on carbon monoxide in lieu of oxygen molecules with disastrous results.

2. *Flexibility.* There are situations in which a protein interacts with a ligand or another protein, and this interaction is accompanied by a change in protein conformation that facilitates the reaction with yet another molecule. Allosteric proteins demonstrate this functionality, and the

overall change in the protein's conformation may involve residues that are quite distant from the binding site that initiates the change.

3. *Control of a correct folding pathway.* In some sense, the folding of a protein involves many of the forces that contribute to the formation of a binding site. However, folding is functionality that must be accomplished *before* the binding site can be established. As such, the tertiary structure must have a level of organization that will ensure the correct fold. It has been experimentally observed that various peptide sequences will simply not fold. For those that do fold, it may be necessary to have a primary sequence with key residues that foster particular intermediate transitory folds acting as transitional conformations. Presumably, these are necessary to guide the folding process so that the final conformation is indeed the lowest energy conformation. The reader may consult [SK00] for an opinion that stresses the importance of folding pathways and [Do03] for a discussion about transition state conformations that act as a quality control mechanism in protein folding. A discussion of ultrafast and downhill protein folding may be found in [Dy07].* Models for folding and misfolding can be found in [Do06]. Another aspect of the correct folding of proteins deals with cotranslational folding. This involves the assumption that proteins start to take on significant aspects of their tertiary structure during translation when they are still attached to the ribosome. Such an assumption also has implications about the folding pathway (see [FB97] and [Ko01]). It should be stressed that there is still considerable controversy about the various phases of the folding process (for example, see [LB98]). Much of this should be settled as more data becomes available.

4. *Stability of the protein.* After the completion of folding, a complex interaction of the residues must ensure that the final structure will be maintained without the possibility of a different conformation being formed. In the ideal situation, a protein folds to a tertiary structure, which is characterized as the lowest energy conformation, and it stays in this conformation except for any flexible changes that are necessary for its functionality. As described earlier, disulfide bridging enhances the preservation of conformation. In less than ideal

---

* Issue 1, vol. 17, of *Current Opinion in Structural Biology* has several papers devoted to protein folding issues.

situations, it is possible that the protein undergoes a refolding that produces a different conformation. The endogenous cellular prion protein PrP^C that can undergo a conformational change to become the scrapie isoform PrP^{Sc} gives a good example of such metastable behavior. The PrP^C conformation contains 43% alpha helix and no beta sheets, whereas the PrP^{Sc} form contains 20% alpha helix and 34% beta sheet. Unfortunately, PrP^{Sc} has a tendency to form aggregates in cerebral tissue, and existing proteolytic mechanisms do not clear these aggregates. With unchecked accumulation, neurological damage occurs, ultimately leading to death.

These points emphasize that tertiary structure is necessarily complex because it must do a lot more than simply provide a foundation for the binding site. The additional responsibilities of tertiary structure include the support of delicate mechanisms that foster the proper behavior of the protein over its entire "life cycle." This includes its creation by means of a guided folding pathway, the stable maintenance of its conformation and binding site, and its final susceptibility to a degradation process that will lead to the reutilization of its amino acids.

Meeting all of these requirements is a complex optimization problem with multiple objectives. Furthermore, it is usually quite difficult to explain the functional significance of any one amino acid in terms of its contribution to one or more of the four issues just described. The proximal and distal histidine residues in myoglobin, various other residues that act as secondary structure "breakers," and particular residues that form hairpin structures (see [IA05]) are exceptions to this statement, but in the general case, a particular residue serves along with several other amino acids to define the various functional aspects of the tertiary structure.

Despite the possible lack of regularity in a tertiary structure, the fold appearance of the structure may be replicated in another protein even though its primary sequence is reasonably different. In Figure 2.26, we can see an obvious similarity between myoglobin and a single chain within hemoglobin, which is also an oxygen carrying protein. This structural similarity is present even though the primary sequences have only 29% of their primary sequences matching identically. As noted in [OM97]: "At 30% sequence identity, proteins will almost certainly have the same overall fold." This is especially significant for the globin family where similar folds have been seen for even lower sequence identities (less than 15%). The alignment (Figure 2.27) was done using the TCoffee server (see [Po03]).

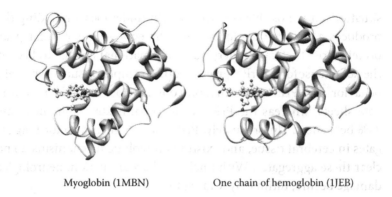

Myoglobin (1MBN)   One chain of hemoglobin (1JEB)

FIGURE 2.26   Structural comparison of myoglobin and a single chain of hemoglobin.

This example underscores the observation that tertiary structure is more conserved than sequence in an evolutionary descent.

Even though there are several amino acids that can be changed while still maintaining the functionality of the globin fold, there is a critical location at which a mutation can produce a significant change in functional behavior. The reader may note that the sixth residue of 1JEB is "E" for glutamic acid. Sickle-cell anemia is an inherited disease that leads to single mutational change in which this glutamic acid is replaced by valine. This produces a small hydrophobic patch on the surface of the protein. Unfortunately, this patch is able to bind with a hydrophobic pocket in the deoxygenated conformation of another hemoglobin molecule. This "stickiness" of the hemoglobin proteins causes a clumping that leads to distortion of the red blood cell, which then takes on a "sickle-cell" appearance. These blood cells have a shorter life span than normal cells, and so the patient has anemia caused by low levels of hemoglobin. Sickle cells can also become trapped in small blood vessels, thus preventing blood flow.

```
1MBN   VLSEGEWQLVLHVWAKVEADVAGHGQDILIRLFKSHPETLEKFDRFKHLKTEA
1JEB   SLTKTERTIIVSMWAKISTQADTIGTETLERLFLSHPQTKTYFPHFDLHPGSA

1MBN   EMKASEDLKKHGVTVLTALGAILKKKGHHEAELKPLAQSHATKHKIPIKYLEFI
1JEB   QLRA------HGSKVVAAVGDAVKSIDDIGGALSKLSELHAYILRVDPVNFKLL

1MBN   SEAIIHVLHSRHPGDFGADAQGAMNKALELFRKDIAAKYKELGYQG
1JEB   SHCLLVTLAARFPADFTAEAHAAWDKFLSVVSSVLTEKYR------
```

FIGURE 2.27   (See color insert following page 72) Sequence comparison of myoglobin and a single chain of hemoglobin.

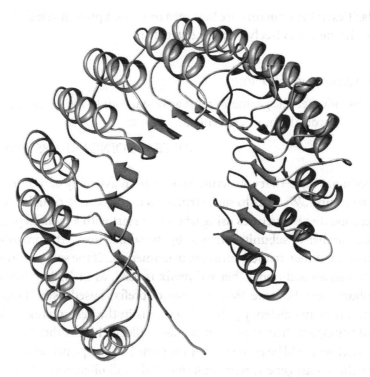

FIGURE 2.28 (See color insert following page 72) A tertiary structure involving repeated secondary structures (chain D of 1A4Y).

As a final note on tertiary structure, Figure 2.28 is presented to illustrate a protein that does have a remarkable amount of regular structure. This is from a human placental ribonuclease inhibitor that binds the blood vessel-inducing protein human angiogenin (see [PR97]). The angiogenin molecule is not displayed in order to show the extraordinary repetition of alpha helix and beta sheet structures that combine to form the large "horseshoe" tertiary structure. Although there is an extensive beta sheet present, it does not have the same twist that is often seen for smaller sheets. The beta strands have collected into a cylindrical formation that almost goes full circle. The regularity of their positions seems to be coordinated by the "hair roller" placement of the helices in the structure.

Many other remarkable folds can be seen in [OM97] and especially [PR04]. Space limits the presentation of these diagrams, but the interested reader may fetch and display the ribbon diagrams for the more picturesque titles: β Barrel (2POR), β 7 Propeller (2BBK chain H), β 2 Solenoid (1TSP), and $\alpha\beta$4-Layer Sandwich (2DNJ chain A). For most of these examples, the regularities in the structure are best seen by using a protein visualizer that allows the viewer to freely rotate the structures.

2.2.4 Quaternary Structure

It has adopted the geometry most advantageous to the species or, in other words, the most convenient.

JULES H. POINCARE (1854–1912)

A protein has quaternary structure when it is an assembly of multiple tertiary structures. We refer to such structures as oligomeric complexes, and we describe these structures using labels that start with a prefix representing the number of subunits followed by the suffix "mer." The progression is monomer, dimer, trimer, tetramer, pentamer, etc. If the subunits are different, then an additional "hetero" prefix is used, as in "heterodimer." If the subunits are the same, then the "homo" prefix is used, as in "homotetramer." From an efficiency point of view, it is in the best interest of a cell to build complex structures from smaller building blocks that are all the same. Each unit will be prescribed by the same DNA sequence, and so there is a parsimonious genetic representation of the full oligomeric structure.

In some cases, the quaternary structure can show a low level of recursive assembly. For example, three heterodimers may be assembled to form a "trimer of dimers." A beautiful example of this construction is the insulin protein illustrated in Figure 2.32. The assembly of this hexamer is a bit complicated, so we will explain it as a series of steps starting with the monomer illustrated in Figure 2.29. This is a single molecule made up of two peptide chains that are joined together using two disulfide bridges. There is a third disulfide bridge, but it connects two cysteine residues within the same chain. To see how the three disulfide bridges interconnect the residues in the three chains, consider Figure 2.30.

Two of these monomers make up the insulin dimer by having their beta strands combine in an antiparallel fashion to make a small beta sheet. This is illustrated in Figure 2.31 where the A and B chains appear in the background with C and D chains in the foreground. Chains C and D correspond to chains A and B in the dimer assembly, and they have the same type of disulfide connectivity. In Figure 2.31, all disulfide bridges have been removed for clarity.

Figure 2.32 shows the full insulin hexamer made up of three dimers each one of the same construction as that shown in Figure 2.31. The

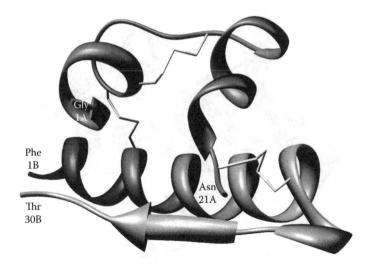

FIGURE 2.29 (See color insert following page 72) The insulin monomer has two chains connected by disulfide bridges.

positions of the three dimers are coordinated by means of two zinc atoms. To see how the zinc atoms are related to a dimer, they have been included in Figure 2.31.

To understand how the dimers combine to make a hexamer, imagine taking the dimer of Figure 2.31 and tilting the entire assembly so that your line of view is such that the darker zinc atom is exactly behind the brighter zinc atom. This will put the dimer into the exact orientation that the upper third of the hexamer shows in Figure 2.32. Notice that in this orientation, the beta sheets are seen on edge, and so they are somewhat difficult to discern.

There are many other examples of quaternary structure, some of which we will see later.

A-Chain   21 amino acids

G I V E Q C C T S I C S L Y Q L E N Y C N

F V N Q H L C G S H L V E A L Y L V C G E R G F F Y T P K T

B-Chain   30 amino acids

FIGURE 2.30 Six cysteine residues of the monomer are interconnected using disulfide bridges.

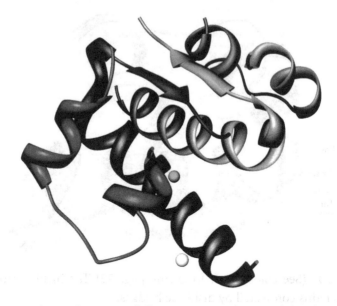

FIGURE 2.31 (See color insert following page 72) One of the three dimers making up the hexamer of an insulin structure.

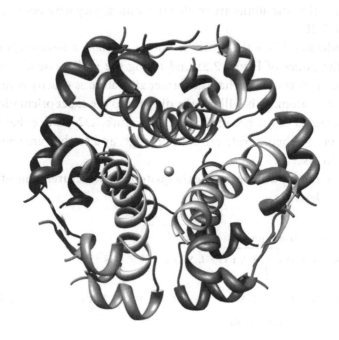

FIGURE 2.32 (See color insert following page 72) Hexamer structure of insulin (PDB code 1ZNJ).

2.2.5 Protein Functionality

As already stated, protein structure facilitates protein functionality. Even though this book does not consider functionality in any significant depth, we now present a very high level overview of the various processes and structural elements that utilize proteins.

The study of protein function includes

- Enzymatic catalysis

  - Almost all chemical reactions in a cell are facilitated by a catalytic mechanism involving a protein.

  - Reactions are accelerated, increasing the rate, sometimes by as much as a factor of 10^{17}.

  - Enzymes may use any of the following strategies:

    - Provide an alternative reaction route

    - Stabilize reaction intermediates

    - Facilitate alignment of reacting groups

- Signal transmission

  - Information can be moved from one part of a biological system to another. This is done, for example, by neurotransmitters, hormones, and cytokines.

    - Cytokines act as signals in immune systems and during embryogenesis.

    - They typically bind to a cell-surface receptor, and this initiates a signal cascade resulting in an upregulation or downregulation of some particular gene expression.

- Regulation

  - PPI (protein–protein interaction): It may enable or inhibit one of the proteins involved in an interaction.

  - Transcription regulation: Proteins may assemble to form transcription factors that regulate gene expression in DNA.

- Immune system

  - The immunoglobulins are examples of proteins that play a central role in the defense mechanisms of the body. They recognize foreign molecules and help the body to destroy viruses and bacteria. Other components of the immune system include MHC (major histocompatibility complexes) and various receptor proteins.

- Molecular transport

  - Some membrane proteins are capable of moving small molecules into or out of a cell. For example, the Na^+/glucose transporter is a membrane protein that brings in both sodium ions and glucose molecules with sodium ions moving down their concentration gradient, whereas the glucose molecules are pumped up their concentration gradients.

- Motor elements

  - Changes in protein conformation produce mechanical movement, be it large muscle movement or the small flagellar motors of a bacterium.

- Structural systems

  - Proteins act as the building blocks of living systems.

  - In large complex systems, they contribute to the formation of large structures such as bone, hair (collagen), nails (keratin), and many other tissues.

  - Smaller structures include strands of silk, viral coats, and frameworks for the movement of molecules within cells.

2.2.6 Protein Domains

We have already seen the progression of structural organization: primary sequence, secondary structure, tertiary structure, and quaternary structure. The protein *domain* is another structural concept that is important in the discussion of protein functionality. Although the definition of a domain varies somewhat in protein research papers, here is a reasonable definition:

> A domain is a region of a protein that has more interactions within itself than with the rest of the chain.

These interactions include the formation of a hydrophobic core. Indeed, it has been observed that protein domains can often fold independently. The observation of independent folding is due to experiments that have shown that the subsequence of amino acids in a domain can often fold into the final conformation even when the rest of the protein is absent.

The following points provide more information about the formation of domains:

- Domains can be structurally similar even though they are in proteins that are otherwise quite different. Such domains usually show a high-sequence similarity.

- Domains tend to be compact and globular. Linkages between domains are often loop structures and hardly ever helices or beta strands.

- Domains have distinct solvent accessible surfaces and are typically separated by water molecules.

- Residues within a domain will contact other residues in the same domain. There is very little contact, if any, between residues in different domains.

- Usually a domain is formed from a contiguous residue sequence of the protein. In rare cases, a domain is made up of two or more regions of protein that are subsequences from one or more polypeptide chains. This is why expert visual inspection is often required to determine domains.

Multiple domains within a single protein often cooperate to accomplish related tasks. For example, the reverse transcriptase protein of HIV has various domains that work together to convert viral RNA to DNA, followed by destruction of the viral RNA. So, by studying the domain structure of a protein, we can strive to understand how design principles facilitate particular types of functionality. Although this is an important goal, the mechanisms of cooperative function are not fully understood in many cases.

Figure 2.33 presents a simple example of the 6PAX protein that uses two domains to interact with a segment of DNA. This protein is part of a family called the *homeodomain proteins*. They interact with DNA to provide transcription regulation, and they are important in development and cell differentiation. When the 6PAX protein interacts with DNA, the event initiates a cascade of protein interactions that leads to the development

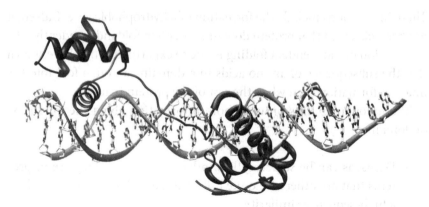

FIGURE 2.33 The two domains of 6PAX interact with DNA.

of various eye structures. In the figure, it is clear that the protein has two hydrophobic cores, each responsible for interacting with a particular DNA segment. The two cores are linked by a rather lengthy loop structure.

2.3 AN OVERVIEW OF RNA STRUCTURE

Ribonucleic acid is another very important macromolecule with a complicated structure that supports its biochemical functionality. Debating the emergence of life on earth, biologists have conjectured that the existence of DNA was preceded by the formation of RNA.* Presumably, this was facilitated by two significant properties of RNA:

- RNA has information storage capability.

- RNA can act as an enzyme.

We currently see these properties utilized in modern cells because RNA is incorporated into ribosomes, nucleases, polymerases, and spliceosomes.

For many years, the importance of RNA was not fully appreciated. It was mostly considered to be a carrier of information, as in messenger RNA (mRNA) or as a carrier of amino acids during protein synthesis, as in transfer RNA (tRNA). Even in ribonucleoprotein complexes, it was considered to be mainly providing a structural framework for the protein. Later, these viewpoints changed when Cech and Altman discovered that

---

* Because all chemical evidence is now lost, this topic is subject to considerable debate. See www.amsci.org/amsci/articles/95articles/cdeduve.html and www.panspermia.org/rnaworld.htm.

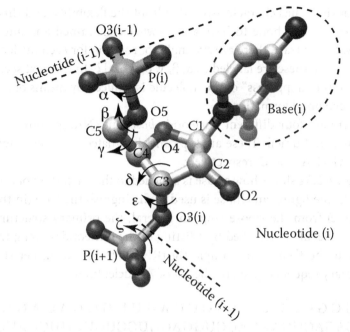

FIGURE 2.34 (See color insert following page 72) A nucleotide and its flexible backbone.

RNA can provide catalytic capabilities. In 1989, they were awarded the Nobel Prize in chemistry for this work.

2.3.1 Nucleotides and RNA Primary Sequence

Just as amino acids are the building blocks of proteins, nucleotides are the units that are used to sequentially construct RNA molecules. As illustrated in Figure 2.34, RNA has a backbone structure that alternates between phosphate and ribose groups. Each nucleotide unit has a backbone comprised of six atoms. Starting just below the uppermost dashed line, we see the phosphorous atom P(i) for the ith nucleotide and just below that an oxygen atom labeled as O5. Continuing along the backbone, we encounter C5 of the ribose group followed by C4, C3 in the ribose ring, and finally O3.

The chain has a direction that is specified by the C5 and C3 atoms of the ribose group. Consequently, moving along the backbone in the direction of nucleotides with increasing index means that we are going in the 3' direction. Moving in the opposite direction of decreasing nucleotide indexes is considered going toward the 5' end.

As we will see in a later chapter, a protein backbone essentially has two dihedral angles for each amino acid. The amide plane that is part of the backbone has characteristics that limit the flexibility, and this causes the peptide backbone to have a somewhat constrained amount of flexibility. By contrast, there are six angles of rotation for each nucleotide. In Figure 2.34, these are labeled as α, β, γ, δ, ε, and ζ. In the next section, we see that the "floppiness" of the molecule is reduced by means of extensive hydrogen bond formation.

There are four different nucleotide *bases* in RNA: guanine, cytosine, adenine, and uracil. These are typically designated with the single letter codes G, C, A, and U, respectively.

Figure 2.34 shows how a base is attached to the C1 carbon of the ribose ring. In the figure, an ellipse is used to distinguish the base (in this case, cytosine) from the ribose ring. In general, the primary structure of an RNA molecule is specified by a listing of the nucleotides going from the 5′ end to the 3′ end. For example, the tRNA with PDB identifier 1EHZ has a primary sequence given by the list of 76 nucleotides:

G C G G A U U U A G C U C A G U U G G G A G A G C G C
C A G A C U G A A G A U C U G G A G G U C C U G U G U U C G A U C C A
C A G A A U U C G C A C C A

2.3.2 RNA Secondary Structure

We have already seen that proteins have a secondary structure that is determined by hydrogen bonding between atoms that are part of the backbone. For RNA, conformational stability is also enhanced by hydrogen bonding, but in this case the hydrogen bonds are established between donor and acceptor atoms that are part of the nucleotide bases. This type of hydrogen bonding is called *base pairing*, and it occurs when the nucleotide chain folds back on itself so that double ring nucleotides (purines) are in close proximity to single ring nucleotides (pyrimidines). Figure 2.35 illustrates GC and AU base pairing. Note that the GC association involves three hydrogen bonds, whereas AU has only two hydrogen bonds.[*] The GC hydrogen bonds of Figure 2.35 also show that bond formation is not necessarily in perfect alignment. The GC example from 1B23 shows a slightly staggered placement of the two residues, undoubtedly due to the positioning constraints imposed by other residues in the stem.

---

[*] Figure 2.35 data sources:
GC: Residues 30 and 40 of 1B23.
AU: Residues 66 and 7 of 1EHZ.
GU: Residues 4 and 69 of 1EHZ.

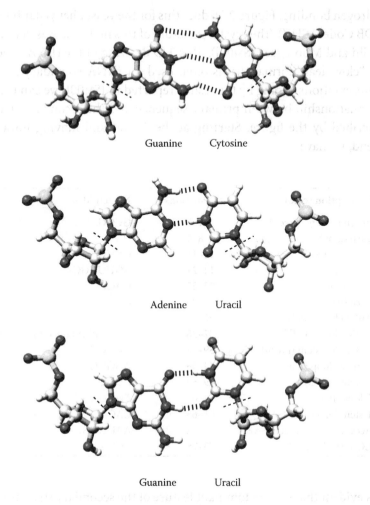

Guanine   Cytosine

Adenine   Uracil

Guanine   Uracil

FIGURE 2.35 (See color insert following page 72) Hydrogen bonding between complementary purine–pyrimidine pairs.

Occasionally, RNA will form GU pairs as shown in the last pairing in Figure 2.35. This is the so-called *wobble* base pairing. The counterpart of this, a GT pairing in DNA, is extremely rare (as one would imagine, because of the information carried in normal complementary base pairs). RNA is more forgiving with respect to GU pairs, and in fact, there is some research being done to demonstrate that wobble pairs can play a particular functional role [Vм00].

In many cases, RNA base pairing can be illustrated using a 2-D diagram that is drawn to show the pairs that are in association due to hydrogen bonding. Figure 2.36 does this for the yeast phenylalanine tRNA (PDB Code: 1EHZ). The crystal structure of this molecule was determined by Shi and Moore (see [SM00]). This 2-D drawing of the tRNA is done in the "cloverleaf" format that is often used for tRNA secondary structure representations. Going by the results reported in [SM00], we can trace out the relationship between primary sequence and the secondary structure described by the figure. Starting at the 5′ end and moving toward the 3′ end, we have

Description (color)	Positions	Nucleotides
Acceptor stem (green)	1–7	GCGGAUU
Turn (gray)	8–9	UA
D stem (green)	10–13	GCUC
D loop (blue)	14–21	AGUUGGGA
D stem (green)	22–25	GAGC
Turn (gray)	26	G
Anticodon stem (green)	27–31	CCAGA
Anticodon loop (blue)	32–38	CUGAAYA anticodons (violet)*
Anticodon stem (green)	39–43	UCUGG
Variable loop (blue)	44–48	AGGUC
T stem (green)	49–53	CUGUG
T loop (blue)	54–60	TψCGAUC
T stem (green)	61–65	CACAG
Acceptor stem (green)	66–72	AAUUCGC
CCA end (yellow)	73–76	ACCA

It is evident that the predominant feature of the secondary structure is the tendency to use base pairing to build "stem" formations, each ending in a "loop." Occasionally the loop structure is responsible for some particular chemical interactions that support its functionality. In the case of 1EHZ, nucleotides 34, 35, and 36 provide the anticodon triplet that is necessary for phenylalanine tRNA activity.

In addition to stems and loops, RNA secondary structure can incorporate internal loops and bulges. These are essentially stem structures that bulge out because there is a lack of base pairing for a short stretch of nucleotides. Internal loops tend to have the same number of unpaired nucleotides

---

* The Y base at position 37 is a modified purine called wybutine. Its presence has a stabilizing effect on codon binding.

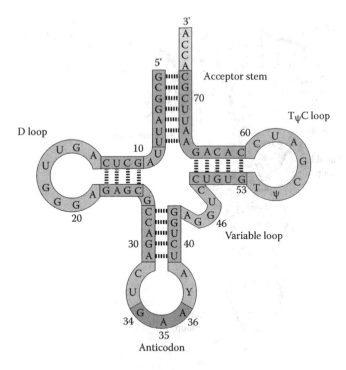

FIGURE 2.36 (See color insert following page 72) Example of RNA secondary structure (From H. SHI AND P. B. MOORE. The crystal structure of yeast phenylalanine tRNA at 1.93 Å resolution: a classic structure revisited. *RNA*, **6** (2000), 1091–1105).

on each side of the stem, whereas a bulge is characterized by an imbalance in the number of unpaired bases. For example, with several unpaired bases on one side and few on the other side, the effect is to get a bulge and the stem typically continues on in an altered direction. Finally, it is possible to have a multiloop, which is a loop having various stems arising from it. This is seen in Figure 2.36, if we consider the center of the cloverleaf to be a multiloop with four stems going off in four different directions.

2.3.3 RNA Tertiary Structure

The stem structures of RNA tend to form short helical structures that are locally similar to the much longer double helix structure of DNA. This gives a lot of stability to the molecule and helps to counteract the floppiness arising from the six torsional angles between consecutive nucleotides mentioned earlier when we introduced the structure of a nucleotide. An example of RNA tertiary structure can be seen in Figure 2.37, which illustrates the

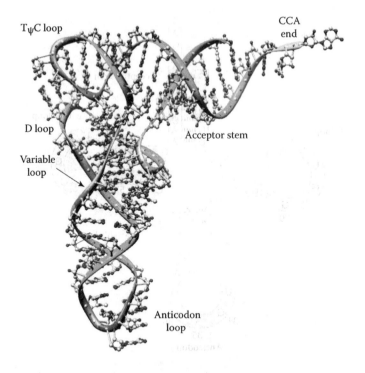

FIGURE 2.37 (See color insert following page 72) An example of RNA tertiary structure.

beautiful 3-D conformation of phenylalanine tRNA (PDB code: 1EHZ). The ribbon backbone has been given the same coloration as that used in the secondary structure of Figure 2.36 so that the reader may see the correspondences in structure. A close inspection of Figure 2.36 allows one to see the base pairings in the various stems. There is also a rather complicated pseudoknot arrangement that defines the tertiary structure. The "arms" of the cloverleaf participate in a *coaxial* stacking of the acceptor stem with the T stem. There is another coaxial stacking of the D stem with the anticodon stem.

We have now covered the fundamentals of RNA structure. This will be useful later when we consider algorithms that try to predict base pairing of RNA when given the primary sequence as an input. However, these algorithms often have limitations because it is very difficult to predict *all* the base pairs without extra information about the 3-D conformation of the molecule. The difficulties arise because of the additional hydrogen bonding that can occur outside of the stem regions. Specifically, it is also possible to have extra base pairing between nucleotides that are in different loop regions. With a careful inspection of Figure 2.37, one can see this occurring between nucleotides in

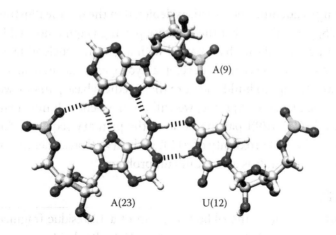

FIGURE 2.38 (See color insert following page 72) Pseudoknots complicate base pairings.

the TψC loop and the D loop. This type of base pairing is referred to as a *pseudoknot*. These pairings are important structural features because they help to increase the conformational stability of the molecule.

For the sake of clarity, these "long links" have not been included in Figure 2.36. We now list them as pairs of nucleotides, complete with positional index:

[U(8)::A(14)]   [G(10)::G(45)]   [G(15)::C(48)]

[G(18)::ψ(55)]*   [G(19)::C(56)]   [G(22)::G(46)]

[A(9)::A(23)]   [G(26)::A(44)]   [T(54)::A(58)]

If you inspect Figure 2.35 to understand which nucleotides are being linked, you will discover that essentially all of these extra hydrogen bond associations involve nucleotides that are outside of the stem regions. A rule of thumb would be that once a nucleotide is based-paired, there is no possibility of any additional hydrogen bonding. However, there are exceptions to this rule: Because of the close compaction of the RNA molecule, it is possible to bring nucleotides into sufficiently close proximity that additional hydrogen bonds can form. A good example of such an exception is given by the pairing of A(23) with A(9). This is remarkable because A(23) is in the D stem and is already base paired with U(12). A diagram of this *ménage a trois* is given in Figure 2.38. If you compare this with the usual

---

* The symbol ψ represents pseudouridine, a posttranscriptionally modified ribonucleotide.

base pairing of adenine and uracil, as depicted in the middle illustration of Figure 2.35, you will see that the AU base pairing is quite normal but the adenine at position 9 (not base paired with any other nucleotide because it is located in a turn) is able to develop a hydrogen bond association with A(23). If we attempt to build software that predicts base pairs by working only from the primary sequence, we will not have enough information to predict the [A(23)::A(9)] pairing because the primary sequence does not tell us about the 3-D proximity of A(23) with A(9). Consequently, predicting all hydrogen bonds is a very difficult problem.

2.4 EXERCISES

1. Figure 2.39 shows the ribbon diagram of a 12-residue fragment of a longer protein. What is the Protein Data Bank identifier of that protein?

2. What is the biological function of myoglobin? List and describe all the structural features of the globin fold that support this functionality.

3. Figure 2.40 shows the ribbon diagram of rice cytochrome C (PDB code 1CCR). There are only two histidine residues in this protein, and they are shown colored in black. One other residue has been made visible. There is a methionine residue with its sulfur atom hovering just above the Fe atom in the heme group. Do some Web searching to discover the biological function of cytochrome C. Compare the structures of myoglobin and cytochrome C by referring to Figures 2.40, 2.21, and 2.25. How is the difference in functionality supported by the differences in structure?

4. How many different topologies can we have for a three-stranded beta sheet? How many for a four-stranded beta sheet? Draw each sheet as a set of arrows. Ignore the connectivity between the strands. Be sure

FIGURE 2.39 (See color insert following page 72) Peptide sequence for Exercise 1.

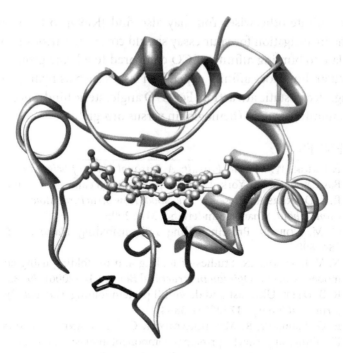

FIGURE 2.40   Rice cytochrome C for Exercise 3.

that you do not count a sheet twice when it is the same sheet already listed but with a different orientation in 3-D space.

5. Essay assignment: Cyclotides are peptides that have a cyclic backbone. They usually have short primary sequences and often fold into loop structures that are stabilized by cysteine bridges. What are the anticipated uses of cyclotides? How would their structure facilitate these applications? How is the loop closure accomplished?

6. Essay assignment: It has been observed that the heme prosthetic group shows an affinity for CO, which is up to 100,000 times higher than the affinity for O_2 in an unhindered model (see [JK95]). However, a distal histidine can provide some steric hindrance (discussed in [Om88]). It should be noted that the geometry of the Fe–C–O configuration has been debated. For example, Lim et al. ([LJ95]) state: "These solution results help to establish how myoglobin discriminates against CO, a controversial issue dominated by the misconception that Fe–C–O is bent." However, a casual inspection of the Fe–C–O triplet in cocrystallizations such as 1VXF would seem

to indicate otherwise. You may also find [KP99] to be of interest. The investigation for your essay should cover the various issues that relate to binding affinity of CO compared to a heme group. In particular, how does affinity depend on issues such as hydrogen bonding, electrostatic attraction, Fe–C–O angle, steric hindrance, and the containing protein (hemoglobin versus myoglobin)?

REFERENCES

[Cr88] F. Crick. *What Mad Pursuit: A Personal View of Scientific Discovery.* Basic Books, New York, 1988 (a reprint edition came out in June 1990).

[Di05] R. E. Dickerson. *Present at the Flood: How Structural Molecular Biology Came About.* Sinauer, Sunderland, MA, 2005.

[Do03] C. M. Dobson. Protein folding and misfolding. *Nature,* **426** (2003), 884–890.

[Do06] N. V. Dokholyan. Studies of folding and misfolding using simplified models. *Current Opinion in Structural Biology,* **16** (2006), 79–85.

[Dy07] R. B. Dyer. Ultrafast and downhill protein folding. *Current Opinion in Structural Biology,* **17** (2007), 38–47.

[Em04] E. G. Emberly, R. Mukhopadhyay, C. Tang, and N. S. Wingreen. Flexibility of β-sheets: principal component analysis of database protein structures. *Proteins: Structure, Function, and Bioinformatics,* **55** (2004), 91–98.

[Fb97] A. N. Fedorov and T. O. Baldwin. Cotranslational protein folding. *The Journal of Biological Chemistry,* **272** (1997), 32715–32718.

[Fr79] J. S. Fruton. Early theories of protein structure. *Annals of the New York Academy of Sciences,* **325** (1979), 1–20.

[Hc02] B. K. Ho and P. M. G. Curmi. Twist and shear in β-sheets and β-ribbons. *Journal of Molecular Biology,* **317** (2002), 291–308.

[Ia05] K. Imai, N. Asakawa, T. Tsuji, M. Sonoyama, and S. Mitaku. Secondary structure breakers and hairpin structures in myoglobin and hemoglobin. *Chem-Bio Informatics Journal,* **5** (2005), 65–77.

[Jk95] P. Jewsbury and T. Kitagawa. Distal residue-CO interaction in carbonmonoxy myoglobins: a molecular dynamics study of three distal mutants. *Biophysical Journal,* **68** (1995), 1283–1294.

[Kp99] G. S. Kachalova, A. N. Popov, and H. D. Bartunik. A steric mechanism for inhibition of CO binding to heme proteins. *Science,* **284** (1999), 473–476.

[Kb58] J. C. Kendrew, G. Bodo, H. M. Dintzis, R. G. Parrish, H. Wyckoff, and D.C. Phillips. A three-dimensional model of the myoglobin molecule obtained by x-ray analysis. *Nature,* **181** (1958), 662–666.

[Ko01] V. A. Kolb. Cotranslational protein folding. *Molecular Biology,* **35** (2001), 584–590.

[Kd99] J. Kuriyan and J. E. Darnell Jr. An SH2 domain in disguise. *Nature,* **398,** March 4 (1999), 22–24.

[LB98]   D. V. LAURENTS AND R. L. BALDWIN. Protein folding: matching theory and experiment. *Biophysical Journal*, **75** (1998), 428–434.

[LJ95]   M. LIM, T. A. JACKSON, AND P. A. ANFINRUD. Binding of CO to myoglobin from a heme pocket docking site to from nearly linear Fe–C–O. *Science*, **269** (1995), 962–966.

[MG00]   T. MIZUTANI, C. GOTO, AND T. TOTSUKA. Mammalian selenocysteine tRNA, its enzymes and selenophosphate. *Journal of Health Science*, **46** (2000), 399–404.

[OM88]   J. S. OLSON, A. J. MATHEWS, R. J. ROHLFS, B. A. SPRINGER, K. D. EGEBERG, S. G. SLIGAR, J. TAME, J.-P. RENAUD, AND K. NAGAI. The role of the distal histidine in myoglobin and haemoglobin. *Nature*, **336,** November 17 (1988), 265–266.

[OM97]   C. A. ORENGO, A. D. MICHIE, S. JONES, D. T. JONES, M. B. SWINDELLS, AND J. M. THORNTON. CATH—a hierarchic classification of protein domain structures. *Structure*, **5** (1997), 1093–1108.

[PH80]   S. E. PHILLIPS. Structure and refinement of oxymyoglobin at 1.6 Å resolution. *Journal of Molecular Biology*, **142** (1980), 531–554.

[PR04]   G. A. PETSKO AND D. RINGE. *Protein Structure and Function*. New Science Press, London, U.K., 2004.

[PR97]   A. C. PAPAGEORGIOU, R. SHAPIRO, AND K. RAVI ACHARYA. Molecular recognition of human angiogenin by placental ribonuclease inhibitor—an x-ray crystallographic study at 2.0 Å resolution. *The EMBO Journal*, **16** (1997), 5162–5177.

[PC53]   L. PAULING AND R. B. COREY. A proposed structure for the nucleic acids. *Proceedings of the National Academy of Sciences of the United States of America*, **39** (1953), 84–97.

[PO03]   O. POIROT, E. O'TOOLE, AND C. NOTREDAME. Tcoffee@igs: A Web server for computing, evaluating and combining multiple sequence alignments. *Nucleic Acids Research*, **31** (2003), 3503–3506.

[SK00]   J. SKOLNICK. Putting the pathway back into protein folding. *Proceedings of the National Academy of Sciences U.S.A.*, **102** (2005), 2265–2266.

[SM00]   H. SHI AND P. B. MOORE. The crystal structure of yeast phenylalanine tRNA at 1.93 Å resolution: a classic structure revisited. *RNA*, **6** (2000), 1091–1105.

[TR01]   C. TANFORD AND J. REYNOLDS. *Nature's Robots: A History of Proteins*. Oxford University Press, New York, 2001.

[TR08]   A. TRAMONTANO. (Private communication).

[VM00]   G. VARANI AND W. H. MCCLAIN. The G.U wobble base pair: a fundamental building block of RNA structure crucial to RNA function in diverse biological systems. *EMBO Reports*, **1** (2000), 18–23.

CHAPTER 3

Data Sources, Formats, and Applications

When the early explorers of America made their first landfall, they had the unforgettable experience of glimpsing a New World that no European had seen before them. Moments such as this—first visions of new worlds—are one of the main attractions of exploration. From time to time scientists are privileged to share excitements of the same kind. Such a moment arrived for my colleagues and me one Sunday morning in 1957, when we looked at something no one before us had seen: a three-dimensional picture of a protein molecule in all its complexity.

JOHN C. KENDREW, 1961

3.1 MOTIVATION

In the previous quotation, one can sense the joy of discovery experienced by Dr. Kendrew [KE61] and his coworkers. After many months of work, their efforts resulted in the 3-D x-ray analysis of myoglobin and the subsequent publication of various diagrams describing this structure. This included a color picture of the protein with atoms and bonds represented in "stick" form.

One can only wonder what Kendrew would have thought had he known that, 50 years later, known protein structures would number in the tens of thousands and be freely accessible. In fact, databases containing these structures have even been suggested as sources for high school science fair projects.*

For people working in the area of structural bioinformatics, these data sources are important† for the realization of various goals, such as the following:

- The design of geometric algorithms for manipulating structural data.

- The design of pattern analysis algorithms for detecting structural patterns in macromolecules.

- The application of these algorithms to further understanding of the cellular processes possibly leading to therapeutic interventions.

Presently, there are several databases dealing with molecular biology. An excellent review of the many databases that are available can be found in a special issue of *Nucleic Acids Research* (Volume 28, Issue 1, 2000). The review covers both sequence and structure databases. Our main focus will be on the latter.

3.2 SOURCES OF STRUCTURAL DATA

The study of protein domains and their functionality requires a reliable source of structural data for both macromolecules and ligands. Because the main purpose of this text is algorithms, we will not spend much time discussing the many data sources that are available, but it is worthwhile to present a quick review of a few sources that are relevant to material discussed in the remaining chapters.

3.2.1 PDB: The Protein Data Bank

The PDB is a large protein, public domain repository maintained by the RCSB (Research Collaboratory for Structural Bioinformatics) [Bw00]. The RCSB describes itself as "a nonprofit consortium dedicated to improving our understanding of the function of biological systems through the study of the 3-D structure of biological macromolecules."

For people studying protein structure, the PDB is a very important source of structural data.† It contains thousands of files, each reporting

---

* Science Buddies home page: http://www.sciencebuddies.org/mentoring/project_ideas/ home_BioChem.shtml.

† According to the RCSB Web site, over 10,000 scientists, students, and educators visit the PDB Web site every day and, on average, 2.2 files are downloaded every second.

on the 3-D coordinates of atoms in various molecular categories including proteins, peptides, nucleic acids, protein–nucleic acid complexes, and carbohydrates.

Most of the atomic coordinates have been determined by x-ray crystallography, but there are also structures that have been determined by NMR techniques. As of June 17, 2008, there were 51,366 entries in all.

Nuclear magnetic resonance (NMR) has been applied to proteins fairly recently [Wu86]. It has the advantage of being able to determine the structure of proteins in solution. The structure of some membrane proteins has been determined using NMR, for example, 2JQY. After processing NMR data, researchers derive a long list of geometric conformational restraints that are subsequently used to calculate structure (see [Ns04]). NMR is also used in drug design studies (see [Hκ01]).

Structures are usually reported in journal publications, and in these circumstances, a PDB file will contain pertinent information such as title of the research paper, names of authors, experimental method, parameters, and unit cell specification. The PDB Contents Guide (available at http://www.wwpdb.org/docs.html) provides a full list of all the various types of records that can be found in a PDB file.

When using the PDB, students should appreciate that this is "real-world" data that has been assembled from a wide selection of labs from several countries. Although there is a sustained effort to maintain high-quality data, there are PDB files that are inconsistent, informal, and not fully checked for errors. Other limitations are a result of the data acquisition process itself: For early x-ray structures, the resolution was not high enough to accurately position the hydrogen atoms of a molecule, and so only the "heavy" atoms were given in the file. More recent PDB files based on very high resolution x-ray studies may contain coordinates for hydrogen atoms.

Furthermore, software such as visualization applications that must determine the bonds between atoms will usually do this by inference. There is a CONECT record that, for each atom in the chemical component, lists how many and to which other atoms that atom is bonded. However, most bonds have to be inferred from the atom name. For example, when a PDB file has ATOM records for the atoms in valine, they will specify the coordinates of atoms labeled CG1 and CG2, but there will be no bond information for these atoms. So, the software would have to have information that specifies the existence of bonds between the beta carbon CB and both CG1 and CG2 (there is no bond between CG1 and CG2).

3.2.2 PDBsum: The PDB Summary

PDBsum is a Web site* that provides summaries and analyses of structures from the PDB. Entries are accessed by their PDB code or by doing a simple text search. This is an excellent starting point for information related to proteins. There are many links to a variety of other sites with more information that is relevant to the protein being summarized, for example, PDB, MSD, SRS, MMDB, JenaLib, OCA, CATH, SCOP, FSSP, HSSP, PQS, ReliBase, CSA, ProSAT, STING, Whatcheck, and EDS. The PDBsum for a particular protein will also include a Procheck diagram (giving Ramachandran plot statistics), an assessment of clefts on the protein surface (useful for binding-site prediction), and a visualization of the protein surface. Other tabs on the page lead to pages that describe any ligands that are bound to the protein, as well as a description of any protein–protein interactions. Enzyme reactions for the protein are also reported.

3.2.3 SCOP: Structural Classification of Proteins

It is well established that proteins have structural similarities. This may be due to common evolutionary origins or possibly convergent evolution from origins that are not really related. The SCOP hierarchy† has been created by visual inspection but with the help of various software tools. It hierarchically organizes proteins according to their structure and evolutionary ancestry, and also provides entry links for coordinates, structure images, literature references, etc. (see [MB95], [LB02], and [AH04]). Release 1.73 (November 2007) contains 97,178 domains related to 34,494 PDB entries.

The unit of categorization in SCOP is the protein domain, not the protein itself. Some small proteins have a single domain, so in this case, the entry will be an entire protein, but in general, a large protein may have more than one domain, leading to as many entries.

There are four hierarchical levels in SCOP, and they can be characterized as follows:

- **Classes**: A class contains domains with similar global characteristics. There is not necessarily any evolutionary relation. For example, release 1.73 contains the following classes:

  1. All alpha proteins

  2. All beta proteins

---

* http://www.ebi.ac.uk/thornton-srv/databases/pdbsum/.
† http://scop.mrc-lmb.cam.ac.uk/scop/.

3. Alpha and beta proteins (α/β)

4. Alpha and beta proteins ($\alpha + \beta$)

5. Multidomain proteins (alpha and beta)

6. Membrane and cell surface proteins and peptides

7. Small proteins

8. Coiled coil proteins

9. Low-resolution protein structures

10. Peptides

11. Designed proteins

Some noteworthy points: Membrane proteins are considered to be a separate class. Small proteins are stabilized by disulfide bridges or by metal ligands in lieu of hydrophobic cores.

- **Folds**: A fold is a particular topological arrangement of alpha helices, beta strands, and loops in 3-D space. Release 1.73 reports on 1086 different folds across the first seven classes. It should be noted that many different protein sequences can produce the same fold. A short description of the most significant structural features is used as the name of the fold (some examples: Globin-like, Long alpha-hairpin, Cytochrome c, Frizzled cysteine-rich domain, etc.).

- **Superfamilies**: A superfamily contains a clear structural similarity. The stress on structural similarity is to emphasize the fact that they might have low sequence identity. There is a probable common evolutionary origin (i.e., they are probably homologous proteins). Superfamilies share a common fold and usually perform similar functions. Release 1.73 contains 1777 superfamilies spread over the first seven classes.

- **Families**: A family contains a clear sequence similarity (i.e., there is an obvious evolutionary relationship). The pair-wise residue identity (sequence similarity) is at least 30%. In some cases, this figure may be lower (say, 15%), but the proteins are still put in the same family because they have very similar functions and structures (an example of this would be the family of globins). Release 1.73 contains 3464 families in the first seven classes.

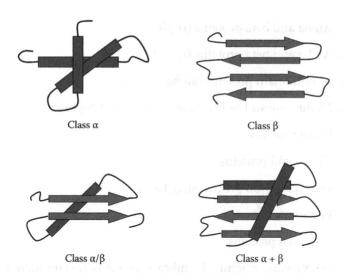

Class α   Class β

Class α/β   Class α + β

FIGURE 3.1   Cartoons illustrating the first four SCOP classes.

Figure 3.1 shows some cartoons that illustrate examples from the first four classes of SCOP. The alpha class is a bundle of α helices connected by loops, and the beta class is a sheet comprised of antiparallel β strands. Note the difference between the classes α/β and α+ β. The former is mainly parallel β strands with intervening α helices, and the latter mainly segregated α helices and antiparallel β strands positioned so that the helices pack against the β sheet. As an exercise, the reader should investigate the various classes of the SCOP hierarchy and use protein display software to get images of representatives from various classes.

It should be emphasized that the utility of SCOP arises from its hierarchical organization as established by structural similarity. Because it is easier to design algorithms that evaluate sequence similarity, such comparisons may lack the effectiveness of structural comparisons, especially when function prediction is desired. The globin family is a good example of this situation because it has very similar physical structures even though sequence similarity is fairly weak.

3.2.4 CATH: The CATH Hierarchy

CATH* is another protein structure classification hierarchy. It is based on the following levels:

---

* http://www.cathdb.info/latest/index.html.

- **Class**: This level has categories for the global characteristics of the secondary structures. The main categories within the class level are

  - **Class 1**: Mainly alpha

  - **Class 2**: Mainly beta

  - **Class 3**: Mixed alpha–beta

  - **Class 4**: Few secondary structures

- **Architecture**: The architecture level contains groups of topologies that share similar structural features as determined by the orientations of the secondary structures. Connectivity between secondary structures is not considered.

- **Topology**: The topology level groups structures into fold families that show similarity due to both overall shape and connectivity of the secondary structures. The grouping is automatically done by the SSAP structure comparison algorithm ([To89a], [To89b]) and CATHEDRAL ([Pb03]). Structures that have an SSAP score of 70, and such that at least 60% of the larger protein matches the smaller protein, are assigned to the same topology level or fold group.

- **Homologous superfamily**: As expected, "homologous" implies a common ancestor. Similarities are determined by sequence comparison and then by structure comparison using SSAP. Two structures are in the same homologous superfamily if any of the following hold

  - Sequence identity > 35%

  - SSAP score > 80 and sequence identity > 20%

  - SSAP score > 80 and 60% of larger structure is equivalent to the smaller structure; the domains have related functions

- **Sequence families**: Structures within an H level are clustered into smaller groups based on sequence identity. Domains clustered in the same sequence family level have sequence identities greater than 35% with at least 60% of the larger domain equivalent to the smaller domain. At this level, we have highly similar structure and function.

As with SCOP, classification is for protein domains, but unlike SCOP, most classification is done automatically by software. When this is in doubt, classification is done by visual inspection.

The CATH levels just described represent the upper or main levels of the hierarchy. Descendent from the H-level, we find the sequence family levels designated as the "SOLID" sequence levels. The S, O, L, and I levels provide further subdivisions that cluster domains on similarities in sequence identity (35% for S, 60% for O, 95% for L, and 100% for I). The D-level is a final "counter" designation that is used to uniquely identify different proteins that have the same CATHSOLI path in the hierarchy (see [GL06]).

Every domain in CATH has a classification code, which is written as a sequence of 9 numbers, one for each level within the CATHSOLID, separated by periods. As there are only 4 class levels, the classification number will always begin with 1 for mainly alpha, 2 for mainly beta, 3 for mixed alpha–beta, or 4 for few secondary structures.

A CATH code with fewer than 9 numbers will represent a group of domains at some level of the hierarchy, for example, 3.40.1280 is the CATH code for a particular topology, namely, topology 1280 (alpha/beta knot) within architecture 40 (3-layer (aba) sandwich), which is within class 3 (mixed alpha–beta).

In Figure 3.2, we see some representatives of the CATH classes. Each domain is surrounded by a dashed line. Domain 1CUK03 is mainly alpha helices and is the third domain within 1CUK. Domain 1PDC00 is mainly beta strands and is the only domain within the protein 1PDC. Domain 1RTHA1 is mixed alpha–beta, and it is the first domain within the A chain of 1RTH. It is noteworthy that, although many domains are comprised of a single contiguous subsequence of residues, 1RTHA1

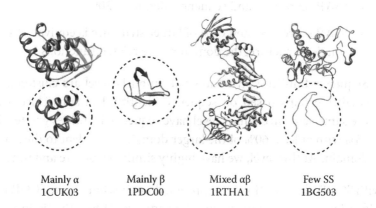

| Mainly α | Mainly β | Mixed αβ | Few SS |
| 1CUK03 | 1PDC00 | 1RTHA1 | 1BG503 |

FIGURE 3.2 Some examples from the CATH classes.

contains two subsequences, namely, residues 1 to 90 and residues 116 to 156. For the sake of clarity, only chain A of 1RTH has been shown. Finally, 1BG503 is the third domain within 1BG5 and is a good example of a domain with essentially no secondary structure in the form of helices or beta strands.

In version 3.1.0 of CATH (released in January 2007), there are 5 architecture entries for mainly alpha, 20 for mainly beta, 14 for mixed alpha–beta, and 1 for few secondary structures.

To illustrate the types of domains that one can observe at the architecture level, let us inspect some of the descendents of the mixed alpha–beta class. Version 3.1.0 lists the following 14 entries at the architecture level:

Roll	4-Layer sandwich
Super roll	Alpha–beta prism
Alpha–beta barrel	Box
2-Layer sandwich	5-Stranded propeller
3-Layer (aba) sandwich	Alpha–beta horseshoe
3-Layer (bba) sandwich	Alpha–beta complex
3-Layer (bab) sandwich	Ribosomal protein

To get an idea of what we might expect at this level, consider Figure 3.3. It shows representatives* of 3 of the 14 entries. The first example shows a 2-layer sandwich with two alpha helices hovering over various beta strands making up a beta sheet. The scene has been adjusted to show a small gap between the two components of the domain. In fact, the domain is comprised of two separate subsequences in the A chain of 1C0P, namely, residues 1077 to 1136 inclusive and residues 1186 to 1288 inclusive. The next example is a 3-layer (aba) sandwich. It has been positioned so that we can see the beta sheet sandwiched between the two groups of alpha helices. Finally, the first domain in the B chain of 1BHT shows the bba structure with a beta sheet positioned between some beta strands above the sheet and an alpha helix below it.

Before making another descent in the hierarchy, it should be mentioned that Figure 2.28 gives a fine example of an alpha–beta horseshoe domain.

If we descend into the 3-layer (aba) sandwich grouping, there are 100 entries below this within the topology level. For example, the domain

---

* These representatives are the same as those seen at http://www.cathdb.info/cgi-bin/cath/GotoCath.pl?cath=3.

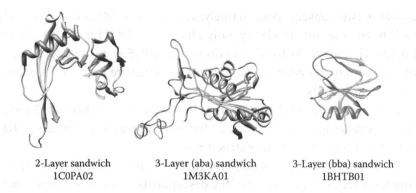

| 2-Layer sandwich | 3-Layer (aba) sandwich | 3-Layer (bba) sandwich |
| 1C0PA02 | 1M3KA01 | 1BHTB01 |

FIGURE 3.3   Some architecture examples from the mixed alpha–beta class.

1IPAA02 has the topology of an alpha/beta knot. The full CATH code for this domain is 3.40.1280.10.9.2.1.1.1. Now let us consider another domain with a different topology but the same class and architecture. If we inspect domain 1PSZA01, we see that it has a CATH code of 3.40.50.1980.14.1.1.1.1. Note that the topology number is 50, which is for the Rossmann fold. Figure 3.4 gives us an illustration of these two domains. Note the similarity in structure. The domains have been positioned so that the 3-layer (aba) sandwich (alpha helices, beta strands, and alpha helices) is easily seen.

As an additional note, it may be mentioned that these two proteins have another significant relationship. It is an example of *circular permutations* in proteins (see [UF01]). Proteins A and B are related by a circular permutation if the N-terminal region of protein A has significant sequence similarity to the C-terminal region of protein B and, vice versa. As noted earlier, the architecture level does not take into account the connectivity between secondary structural elements (alpha helices and beta strands), so these two proteins would be expected to occupy the same architectural level while existing in different topologies because they do differ in the connectivity of their secondary structural elements.

3.2.5 PubChem

PubChem* is a database-holding structural information for millions of molecules. It is maintained by the NCBI (National Center for Biotechnology Information) within the National Library of Medicine, NIH (National

---

* http://pubchem.ncbi.nlm.nih.gov/.

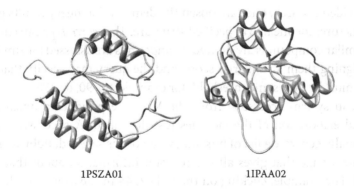

1PSZA01   1IPAA02

FIGURE 3.4 Some examples of different topologies within the same architecture.

Institute of Health). Data is free and can be downloaded via FTP. Currently (June 2008), PubChem has over 19 million compounds.

The PubChem Web site supports three linked databases:

- PubChem Compound: Searches for compounds using names, synonyms, or keywords

- PubChem Substance: Supports searching for chemical substances

- PubChem BioAssay: Does searching on bioassay results

Search facilities are supported using sophisticated queries that can involve chemical structure specifications. A structure in a query may be specified by a SMILES string, InChI string, or through a MOL file (to name just three examples). Any compound in the PubChem database will have been assigned a unique compound identifier (CID), and this can also be used in a query to make a specific request.

Compounds can also be specified using common names* or synonyms. For example, a search for "aspirin" will produce 36 hits, each with a different CID but all related in some way to the usual 2-acetyloxybenzoic acid.

Various search filters can be used to narrow the results of a query that might otherwise produce too many hits. For example, using the expression 150:300[mw], as part of a compound text search, will restrict the results of the search so that compounds have a molecular weight between 150 and 300.

---

* PubChem street cred: synonyms for cocaine include "sugar," "sleighride," "blow," and "blizzard."

To widen a search, one can loosen the demand that query results generate structures identical to a specified structure. The *search type* can stipulate that similar compounds are desired. Compounds are assessed for similarity by assigning them a similarity score, and the query may specify that such scores must be over some threshold, for example, 80, 90, or 95%.

A help system accessible from the Web site provides a much more detailed elaboration of the facilities just described. A successful search will usually generate a list of hits and, when one is selected, PubChem will generate a page that gives all the relevant information about that compound. For example, clicking on the CID 2244 in the aspirin hit list will produce a page with several other links to pages that tell you more about this compound including pharmacological action and a list of proteins that bind with this molecule. The top page for CID 2244 shows a drawing of the compound, and gives a great deal of information about this compound, including molecular weight, molecular formula, hydrogen bond donor count, hydrogen bond acceptor count, rotatable bond count, and topological polar surface area.

Structure files give the coordinates of the "flattened" molecule. The coordinates allow visualization software to draw the molecule as a 2-D graphic. Bond connectivity information can be used by energy minimization algorithms to derive the 3-D coordinates of the molecule. This can be a complicated procedure, and because there may be several conformations, each corresponding to an energy value that is a local minimum, PubChem avoids the specification of any particular 3-D conformation. For example, the x, y, and z coordinates given for methane are

```
2.5369      0.0000      0.0000      C
3.0739      0.3100      0.0000      H
2.0000     -0.3100      0.0000      H
2.2269      0.5369      0.0000      H
2.8469     -0.5369      0.0000      H
```

Structure files are available in three different formats: ASN1, XML, and SDF.

3.2.6 DrugBank

On November 29, 2007, the PubChem Announcements Web page noted that structures from DrugBank had been made available in PubChem. As of January 1, 2008, this included 4338 drugs. To get drug information that goes well beyond structural data, the reader may visit the DrugBank home page* (see [WK06]). Quoting from the abstract:

> DrugBank is a unique bioinformatics/cheminformatics resource that combines detailed drug (i.e., chemical) data with comprehensive drug target (i.e., protein) information.

In addition to this, there are more than 14,000 protein or drug target sequences linked to these drug entries. Each DrugCard entry contains more than 80 data fields including drug/chemical data and drug target or protein data. Various data fields contain links to other databases such as KEGG, PubChem, ChEBI, PDB, Swiss-Prot, and GenBank.

The DrugBank home page has tabs that link to:

- **Browse**: the main DrugBank browser

- **PharmaBrowse**: a category browser that allows the user to explore the database by starting with physiological category, for example, alimentary tract and metabolism, blood and blood-forming organs, cardiovascular system, dermatologicals, genito-urinary system, sex hormones, etc.

- **ChemQuery**: a structure query tool that allows the user to draw the structure, convert it to a MOL file, and then submit it as a query

- **Text Query**: a text search facility that allows for partial matches and misspellings.

- **SeqSearch**: supports a BLAST search

- **Data Extractor**: a high-level data search engine that allows users to construct complex or constrained queries

- **Download**: allows users to download protein sequences, DNA sequences, DrugCards, and structure files (SDF)

3.3 PDB FILE FORMAT

There are dozens of file formats for chemical data. This section will provide a brief overview of the PDB file format because this is useful for material in the remainder of the book. A full specification of the PDB file format

---

* URL for the DrugBank home page: http://redpoll.pharmacy.ualberta.ca/drugbank/.

is rather extensive, but there is online documentation available.* We will cover only a few of the more relevant entries.

A PDB file is comprised of text lines, each 80 columns† in length and terminating in an end-of-line character. Each line in the file starts with a left-justified record name that occupies the first six columns of the record. The record name specifies the record type, and this dictates the format of the remaining columns, which may be subdivided into a sequence of fields. Some examples follow:

Record Name	Contents
HEADER	First line of the file, it contains PDB code, classification, and date of deposition
COMPND	A description of the macromolecular contents
SOURCE	Biological source of the macromolecule
REVDAT	Revision dates
JRNL	Reference to a journal article dealing with this structure
AUTHOR	List of contributors
REMARK	General remarks—some structured, some free form
SEQRES	Primary sequence of the residues on a protein backbone
ATOM	Contains the coordinates of a particular atom

The first seven record types in this list of examples are reasonably self-explanatory. Let us take a closer look at the SEQRES record. The next four lines show the SEQRES records for crambin (1CRN), a protein that is often used as an example because it has only 46 residues. An additional two lines have been added to indicate column positions.

```
SEQRES   1 A   46  THR THR CYS CYS PRO SER ILE VAL ALA ARG SER ASN PHE
SEQRES   2 A   46  ASN VAL CYS ARG LEU PRO GLY THR PRO GLU ALA ILE CYS
SEQRES   3 A   46  ALA THR TYR THR GLY CYS ILE ILE ILE PRO GLY ALA THR
SEQRES   4 A   46  CYS PRO GLY ASP TYR ALA ASN
0        1        2         3         4         5         6         7
1 2 3 4 5 6 7 8 9 0 1 2 3 4 5 6 7 8 9 0 1 2 3 4 5 6 7 8 9 0 1 2 3 4 5 6 7 8 9 0 1 2 3 4 5 6 7 8 9 0 1 2 3 4 5 6 7 8 9 0 1 2 3 4 5 6 7 8 9 0
```

The table below gives an explanation for each field in the record.

---

* A tutorial on the topic can be found at http://www.wwpdb.org/documentation/format23/v2.3.html.

† The 80-character format is undoubtedly a legacy from the days when data was often stored on 80-column punch cards.

Field Columns	Contents
1 to 6	SEQRES record name
9 to 10	Serial number (line number within the SEQRES group of lines)
12	Chain identifier
14 to 17	Total number of residues
$20 + 4^*n$ to $22 + 4^*n$	Residue names (each three characters)

The ATOM records contain the coordinates of all atoms in the macromolecule. To illustrate the typical ATOM record, we give the first four ATOM records of the crambin file:

```
ATOM       1   N    THR A   1       17.047  14.099   3.625  1.00  13.79           N
ATOM       2   CA   THR A   1       16.967  12.784   4.338  1.00  10.80           C
ATOM       3   C    THR A   1       15.685  12.755   5.133  1.00   9.19           C
ATOM       4   O    THR A   1       15.268  13.825   5.594  1.00   9.85           O
0          1         2         3         4         5         6         7
1 2 3 4 5 6 7 8 9 0 1 2 3 4 5 6 7 8 9 0 1 2 3 4 5 6 7 8 9 0 1 2 3 4 5 6 7 8 9 0 1 2 3 4 5 6 7 8 9 0 1 2 3 4 5 6 7 8 9 0 1 2 3 4 5 6 7 8 9 0 1 2 3 4 5 6 7 8 9
```

The next table gives an explanation for each field in the record.

Field Columns	Contents
1 to 6	ATOM record name
7 to 11	Serial number (line number within the ATOM group of lines)
13 to 16	Atom name
18 to 20	Residue name
22	Chain identifier
23 to 26	Residue sequence number
31 to 38	Atomic x coordinate in Angstroms
39 to 46	Atomic y coordinate in Angstroms
47 to 54	Atomic z coordinate in Angstroms
55 to 60	Occupancy
61 to 66	Temperature factor
77 to 78	Atomic symbol (right justified)
79 to 80	Atomic charge

For proteins, ATOM records have a listing order that goes from the amino end to the carboxyl end. The sequence of ATOM records in a chain is terminated by a TER record that contains TER in columns 1 to 3 inclusive.

Other record types that may be of interest are listed in the following table.

Record Name	Contents
HET	Names of hetero-atoms (usually designates atoms and molecules that are separated from the protein)
FORMUL	Chemical formula for the hetero-atoms
HELIX	Location of helices (designated by depositor)
SHEET	Location of beta sheets (designated by depositor)
SSBOND	Location of disulfide bonds (if they exist)
ORIGn	Scaling factors to transform from orthogonal coordinates
SCALEn	Scaling factors transforming to fractional crystal coordinates

3.4 VISUALIZATION OF MOLECULAR DATA

Many molecular visualization tools are available free for student use. These could be described with many colorful screen shots, but we will not do this because it is much more rewarding to go out and do the actual interactive exploration on your own. Consequently, we limit this section to a quick overview of some protein visualization software that is currently available.

The following is a partial list of the many visualization software packages that are currently available.

Visualization Software	Available from
QuickPDB	http://www.sdsc.edu/pb/Software.html
KiNG (Kinemage)	http://kinemage.biochem.duke.edu/software/king.php
WebMol	http://www.cmpharm.ucsf.edu/cgi-bin/webmol.pl
Jmol	http://jmol.sourceforge.net/
Swiss PdbViewer	http://ca.expasy.org/spdbv/
MOLMOL	http://hugin.ethz.ch/wuthrich/software/molmol/
BioDesigner	http://pirx.com/biodesigner/
PyMOL	http://pymol.sourceforge.net/
BallView	http://www.ballview.org/
UCSF Chimera	http://www.rbvi.ucsf.edu/chimera/

These will act as examples for the discussions following.

Visualization software can be compared with respect to several attributes:

- Is it a stand-alone application or a browser plug-in?

- What does it offer for changing the viewing perspective?

- What graphical representations does it support?

- Are there any special visual effects?

- What selection abilities are supported?

- Does it make any computational tools available?

- Does it have the ability to label atoms and residues?

- Are there any "extra" features?

We now consider each of these in turn.

3.4.1 Plug-In versus Stand-Alone

Plug-ins and applets are visualizer applications that execute within a browser or in another applet window. Examples of these are QuickPDB, KiNG, WebMol, and Jmol. Stand-alone applications are usually downloaded as applications that are then executed as separate software without the need for a browser. Examples of these are Swiss PdbViewer, MOLMOL, BioDesigner, PyMOL, BallView, and UCSF Chimera.

3.4.2 Change of Viewing Perspective

When a molecule is being displayed, we will often need to change our perspective or point of view. So, the visualization software should provide mouse actions to do

- Rotations

- Translations

- Zooms

Although most visualization packages offer this functionality, they may differ somewhat in terms of the ease with which such operations are accomplished. It is important that various mouse buttons are used in a natural way without requiring the user to do a menu selection to accomplish a change in viewing perspective. Some visualizers offer stereo views of a molecule.

3.4.3 Graphical Representation

There are a variety of ways to display a protein molecule. For example:

- Atoms and bonds may be shown in different graphical representations:
  - Wire-frame

- Stick

- Ball and stick (radii may be changeable)

- Spheres

- The overall structure may appear in "cartoon" form:

  - Secondary structure appearing in a ribbon representation (flat, edged, or rounded)

- The software may illustrate a molecular surface representation of a

  - Van der Waals surface

  - Solvent accessible surface

  - Connolly surface

The use of color in a molecular scene is very important when it is necessary to distinguish various components of the illustration. Color may be applied to atoms, bonds, cartoons, or surfaces.

Typical choices for atom coloring include

- A single color applied to all atoms

- The CPK colors palette to distinguish atom types

- Editable colors of the user's choosing

It may be possible to choose atom "material" so that the atom sphere has the appearance of plastic, glass, dull, or shiny metal.

Ribbons that represent secondary structure may be colored

- Using a single color applied uniformly to all ribbons

- By residue

- With respect to hydrophobicity

- With respect to pKa value

- By secondary structure type

- By molecule

- By chain

When software computes a molecular surface, it is often desirable to color the surface so that it imparts more information about the atoms just below the surface. Typical functionality includes the ability to change color to illustrate hydrophobicity and charge polarity. The appearance of the surface itself and whether or not it hides the underlying atoms can often be specified. For example, surface construction may have the appearance of being solid, mesh like, or comprised of dots. Some applications allow surface transparency to be modified across various percentages. The versatile graphics of BioDesigner are especially noteworthy.

3.4.4 Visual Effects

These are changes to the scene that affect overall appearance. Possibilities include

- Color change of background

- Changing the lighting of the scene by adjusting ambient light, applying spot lighting, or by adding extra lights

- Introducing scene perspective

- Introducing "fog" to get atmospheric perspective

3.4.5 Selection Abilities

Selection allows you to designate parts of the molecule on which future commands will act. For example, a selection may determine the parts of a molecule that are subjected to a change of color, graphical representation, etc.

A selection may be done by specifying a

- Protein chain

- Element (C, H, N, O, P, S)

- Functional group

- Amino acid name or amino acid category

- Structure

- Subsequence (usually done via a separate dialog box that lists the residues)

- Prosite pattern

- Zone (atoms within some distance of a designated atom)

For example, UCSF Chimera allows functional groups such as the following to be selected: acyl halide, aldehyde, akyl halide, amide, amine (aliphatics and aromatics), aromatic ring, carbonyl, carboxylate, disulfide, ester, ether, hydroxyl, imine, methyl, nitrile, nitro, nucleoside base (adenine, cytosine, guanine, thymine, and uracil), phosphate, phosphinyl, purine, pyrimidine, sulfide, sulfonate, sulfone, thiocarbonyl, and thiol.

In Chimera, amino acid categories include aliphatic, aromatic, charged, hydrophobic, negative, polar, positive, small, and tiny. Structure selection allows one to specify backbone, ions, ligand, main, secondary structure (helix, strand, or turn), side chain/base, or solvent.

In some cases, selections of atoms or residues are made using a mouse click with the cursor hovering over a residue in the scene. Once a selection is made, functionality may be provided to modify that selection by

- Clearing the selection

- Inverting or complementing a selection (selecting everything but the items previously selected)

- Selecting all of the molecule

- Undoing a selection

- Naming and storing selections for later restoration

3.4.6 Computational Tools

Some visualization software can perform various computations on the molecule being displayed. This may include computations dealing with molecular geometry (calculating bond lengths, bond angles, dihedral angles, or the distances between two arbitrary atoms). There are various applications such as drug design where it is very beneficial to calculate and display hydrogen bonds. Other useful calculations include the evaluation of force fields as described earlier in Chapter 1. As an example of these capabilities, we mention that Swiss PdbViewer has several computational tools including hydrogen bond calculations, surfaces, electrostatic potentials, electron density maps, and energy minimization.

3.4.7 Extras

These are features that are seldom seen but can be very useful:

- Session save and restore

- Saving images

- Collaboratory sessions with other people

- Structure comparisons and structural alignment

- Molecular animations

- Computation of Ramachandran plots

3.5 SOFTWARE FOR STRUCTURAL BIOINFORMATICS

We finish this chapter with a short list of software packages that are useful for structural bioinformatics with a concise overview of the functionality provided. There are many packages being sold by companies for the pharmaceutical industry. Some are quite expensive. We will avoid any product promotion and will concentrate on software packages that are freely available (including open source packages).

3.5.1 PyMOL

As stated on the PyMOL Web site:

> PyMOL is a molecular graphics system with an embedded Python interpreter designed for real-time visualization and rapid generation of high-quality molecular graphics images and animations. It can also perform many other valuable tasks (such as editing PDB files) to assist you in your research.

PyMOL has already been mentioned in the previous section as visualization software, but it is also included in this section because it has the objective of providing functionality for computational chemistry, modeling, and informatics. More importantly, the software (available from http://pymol.sourceforge.net/) is an open source.

PyMOL has good quality graphics and Warren DeLano, the principal at DeLano Scientific, takes great pride in the number of times PyMOL is used to generate molecular scenes for the covers of various journals. The home page has a Wiki and a FAQ section.

3.5.2 Eclipse

Eclipse and its associated Rich Client Platform is an open development system that is freely available from http://www.eclipse.org/. Although not directly intended for bioinformatics, it is a good starting point for rapid Java development with the ability to provide the design of sophisticated user interfaces. In particular, the Rich Client Platform helps the developer in the building of platform-independent applications. For some developers, there might be an issue related to somewhat slow Java execution times when applications are calculation intensive. There is a lot of documentation support (although some of it might be out of date because the development platform has evolved quite rapidly).

Bioclipse (http://www.bioclipse.net/) is one of the first bioinformatics applications to go along this development route.

3.5.3 MarvinSketch

There are applications that allow a user to build small molecules using an interface that facilitates the construction of a 2-D representation of a molecule. This is then converted to a file representation in a conventional format, for example, a structure-data file (SDF). If you need a 3-D structure, there are packages that will accept the 2-D version and will do an energy minimization analysis to generate a low energy 3-D conformation.

A good example of a 2-D drawing package is MarvinSketch available from ChemAxon: www.chemaxon.com/marvin/. There is a Java applet that is freely available. You can also get a download to operate as a standalone. The interface allows you to build molecules, and it will also enable seeing your molecule as a 3-D atom/bond representation. This can be saved as an image or as a structure file. Some calculation tools are available (others may require the purchase of a license).

3.5.4 ACD/ChemSketch

ChemSketch is available from ACD/Labs: http://www.acdlabs.com/. ACD/Labs is a company that also offers other "free stuff," and they take pride in being kind to academia: http://www.acdlabs.com/educators/.

ChemSketch has many sophisticated and useful features:

- Runs in structure or draw mode

- Can build animations

- Can export to PDF

- Converts InChI codes into chemical structures

- Can search structure in PubChem and eMolecules Web databases

Some capabilities are only available for the commercial version.

3.5.5 JOELib2

JOELib2 is a Java-based cheminformatics algorithm library for computational chemistry. As such, it does not really have a graphical–user interface. The software is available from http://www-ra.informatik.uni-tuebingen.de/software/joelib/.

It provides algorithm support for

- Structured data mining of molecular databases

- Feature calculation for QSAR analysis

- Fragmenting and generating molecules

- Many computational chemistry calculations that were originally in the OELib library from OpenEye Scientific Software*

3.5.6 Chemistry Development Kit (CDK)

CDK is a Java library for structural cheminformatics and bioinformatics. It is available from http://sourceforge.net/projects/cdk/. CDK is used by other software to provide functions that are needed for their calculations. Examples include Bioclipse, JChemPaint, Jmol, and others.

CDK facilitates many calculations. Here is a very partial list:

- QSAR descriptor calculations

- Surface areas

- Geometric transformations

- Writing molecule objects in various output formats

- Determination of maximum common substructures (for drug design)

- Fragmenting of molecules

3.5.7 BioPython

The BioPython project (see http://biopython.org/wiki/Main_Page) is spearheaded by a group of developers who are making Python-based tools freely available for computational molecular biology. BioPython is useful for

- Working with biological sequences

- Parsing biological file formats (such as PDB files)

---

* http://www.eyesopen.com/products/toolkits/oelib.html.

- Connecting with biological databases such as SCOP, PubMed, and Entrez

- Developing classification and regression algorithms in bioinformatics

- Deriving a position-specific scoring matrix (PSSM)

If you do not use BioPython, consider BioPerl: http://www.bioperl.org/wiki/Main_Page.

3.6 EXERCISES

1. Write a program that extracts alpha carbon coordinates from a PDB file.

   a. Input: the text of a PDB file.

   b. Output: a file with as many lines as there are alpha carbon atoms in the file. Each line should have six fields containing chain identifier, residue number, residue name, x-coordinate, y-coordinate, and z-coordinate.

2. Download and install the UCSF Chimera Extensible Molecular Modeling System available from http://www.cgl.ucsf.edu/chimera/. Use the software to recreate various figures from Chapter 2. In the following list, we provide a sentence that describes the reason for the exercise:

   a. Figure 2.15: Learn how to use menus to select residues in a beta sheet and display it (in round ribbon form) along with the stick form of the residues.

   b. Figure 2.39: Select the twelve residues from the protein (after doing the relevant exercise), and then do the coloring of the atoms and bonds after generating the ribbon display.

   c. Figure 2.14: Get the beta sheet and display the backbone atoms in the strands.

   d. Figure 2.22: Select the heme group and do the appropriate coloration.

   e. Figure 2.6: Select the cysteine residues and show them in a ribbon diagram.

   f. Figure 2.32: Develop skills in the selection of chains and peptide sequences connected by cysteine bridges followed by application of colors taken from the "all colors" palette.

g. Figure 2.7: Get the SDF file for valine from PubChem and then open the file with Chimera. Use the "Color/by element" menu to color the atoms and bonds.

h. Figures 2.9 and 2.12: Get familiar with menu descents that do a hydrogen bond display. You will also need to select the backbone atoms that are confined to the alpha helix of Figure 2.9 or the beta sheet of Figure 2.12.

i. Figure 2.37 is a good exercise for the selection of nucleotide subsequences and the application of different colors to each selection.

j. Figure 2.23 will give you some experience in the generation of molecular surfaces.

3. Use Chimera to display the surface of the actin-binding region of the dystrophin homologue utrophin (1QAG). Note that this structural protein has two identical chains, each comprised of helices and loops. Display a surface over the entire molecule with the surface over each chain rendered in a different color.

4. Use Chimera to calculate surface area and volume of the following proteins: crambin, pepsin, and the capsid of the PhiX174 virus (1CD3).

5. Use Chimera to do the following: Access the protein structure of baker's yeast topoisomerase II (PDB code 1BJT). Note that this structure only deals with the C-terminal fragment (residues 409–1201). Working with Chimera do the following:

a. Count the number beta strands in this structure.

b. List the number and name of every residue in each beta strand.

c. Determine the number of parallel beta sheets.

d. Determine the number of antiparallel beta sheets.

Repeat steps (a), (b), (c), and (d) using PDBsum. Do you get the same results?

6. The crystal structure for the enzyme ribonuclease T1 (PDB ID 1RGA) includes 142 water molecules and a calcium atom. Work with Chimera to do the following:

     a. Fetch the structure with PDB ID 1RGA

     b. Use Chimera to display hydrogen bonds among the water molecules and between the water molecules and the protein.

     c. Generate a surface over the protein. Make the surface somewhat transparent so that one can see how hydrogen bonds go from various water molecules through the surface and to the partially visible atoms of the protein below the surface.

7. Use SCOP to determine the class, fold, superfamily, and family of chain A in calmodulin (1CM1). How does CATH categorize chain A of calmodulin?

8. How many domains are there in the catalytic fragment of DNA polymerase beta from rat (PDB code 1RPL)?

9. Defects in the human PAX6 gene cause aniridia,* a malformation of the eye. Do an Internet search to discover the protein that is expressed by PAX6. Describe its function along with a diagram of the protein bound to DNA.

10. Look up aspirin in PubChem to get a diagram of its 2-D structure. Go to the DrugBank home page and learn to use the ChemQuery facility. Use the Marvin Sketch application to draw the 2-D structure for aspirin and then convert it to a MOL file that is subsequently submitted to DrugBank. You should get the DrugCard for aspirin. Need more experience with Marvin Sketch? Repeat this exercise for nicotine (described by PubChem as a highly toxic alkaloid).

11. Use Chimera to display calmodulin (1CM1). Look at the sequence for chain A. Using another Chimera window, access the protein 1K93. Look at the sequences for chains D, E, and F. How do they compare with chain A of 1CM1? Do some Internet searching to explain what is going on in 1K93.

12. If you have experienced "intestinal problems" when traveling, it was probably an enterotoxin that was given to you by a slightly† aggressive strain of *E. coli*. Do some Internet searching to discover what is

---

* http://nobelprize.org/nobel_prizes/medicine/laureates/1995/illpres/more-anaridia.html.

† Your gut contains huge quantities of *E. coli* but they are typically nonvirulent. Virulent *E. coli* will cause either noninflammatory diarrhea (watery diarrhea) or inflammatory diarrhea (dysentery with stools that usually contain blood and mucus).

going on at the molecular level. This should involve an explanation of the activities associated with a particular protein (PDB code 1LTT). Use Chimera to illustrate functionality at the molecular level. Cholera utilizes very aggressive protein toxins. What is more deadly for a cell, one molecule of cyanide or one molecule of a cholera toxin?

REFERENCES

[Ah04] A. Andreeva, D. Howorth, S. E. Brenner, T. J. P. Hubbard, C. Chothia, and A. G. Murzin. SCOP database in 2004: refinements integrate structure and sequence family data. *Nucl. Acid Res.*, **32** (2004), D226–D229.

[Bw00] H. M. Berman, J. Westbrook, Z. Feng, G. Gilliland, T. N. Bhat, H. Weissig, I. N. Shindyalov, and P. E. Bourne. The Protein Data Bank. *Nucl. Acid Res.*, **28** (2000), 235–242.

[Gl06] L. H. Greene, T. E. Lewis, S. Addou, A. Cuff, T. Dallman, M. Dibley, O. Redfern, F. Pearl, R. Nambudiry, A. Reid, I. Sillitoe, C. Yeats, J. M. Thornton, and C. A. Orengo. The CATH domain structure database: new protocols and classification levels give a more comprehensive resource for exploring evolution. *Nucl. Acid Res.*, **35** (2007), D291–D297.

[Hk01] M. Heller and H. Kessler. NMR spectroscopy in drug design. *Pure Appl. Chem.*, **73** (2001), 1429–1436.

[Ke61] J. C. Kendrew. The three-dimensional structure of a protein molecule. *Sci. Am.*, **205**(6) (1961), 96–110.

[Lb02] L. Lo Conte, S.E. Brenner, T.J.P Hubbard, C. Chothia, and A. Murzin. SCOP database in 2002: refinements accommodate structural genomics. *Nucl. Acid Res.*, **30** (2002), 264–267.

[Mb95] A. G. Murzin, S. E. Brenner, T. Hubbard, and C. Chothia. SCOP: a structural classification of proteins database for the investigation of sequences and structures. *J. Mol. Biol.*, **247** (1995), 536–540.

[Ns04] S. B. Nabuurs, C. A. E. M. Spronk, G. Vriend, and G. W. Vuister. Concepts and tools for NMR restraint analysis and validation. *Concepts in Magnetic Resonance Part A*, **22A** (1995), 90–105.

[Om97] C. Orengo, A. Michie, S. Jones, D. Jones, M. Swindells, and J. Thornton. CATH—a hierarchic classification of protein domain structures. Structure, 5 (1997), 1093–1108.

[Pb03] F. Pearl, C. Bennett, J. Bray, A. Harrison, N. Martin, A. Shepherd, I. Sillitoe, J. Thornton, and C. Orengo. The CATH database: an extended protein family resource for structural and functional genomics. *Nucl. Acid Res.*, **31** (2003), 452–455.

[Pt05] F. Pearl, A. Todd, I. Sillitoe, M. Dibley, O. Redfern, T. Lewis, C. Bennett, R. Marsden, A. Grant, D. Lee, A. Akpor, M. Maibaum, A. Harrison, T. Dallman, G. Reeves, I. Diboun, S. Addou, S. Lise, C. Johnston, A. Sillero, J. Thornton, and C. Orengo. The CATH

Domain Structure Database and related resources Gene3D and DHS provide comprehensive domain family information for genome analysis. *Nucl. Acid Res.*, **33** (2005), Database Issue D247–D251.

[To89a] W. R. TAYLOR AND C. A. ORENGO. Protein structure alignment. *J. Mol. Biol.* **208** (1989), Vol. 208, 1.

[To89b] W. R. TAYLOR AND C. A. ORENGO. A holistic approach to protein structure alignment. *Prot. Eng.* **2** (1989), 505–519.

[Uf01] S. ULIEL, A. FLIESS, AND R. UNGER. Naturally occurring circular permutations in proteins. *Prot. Eng.* **14** (2001), 533–542.

[Wk06] D. S. WISHART, C. KNOX, A. C. GUO, S. SHRIVASTAVA, M. HASSANALI, P. STOTHARD, Z. CHANG, AND J. WOOLSEY. DrugBank: a comprehensive resource for *in silico* drug discovery and exploration. *Nucl. Acid Res.*, **34** (2006), Database Issue D668–D672.

[Wu86] K. WUTHRICH. *NMR of Proteins and Nucleic Acids*. John Wiley, New York, 1986.

CHAPTER 4

Dynamic Programming

What title, what name, could I choose? In the first place I was interested in planning, in decision making, in thinking. But planning is not a good word for various reasons. I decided therefore to use the word, "programming." I wanted to get across the idea that this was dynamic, this was multistage, this was time-varying—I thought, let's kill two birds with one stone. Let's take a word that has an absolutely precise meaning, namely *dynamic*, in the classical physical sense. It also has a very interesting property as an adjective, and that is it's impossible to use the word *dynamic* in a pejorative sense. Try thinking of some combination that will possibly give it a pejorative meaning. It's impossible. Thus, I thought "dynamic programming" was a good name. It was something not even a congressman could object to. So I used it as an umbrella for my activities.

RICHARD BELLMAN, EXPLAINING HOW HE CHOSE
THE NAME "DYNAMIC PROGRAMMING" IN THE BOOK
EYE OF THE HURRICANE
(*World Scientific Publishing, Singapore, 1984*)

4.1 MOTIVATION

"Dynamic programming" is an algorithm originally developed by Richard Bellman in the early 1950s. At that time, he was working on applied problems dealing with multistage decision processes. As noted in the previous quote, he decided to use the term *dynamic programming* instead of "multistage decision processes" as a name for the algorithm that we are about to describe in this chapter. The "multistage decision" nature of the algorithm will eventually become clear.

For bioinformatics in general, dynamic programming is undoubtedly the most useful paradigm for the design of algorithms. Most students will likely have covered this topic in an earlier year, but it is so important that we review it once again in detail. Dynamic programming is presented at an abstract level in order to present the general steps involved with the strategy.

4.2 INTRODUCTION

Dynamic programs solve optimization problems. The overall strategy underlying the algorithm is that the given problem can be solved by first solving smaller subproblems that have the same structure. Two features characterize these subproblems:

1. Unlike the divide-and-conquer algorithm, which relies on subproblems being independent, dynamic programming handles problems that have overlapping subproblems.

2. We can apply a recursive strategy: We find an optimal solution of each subproblem by first finding optimal solutions for its contained subproblems.

4.3 A DP EXAMPLE: THE AL GORE RHYTHM FOR GIVING TALKS

We start with a simple, dynamic programming example. The bioinformatics content is nonexistent, but it is simple enough to illustrate the main ideas underlying the strategy of dynamic programming.

4.3.1 Problem Statement

You have been hired as a consultant to help Al Gore promote his campaign against global warming. Gore has already decided to tour various cities in the United States in order to lecture on this subject. The tour will take place over n consecutive days. For each of these days, Gore has prescheduled each meeting site, and he knows exactly how many people will show up. We will assume that the ith day has a meeting with $u(i)$ attendees. Although Gore would like to attend all these meetings, the demands of his campaign have weakened him to such an extent that he cannot attend meetings on two consecutive days, and so some meetings must be skipped. (When he does not attend a meeting, there will be a free showing of the movie *An Inconvenient Truth*).

Design an algorithm that will select the days when Mr. Gore will meet the attendees with the objective that, for the entire tour, he will meet the maximum number of these people. Rephrasing the meeting constraint, if he meets attendees on day $i - 1$, then he cannot meet attendees on day i. We will assume there is no meeting planned on the day prior to day 1.

4.3.2 Terminology: Configurations and Scores

We should always distinguish two aspects of optimization problems that are amenable to a dynamic programming solution. We ultimately want some type of *configuration* that provides an answer to the problem. In this problem, a configuration is simply some selection of days chosen from the available n days. The definition of a configuration will change from one problem to the next. The other aspect of the optimization problem is that we can assign a *score* to each configuration. In this problem, the score for a configuration will be the sum of all $u(i)$ values such that i ranges across the selected days that make up the configuration.

With this terminology in place, we can see that the solution of the optimization problem is the derivation of a configuration that provides the optimal score. For this problem, optimal means maximal. For other problems, we might want to find the configuration with the minimal score.

4.3.3 Analysis of Our Given Problem

As stated earlier, we find an optimal solution of our given problem by first finding an optimal solution for various subproblems. We consider the following definition of a subproblem: Find the configuration that gives the optimal score for a problem that is limited to the first i days of the tour. We let this optimal score be represented by the function $S(i)$. Now, we need an analysis of the situation: For day i, Mr. Gore either meets attendees or does not.

- If he does attend the meeting, then for day i the sum $S(i)$ includes $u(i)$, day $i - 1$ is skipped, and moreover, the selection from the first $i - 2$ days must be an optimal selection. If it was not optimal, then we could replace this nonoptimal configuration of days with a better selection having a larger sum of attendees. This sum plus the value of $u(i)$ for day i would lead to a larger score for $S(i)$ in contradiction of the fact that $S(i)$ is already assumed to be optimal. So, Gore meeting attendees on day i implies $S(i) = u(i) + S(i - 2)$ where $S(i - 2)$ is the optimal score for the first $i - 2$ days.

- If Gore does not meet attendees on day i, then nothing has changed since the previous day. In other words, $u(i)$ is not part of $S(i)$, and $S(i) = S(i - 1)$.

Because we want to choose the scenario that yields the larger score $S(i)$, we simply insist that $S(i)$ is the maximum of these two possibilities, in other words:

$$S(i) = \max\{u(i) + S(i - 2), S(i - 1)\}.$$

Pseudocode for the algorithm could be written as

```
S[0] := 0; S[1] := u[1];
for i := 2 to n do
     if (S[i - 1] < S[i - 2] + u[i]) then
          S[i] := S[i - 2] + u[i];
     else
          S[i] := S[i - 1];
return S[n];
```

Note that the value of $S(n)$ will be the best score that is achievable for the entire n days. With a bit of thought, you should also realize that computation of the $S(i)$ values must be completed for all $i = 1, 2, ..., n$ before we can attempt to derive the configuration that gives the solution of the problem. In fact, we must inspect both $S(n)$ and $S(n - 1)$ before discovering whether or not Mr. Gore will attend the meeting on day n. Once that is established, we can then determine whether the meeting on day $n - 1$ is skipped or attended, then we consider $S(n - 2)$, etc. The pseudocode for this algorithm is

```
i := n;
while i > 0
     if (S[i - 1] <> S[i]) then
          write("Day" i "chosen");
          i := i - 2;
     else
          i := i - 1;
```

This completes the solution of the given problem. Now, let us extract from this solution some procedural steps that can be applied to all dynamic programming solutions.

The first step was the most critical one because it identified the subproblems that were to be solved in order to get the solution of the full problem. This means that the optimal configuration for the given problem will be constructed from the optimal configurations of subproblems. These subproblems will have optimal configurations that are in some sense "smaller." The challenge of designing a dynamic programming algorithm will be to understand how these smaller optimal configurations relate to larger optimal configurations and eventually to the largest optimal configuration that we ultimately need. However, if you look back at the analysis of the problem, you will see that this relation is formulated in terms of the scores derived from the optimal configurations. This produces the recursion that is at the heart of the solution.

In general, the score value in the recursion is dependent on various parameters that characterize the extent of the subproblems. In our given problem, S depended only on i. In more complex dynamic programming problems, there may be two (perhaps even three) parameters that act as independent variables for the score function. We will see that it is often advantageous to consider the values of the scoring function to be stored in an array that has as many dimensions as there are parameters.

Using this array, we can visualize the computation of the score function recurrence by stipulating where the initial values are stored and the order of computation when filling the cells of the array. The analysis of the problem will also specify the array cell containing the optimal score for the given problem. Finally, the analysis of the problem will dictate how the solution configuration is recovered. This final step is usually referred to as the trace-back or solution recovery step.

Be sure to understand the significance of the word "back" in the trace-back step. The optimal scores for the subproblems are typically generated by having one or more parameters of the scoring function progressing through values in a direction that we characterize as the "forward" direction. For example, in the Al Gore problem, we have the index i going from 1 to n when the score values are evaluated. After all these computations are completed, we recover the configuration for the optimal solution by reversing direction and letting the parameter (or parameters) progress through values in the opposite direction. In our given problem, i progresses from n down to 1 in the trace-back part of the algorithm.

For students who are just learning about dynamic programming, it is crucial to stress that the formulation of a dynamic programming algorithm should deal first with the scoring function, and then later, with the derivation of the configuration. When faced with a new and complicated dynamic programming problem, inexperienced students will sometimes attempt to derive a final configuration much too soon—immediately trying to understand how the configuration is computed while treating the scoring function as if it is an evaluation to be done after configurations are somehow calculated. This often leads to confusion and failure.

4.4 A RECIPE FOR DYNAMIC PROGRAMMING

To improve the chances of success when designing a dynamic programming algorithm, here is a summary of the steps that are usually taken.

1. **Identify the subproblems and provide a supporting analysis that produces a recursion for the evaluation of the scoring function.**

   We need to specify how the optimal score of a subproblem contributes to the optimal score of a larger subproblem.

2. **Provide details that stipulate how the score values are evaluated.**

   Typically, we store the optimal scores of the subproblems in an array (possibly a big, multidimensional array). We need to

   - Specify the matrix dimensions

   - Describe how the value in a cell of the array depends on the values of other cells in the array (thus specifying the order of computation)

   - Specify where the final optimal score for the full problem will be found in the array

3. **Set values for the base cases.**

   There will be cells that require initial values because the recursion equation alone cannot define all the score values in the array.

4. **Recover the solution configuration.**

   - Keep track of the configurations that provide the optimal values

   - Use a trace-back strategy to determine the full configuration that gives the optimal value for the full problem

4.5 LONGEST COMMON SUBSEQUENCE

Now that we have some dynamic programming methodology established, we apply it to a problem with some extra complexity. The longest common subsequence (LCS) problem usually arises from methods that need to compare strings of nucleotides. The solution of the problem serves to provide a measure of the amount of similarity between two nucleotide sequences. We start with some abstractions that define the problem:

> *Definition:* A **subsequence** of a string X is a string that can be obtained by extracting some or all of the characters from X while still maintaining their sequence order.

For example, if X = "supercalifragilisticexpialidocious," then the characters T = "cialis" would be a subsequence. There are approximately 16 billion subsequences that can be extracted from X.

The reader should not confuse this with the notion of a *substring*, which is a more specialized extraction of characters. A **substring** is a subsequence of *consecutive* characters that are extracted from X.

4.5.1 Problem Statement

We are given two strings: $X = x_1, x_2, ..., x_m$, and $Y = y_1, y_2, ..., y_m$. The problem is to find the longest string Z that is a subsequence of both X and Y.

For example, suppose

X = "supercalifragilisticexpialidocious" and

Y = "supercalafajalistickespeealadojus."

Then, the LCS of X and Y is Z = "supercalfalisticepaldous."*

A small issue of concern is the presentation of strings X, Y, and Z so that a reader can easily verify that Z is indeed the LCS of X and Y. One could write Y below X and then draw lines between the matching characters that produce Z. This is somewhat cumbersome and is typically not done. The usual technique is to insert dash characters into both X and Y strings, so that they have the same length and written out so that matching characters are in the same column. For example:

X: **supercalifragilistic-expi-ali docious**

Y: **supercalaf-aja listickespeealado--jus**

Z: **supercal f a listic e p al do us**

Characters that produce the LCS have been rendered in a bold font.

---

* There is some history associated with this particular Y string. It is related to a legal action brought against Walt Disney Company. See http://en.wikipedia.org/wiki/Supercalifragilisticexpialidocious. Incidentally, the string Y is flagged by Microsoft Word as a misspelling, but X is accepted.

The LCS is not necessarily unique. For example, let X = "ATCTGAT," Y = "TGCATA." Then, both Z = "TCTA" and Z = "TGAT" are legitimate as an LCS. Naturally, both possibilities have the same optimal string length.

4.5.2 Prefixes

We compute the LCS using dynamic programming. Our strategy is to derive the LCS of the given strings X and Y, by first solving various subproblems. Each subproblem involves the computation of an LCS for some pair of *prefixes* of X and Y.

> *Definition:* The ith prefix of a string X, denoted by $X(i)$, will be the substring of X made up from the first i characters of X.

We use lower case, subscripted letters to denote the characters in X. Formally, we denote the full string X as $X = x_1 x_2 \ldots x_m$. Then, the prefix $X(i) = x_1 x_2 \ldots x_i$. Assuming X has m characters, we have $X = X(m)$.

Given input strings $X = x_1 x_2 \ldots x_m$ and $Y = y_1 y_2 \ldots y_n$, our subproblem will be to find the LCS of $X(i)$ and $Y(j)$. Note that the subproblem is parameterized using two indexes i and j with $1 \le i \le m$ and $1 \le j \le n$. The score value for such a subproblem, denoted by $S[i, j]$, is the length of the LCS of the prefixes. The configuration for this subproblem is the LCS itself.

Having defined this notation and the significance of the score function, we are about halfway through the first step of our recipe for dynamic programming introduced in the previous section. It is now necessary to provide the supporting analysis that leads to a recursion defining $S[i,j]$. Because prefixes are essentially defined by their last position, we will be interested in the following "clever observations" about the last character of an LCS.

4.5.3 Relations Among Subproblems

Suppose we consider $X(i) = x_1 x_2 \ldots x_i$ to be a prefix of X, and $Y(j) = y_1 y_2 \ldots y_j$ to be a prefix of Y. Let $Z(k) = z_1 z_2 \ldots z_k$ be the LCS of $X(i)$ and $Y(j)$. There are two possibilities for the last characters of $X(i)$ and $Y(j)$. Either $x_i = y_j$ or $x_i \ne y_j$. Then, the following observations can be made:

1. If $x_i = y_j$, then $z_k = x_i = y_j$, and $Z(k - 1)$ is an LCS of $X(i - 1)$ and $Y(j - 1)$.

2. If $x_i \ne y_j$, then we have the negation of ($z_k = x_i$ and $z_k = y_j$). This means that $z_k \ne x_i$ or $z_k \ne y_i$. This gives us two subcases:

a. $z_k \neq x_i \Rightarrow Z(k)$ is an LCS of $X(i-1)$ and $Y(j)$.

b. $z_k \neq y_i \Rightarrow Z(k)$ is an LCS of $X(i)$ and $Y(j-1)$.

Let us consider Case 1: If $z_k \neq x_i$, then we could simply append x_i to $Z(k)$ to obtain an LCS of $X(i)$ and $Y(j)$ with longer length than k, contradicting the assumption that $Z(k)$ was the *longest* common subsequence. So, $z_k = x_i = y_j$. What about $Z(k-1)$, the characters of $Z(k)$ prior to z_k? We know that they are a common subsequence of $X(i-1)$ and $Y(j-1)$, but can we say that $Z(k-1)$ is the *optimal* common subsequence of $X(i-1)$ and $Y(j-1)$? In other words, has our claim that $z_k = x_i = y_j$ somehow compromised $Z(k-1)$ and it is not optimal? No. If there was an LCS W of $X(i-1)$ and $Y(j-1)$ that had k or more characters, then we would simply append x_i to it to get an LCS of $X(i)$ and $Y(j)$ longer than $Z(k)$, and this would be a contradiction. Consequently, we have the fact that $Z(k-1)$ is an LCS of $X(i-1)$ and $Y(j-1)$.

Now consider Case 2(a): If $z_k \neq x_i$, then $Z(k)$ must be a common subsequence of $X(i-1)$ and $Y(j)$. Using reasoning similar to that of Case 1, we see that $Z(k)$ must be the *longest* common subsequence of $X(i-1)$ and $Y(j)$. If there was a longer common subsequence of $X(i-1)$ and $Y(j)$ (say W with length greater than that of $Z(k)$), then, although W is also a subsequence of both $X(i)$ and $Y(j)$, we would have a common subsequence of $X(i)$ and $Y(j)$ longer than Z, and so derive a contradiction.

Reasoning for Case 2(b) is symmetric to that of Case 2(a).

4.5.4 A Recurrence for the LCS

The different cases just described have implications that define the score function $S[i, j]$. Recall that $S[i, j]$ is the length of the LCS of $X(i)$ and $Y(j)$, in other words, it is the length of $Z(k)$.

1. Case 1: The LCS of $X(i)$ and $Y(j)$ is the LCS of $X(i-1)$ and $Y(j-1)$ extended by one character. This implies that $S[i, j] = S[i-1, j-1] + 1$.

2. Case 2(a): The LCS of $X(i-1)$ and $Y(j)$ is the same as the LCS of $X(i)$ and $Y(j)$. This implies that $S[i, j] = S[i-1, j]$. A similar analysis for Case 2(b) gives us $S[i, j] = S[i, j-1]$.

We want to extract the longest common sequence, and this means that for Case 2 our computation for $S[i, j]$ will be evaluated by selecting the maximum of the two options described. Consequently, the main recurrence is

$$S[i,j] = \begin{cases} S[i-1, j-1]+1 & \text{if } x_i = y_j \\ \max\{S[i-1, j], S[i, j-1]\} & \text{if } x_i \neq y_j. \end{cases} \tag{4.1}$$

This completes Step 1 of our recipe for dynamic programming. For Step 2, we describe the data structure that retains the scores and the computational procedure to evaluate them. The $S[i, j]$ scores will be stored in an n by m array. Considering the recursion, we see that score $S[i, j]$ held in cell $[i, j]$ will depend on score values stored in cells $[i-1, j-1]$, $[i-1, j]$, and $[i, j-1]$. In other words, $S[i, j]$ depends on score entries in a previous row or column. Consequently, it is reasonable to have an order of evaluation that fills in the matrix row by row, starting with the first row. The very last computation will be that of $S[m, n]$, which represents the length of the LCS for the full strings X and Y. This completes Step 2.

Step 3 handles the necessary initialization needed to begin the evaluation of the recursion. The recursion formula claims that entries in row $i = 1$ would require prior values to be set up in row $i = 0$. This can be done by supplying an extra row 0 and column 0 with all entries initialized to 0. These values are independent of the characters in the given sequences. We are really saying that the length of LCS is 0 when one or more of the strings is null (zero length).

Step 4 deals with the recovery of the solution configuration. Filling the score matrix will only compute the length of the LCS. To get the actual LCS, we have to keep track of the cell that is responsible for producing the $S[i, j]$ value. For each $[i, j]$ cell, we keep $D[i,j]$ a back-pointer to one of $[i-1, j-1]$, $[i-1, j]$, or $[i, j-1]$, depending on which cell was used to determine $S[i, j]$. We use D for "direction" and give each direction a symbol that would be represented in the program using some type of encoding:

$$D[i,j] = \begin{cases} \text{up-left} & \text{if} & S[i,j] = S[i-1, j-1]+1 \\ \text{up} & \text{if} & S[i,j] = S[i-1, j] \\ \text{left} & \text{if} & S[i,j] = S[i, j-1]. \end{cases} \tag{4.2}$$

This completes all four steps of the recipe.

Here is a pseudocode version of the algorithm:

```
// Base cases
for i := 0 to m do S[i,0] := 0;
for j := 0 to n do S[0,j] := 0;

// Filling the S[] and D[] matrices
for i := 1 to m do
  for j := 1 to n do
    if (x[i] == y[j]) then
      S[i,j] := S[i - 1, j - 1] + 1;
      D[i,j] := "up-left";
    else S[i,j] := S[i - 1,j];
         D[i,j] := "up";
         if (S[i,j - 1] > S[i - 1,j]) then
             S[i,j] := S[i,j - 1];
             D[i,j] := "left";
return S[m,n];
```

Figure 4.1 presents an example of the score matrix with the direction matrix superimposed. The bottom rightmost cell gives the length of the LCS when the input strings are X = "LOGARITHM" and Y = "ALGORITHM."

To use the back-pointers, we simply start in the final location containing the highest score and then follow the pointers. This produces a path that defines the LCS. All other back pointers are ignored. Figure 4.2

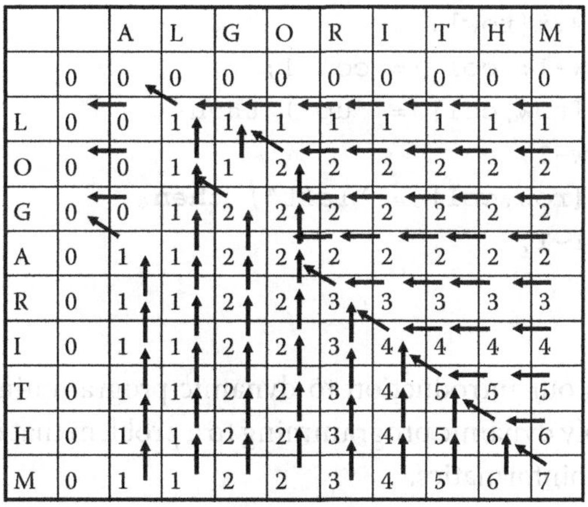

		A	L	G	O	R	I	T	H	M
	0	0	0	0	0	0	0	0	0	0
L	0	0	1	1	1	1	1	1	1	1
O	0	0	1	1	2	2	2	2	2	2
G	0	0	1	2	2	2	2	2	2	2
A	0	1	1	2	2	2	2	2	2	2
R	0	1	1	2	2	3	3	3	3	3
I	0	1	1	2	2	3	4	4	4	4
T	0	1	1	2	2	3	4	5	5	5
H	0	1	1	2	2	3	4	5	6	6
M	0	1	1	2	2	3	4	5	6	7

FIGURE 4.1 Score matrix with superimposed direction matrix.

		A	L	G	O	R	I	T	H	M
	0	0	0	0	0	0	0	0	0	0
L	0	0	1	1	1	1	1	1	1	1
O	0	0	1	1	2	2	2	2	2	2
G	0	0	1	2	2	2	2	2	2	2
A	0	1	1	2	2	2	2	2	2	2
R	0	1	1	2	2	3	3	3	3	3
I	0	1	1	2	2	3	4	4	4	4
T	0	1	1	2	2	3	4	5	5	5
H	0	1	1	2	2	3	4	5	6	6
M	0	1	1	2	2	3	4	5	6	7

FIGURE 4.2   Producing the trace-back path.

shows the path extraction. The main idea behind the path extraction is to go through the optimal configurations and, while doing so, keep track of the occurrences when an "up-left" pointer is used. The origin of the pointer will be a cell that is in a row and column such that x_i and y_j are the same. Considering our previous analysis, this is the situation that causes a character to be appended to the LCS. The code has been written so that the LCS is generated in reverse order. It is then reversed in the final step. The pseudocode to do this trace-back would be as follows:

```
row := m; col := n; lcs := "";
while (row > 0 and col > 0) do
 if (D[row,col] == "upleft") then // x[row] = y[col]
  lcs := lcs.x[row];
  row := row-1; col := col-1;
 else if (D[row,col] = "up") then
  row := row-1;
 else if (D[row,col] = "left") then
  col := col-1;
reverse lcs;
return lcs;
```

This completes our introduction to dynamic programming. In the next chapter, we apply dynamic programming to a problem that is more related to structural bioinformatics.

4.6 EXERCISES

1. Assume you have an n-by-n checkerboard. Each square of the checkerboard is labeled with a coordinate pair so that the square in the ith row and jth column is labeled with the pair (i, j). You must move a checker from the top left corner square $(1, 1)$ to the bottom right corner square (n, n) using at most $2(n - 1)$ moves. Legal moves are as follows:

- A move of the checker down one square

- A move of the checker right one square

- A diagonal move of the checker down one square and to the right one square

In this problem, you will assume that on square $(1, 1)$ you have a purse with zero dollars, and this value may go up or down as you make moves on the checkerboard according to the following *payoff* scheme that dictates changes to the sum in your purse:

- A down move that leaves square (i, j) changes your sum by d_{ij} dollars

- A right move that leaves square (i, j) changes your sum by r_{ij} dollars

- A diagonal move that leaves square (i, j) changes your sum by g_{ij} dollars

You are given all the d_{ij}, r_{ij}, and g_{ij} values as input data. These values may be negative, zero, or positive, so it may be possible to have a negative sum in your purse at any time. You want to find a path from $(1, 1)$ to (n, n) using moves that get as much money as possible, or to end up with the least negative value in your purse, depending on the data.

By providing an efficient algorithm based on dynamic programming, your answer should follow the "four steps recipe" discussed in this chapter.

2. Suppose you are given a sequence of real numbers $x_1, x_2, x_3, \ldots, x_n$. Use a dynamic programming strategy to find a sequence of **consecutive**

numbers with the maximum sum.* In other words, we want to find i, and j such that the following sum is maximized:

$$\sum_{k=i}^{j} x_k.$$

If all the numbers are positive, we simply set $i = 1$, and $j = n$. To make the problem nontrivial, we assume that the sequence of real numbers contains both positive and negative numbers. The algorithm should have execution time $\Theta(n)$. Your answer should follow the "four steps recipe" discussed in this chapter. You should supply a short explanation that justifies the formulation of the recursion.

3. Repeat Exercise 1, subject to the following conditions:

   a. As before, you are working with an n-by-n checkerboard and are only allowed to make the legal moves described by Exercise 1.

   b. The input data is the same as that described by Exercise 1, and the rules for payoff are the same.

   c. You are allowed to start in any square with a purse of zero dollars.

   d. You may finish in any square as long as you follow the rules for legal moves.

   e. You should assume that at least one of the d_{ij}, r_{ij}, and g_{ij} values is positive.

As before, you want to find a path that will give you the most benefit.

---

* In algorithm design courses, this problem is usually known as "Bentley's problem."

CHAPTER 5

RNA Secondary Structure Prediction

The role of proteins as biochemical catalysts and the role of DNA in storage of genetic information have long been recognized. RNA has sometimes been considered as merely an intermediary between DNA and proteins. However, an increasing number of functions of RNA are now becoming apparent, and RNA is coming to be seen as an important and versatile molecule in its own right.

PAUL G. HIGGS
(Quarterly Reviews of Biophysics, 33, 2000, p. 199)

The idea that a class of genes might have remained essentially undetected is provocative, if not heretical. It is perhaps worth beginning with some historical context of how ncRNAs have so far been discovered. Gene discovery has been biased towards mRNAs and proteins for a long time.

SEAN R. EDDY
(Nucleic Acids Research, 27, 2001, p. 919)

5.1 MOTIVATION

The functionality of RNA is determined by its 3-D structure. As described earlier, this structure mainly comprises helical stems and various loops that cooperate to undertake a variety of cellular functions:

- mRNA

  Messenger RNA (mRNA) carries genetic information from DNA to the ribosome, where it directs the biosynthesis of polypeptides.

- tRNA

  Transfer RNA (tRNA) facilitates the transfer of a particular amino acid to the growing polypeptide when its anticodon region recognizes a codon of the mRNA.

- rRNA

  Ribosomal RNA (rRNA) is found in ribosomes and can act as the catalytic agent in protein synthesis.

- ncRNA

  Noncoding RNA (ncRNA) molecules are active in biological processes such as regulation of transcription and translation, replication of eukaryotic chromosomes, RNA modification and editing, mRNA stability and degradation, and protein translocation (excellent introductions to this exciting topic may be found in [ED01] and [ST02]). Strictly speaking, the ncRNA category includes tRNA and rRNA, but these are usually given their own separate categories. More examples of ncRNAs include the following:

  - tmRNA

    Transfer-messenger RNA (tmRNA) is best described as a combination of tRNA and mRNA (see [Zw99]).

  - snRNA

    Small nuclear RNA (snRNA) molecules are part of nuclear ribonucleoprotein particles responsible for splicing of eukaryotic mRNAs.

  - snoRNA

    Small nucleolar RNA (snoRNA) molecules are involved in rRNA modification (see the following discussion).

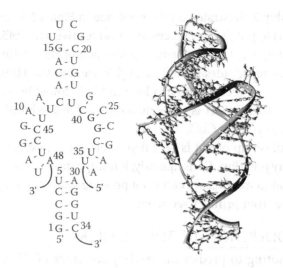

FIGURE 5.1 (See color insert following page 72). RNA–RNA interaction for pseudouridylation. (Adapted from [WF07]).

Recent studies in ncRNAs have uncovered new territory in the exploration of relationships between RNA structure and biological function. Although rRNAs certainly provide interesting functionality, their large size is somewhat daunting.* Studies dealing with smaller RNAs have data sets that are more computationally manageable, thus allowing the detailed analysis of intricate functional mechanisms.

A fine example of noncoding RNA functionality is the interaction between the H/ACA snoRNP and its RNA substrate (see [ED01], [ST02] and [WF07]). The H/ACA snoRNA has a "hairpin-hinge-hairpin-tail" arrangement that involves an H box consensus sequence (5′-ANANNA-3′, where N represents any nucleotide) in the hinge link between the two hairpins. The significance of the "ACA" suffix in the name is also related to a nucleotide sequence, this time in the tail section. The second hairpin has the secondary structure illustrated on the left side in Figure 5.1. The significant feature is a bulge in the stem that forms the so-called "ψ pocket." This pocket combines with a loop from a different RNA, in this case, S14. This is done by establishing 12 base pairs between S14 and the ψ pocket. The 3-D structure

---

* The PDB file 1FFK for the large ribosomal subunit contains 2833 of the subunit's 3045 nucleotides and 27 of its 31 proteins (64,281 atoms, 6497 residues in a 6.5 MB file). If you need to test protein visualization software to see if it can handle large files, 1FFK is highly recommended.

of this assembly is illustrated on the right side in Figure 5.1. The reason for this binding is to put S14 in a precise position relative to a Cbf5 protein (not shown in the figure) that performs a conversion of the uridine at position 41 to pseudouridine* (designated as ψ). We see that the H/ACA snoRNA performs sequence recognition of a loop in S14 followed by a careful setup of the loop so that its U41 is properly positioned in the binding site of the pseudouridine synthase Cbf5.

The careful set up of S14 is very dependent on the overall structure of this three-way junction. Consequently, it is important to understand RNA structure, and in many cases carry out procedures that attempt to predict 3-D structure from primary sequence.

5.2 INTRODUCTION TO THE PROBLEM

Before attempting to predict the tertiary structure of RNA, one would normally try to predict its secondary structure ([Wa78], [Ws78]). We start by describing the problem in terms that are more precise. The problem input will be the primary sequence of nucleotides in the RNA molecule. The objective is to determine where hydrogen bonding takes place among the various nucleotide bases in this primary sequence.

In effect, we want a function $h(i, j)$ defined as:

$$h(i,j) = \begin{cases} 1 & \text{if hydrogen bonding occurs between base}(i) \text{ and base}(j) \\ 0 & \text{if not.} \end{cases} \tag{5.1}$$

By knowing the secondary structure, we would have gained the first step in predicting the tertiary structure. As we will see, not all hydrogen bond interactions can be easily predicted. The predictive process will be some-what iterative, with the goal of steadily refining our assessment of the structure. For example, by determining the "easy" base pairs, we can hope to determine the loops and stems of the RNA molecule. Once these are established, we can try folding the stem structures to predict the tertiary structure, which may lead to the prediction of more base pairs.

Figure 5.2 will give some idea of the complexity involved. In this diagram, we see a drawing of the secondary structure of the nuclear RNase P RNA from *Bufo bufo* (the common toad). This structure was copied from the RNase P database [Br99]. This 397-nt RNA has many base pairs and a secondary structure that reveals the loops, bulges, stems, and multi-loops

---

* Recall from Chapter 2 that tRNA also contained a ψ at its position 55.

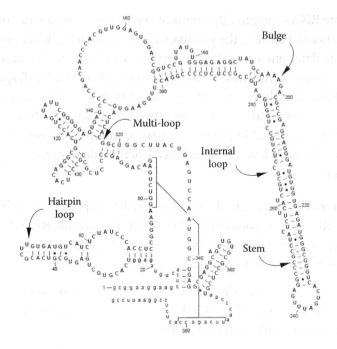

FIGURE 5.2   RNA secondary structure.

that are typical of more lengthy RNA molecules. Note the very sharp U-turn at positions 355, 356, and 357. This is the shortest loop possible. In general, we will assume that a hairpin loop contains at least three nucleotides.

As usual, we progress through the NATURE, SCIENCE, and COMPUTATION analysis to address the main issues:

- NATURE: What are the empirical observations associated with the problem?

- SCIENCE: What are the options for mathematical modeling to frame the problem?

- COMPUTATION: How do we compute answers for the problem?

5.2.1 NATURE

5.2.1.1 Where Do Hydrogen Bonds Form?

Hydrogen bonds between nucleotides will form when these nucleotides have a relative position and orientation that allows hydrogen bond donors to interact with hydrogen bond acceptors. The position and orientation can be established from observational data derived from experiments that investigate RNA structure. By means of crystallization and x-ray analysis, researchers can calculate the positions of all atoms in the RNA molecule. Using this data, one can drive the tertiary structure and the secondary structure that underlies it. Various other studies deal with NMR spectroscopy of RNA, and these experiments also provide conformational data.

These results lead to the following observations about hydrogen bonding. They are arranged in order of increasing complexity (perhaps better stated as increasing levels of annoyance when we do our modeling in the next section).

- Canonical base pairs: The Watson–Crick base pairing, A with U and G with C, corresponds to the AT and GC pairing that is seen in DNA. These are the base pairings seen earlier in Chapter 2, Figure 2.35 and typically occur in stem formations.

- Wobble base pairs: The base pairing G with U is occasionally seen, for example, [4::69] of tRNA[Phe], but substantially less often than either of the canonical base pairings.

- Pseudo-knot pairings: These hydrogen bonds are not due to stem formation but instead are formed when unpaired nucleotides, usually in loops or bulges, are able to establish hydrogen bonding when various stem structures are brought into proximity with one another. RNA researchers have described various types of pseudoknots as "kissing" hairpins, or hairpin-bulge contact.

- Triplets: As illustrated in Chapter 2, Figure 2.38, we cannot simply dismiss two nucleotides from further consideration after they have formed canonical base pairs in a stem region. It is still possible for them to become involved in other hydrogen bond formations to form a triplet.

5.2.1.2 Thermodynamic Issues

Achieving the correct position and orientation for hydrogen bond formation is accomplished when the molecule assumes a 3-D conformation that has minimal free energy [Mt06]. At the risk of oversimplification, it has been established that free energy is lowered by the establishment of base pairs in the stems and raised by other structural features such as loops and bulges (see [Ms99] for a more advanced analysis). For example, a short tight loop will be accompanied by loop strain, and this will raise the energy. The extra considerations that deal with energy-related matters that are different from stem formation help us to understand that an accurate description of the driving forces underlying structure must deal with energy issues and not just the counting of hydrogen bonds.

Thermodynamic theory tells us that the measure of free energy (ΔG) is represented by the equation $\Delta G = \Delta H - T\Delta S$, where ΔH is the enthalpy, T is the temperature, and ΔS is the entropy. However, to provide empirical data that is more easily applied to mathematical modeling, researchers have tried to evaluate free energy contributions in terms of *stacking energy*. In stacking-energy experiments, we assume that energy contributions are dependent on the four bases that make up two consecutive base pairs. Various "melting" experiments have been performed to derive the thermodynamic parameters for these adjacent base pairs ([FB83] is a typical example of such research). The assumption is that energy is lowered owing to base pair *stacking* and *not* from the mere formation of base pairs. The rationale is that unpaired bases will have formed hydrogen bonds with water, and later, when they form base pairs, the net change in energy due to hydrogen bonding is not very significant.

Consecutive base pairs are usually coplanar and tend to be stacked. In fact, several adjacent base pairs will stack to form a stem with a helical conformation. Figure 5.3 shows the base pairs [1::72], [2::71], [3::70], [4::69], and [5::68] of tRNA^Phe.

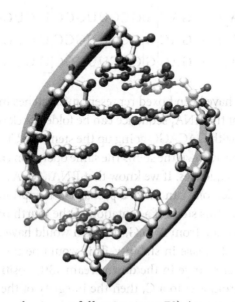

FIGURE 5.3 (See color insert following page 72) An example of base stacking in tRNA^Phe.

5.2.1.3 Consensus Sequence Patterns

In addition to observing structure and assessing thermodynamics, we can simply keep track of frequent patterns that can be observed in nature. For example, if a sequence of four nucleotides is consistent with a pattern that is often seen for "tetraloops," then this would enhance the likelihood that such a pattern is not in a stem; this statistic can be used by computational software that predicts secondary structure. What are tetraloops? Tetraloops are loops with only four nucleotides. These are quite common. For example, in the 16S rRNA secondary structure of *E. coli*, tetraloops account for about 55% of all hairpin loops in this structure [Ww90]. Tetraloops have been roughly categorized by sequence into three classes, which are GNRA, UNCG, and CUUG (see [Hm06]). In this notation, N can be any nucleotide and R is either G or A. In some cases, a particular consensus sequence (GNRA) has been observed to have an extra stability (reported in [Hp91]).

When observing frequent patterns, it should be noted that sequence alignments might seem to be disrupted by characters that do not match, but on closer inspection it becomes clear that the structure associated with each sequence is really the same. Here is a simplified version of the situation. Suppose we have the following aligned sequences:

$$\text{RNA}_1: \quad \cdots\text{GAC} \boxed{\text{A}} \text{GCCUUGGC} \boxed{\text{U}} \text{GUC} \cdots$$
$$\text{RNA}_2: \quad \cdots\text{GAC} \boxed{\text{G}} \text{GCCUUGGC} \boxed{\text{U}} \text{GUC} \cdots$$
$$\text{RNA}_3: \quad \cdots\text{GAC} \boxed{\text{G}} \text{GCCUUGGC} \boxed{\text{C}} \text{GUC} \cdots$$

The columns that have been boxed represent mismatches in the alignment. However, note that the RNA$_1$ sequence can be folded back on itself to make a stem formation with GACAGC going up the stem, CUUG in the loop and GCUGUC going back down the stem. The same operation can be performed for the other two sequences. If we know that RNA$_1$, RNA$_2$, and RNA$_3$ have evolved in that order, one can make a probable assumption about the evolutionary history of this stem. If a mutation in the fourth position of RNA$_1$ changed the nucleotide from A to G, the stem would have a small internal loop with a small decrease in stability. This would be the case for RNA$_2$. If there were a later change in the downstream 13th position causing that nucleotide to go from a U to a C, then the integrity of the stem would be restored to give RNA$_3$ with a stem that has the same hydrogen-bonding pattern as RNA$_1$. This last mutation is called a *compensatory mutation* [Mr04] because it restores the original secondary structure of the RNA. The scenario has also been described by saying that the nucleotides in the aforementioned positions show *covariance*. The main point of this discussion is that when RNA$_1$ and RNA$_3$ are compared there are two mismatches, but the similarity of structure has been preserved. Consequently, in some sense, we should not consider the two mutations to have as much impact as two mutations that are completely independent of each other.

5.2.1.4 Complications

Before leaving the NATURE discussion, we mention three other issues that may be addressed when dealing with secondary structure:

- RNA structure is dynamic and may fluctuate between different fold configurations of nearly equal energy.

- Fold configurations may be sensitive to the salinity and temperature of the ambient solution.

- The most biologically important fold may *not* have the lowest predicted free energy. However, we usually assume that it will be quite close to the minimum.

We now go on to the modeling discussion.

5.2.2 SCIENCE

5.2.2.1 Modeling Secondary Structure

Although we can use 3-D conformation studies to test the success of a secondary structure predictor, the workings of the predictor should be based on thermodynamic theories.* As noted earlier, the RNA molecule assumes a 3-D conformation that has minimal free energy. Although theory informs us that this is true, it is difficult to describe exactly *how* this minimum is achieved. The dependency between free energy and the relative positions of atoms in the molecule is very complicated. As for any macromolecule, the energy surface is characterized as a "rough landscape."

The progress made in RNA secondary structure prediction has been a steady improvement in accuracy at the expense of increasing complexity

---

* Recall the consensus pattern mentioned in the NATURE subsection. Some researchers have used these alignments alone, or in combination with thermodynamics, to generate predictors (e.g., [Bs06]).

in the model. We now consider some of these models, treating them in order of increasing accuracy.

5.2.2.2 Single Base Pairs

In this simple approach, we use a table to evaluate the ΔG decrease when given two consecutive base pairs in a stem. The total free energy is evaluated as:

$$E(S) = \sum_{[i::j] \in S} e(r_i, r_j).$$

(5.2)

In this calculation, $E(S)$ is the decrease in free energy due to stem formation. The variable S represents the set of all base pairings. Each base pair is specified by the indices i and j of the two residues involved. The base pair is written as $[i::j]$ with $i < j$. The notation r_i is used to designate the type of nucleotide at position i in the primary sequence. The "value" of r_i is a symbol such as A, U, C, or G. The term $e(r_i, r_j)$ designates the decrease in free energy that is due to the formation of a single base pair involving residues r_i and r_j. Note that all contributions are assumed independent of one another and the total energy is simply the sum of ΔG values corresponding to all the base pairs. The values for the $e(.,.)$ function are usually specified by consulting a table such as Table 5.1.

These are rough estimates and assumed to be representative of a system with an ambient temperature of 37°C. Note that the −3 and −2 values correspond to the presence of three hydrogen bonds in a GC pairing and the two hydrogen bonds in an AU pairing, respectively. The GU pairing has two hydrogen bonds, but the ΔG value was set to −1 due to the decreased stability that is consistent with the "wobble" behavior. All other pairings are assumed to be zero, for example, $\Delta G(GA) = 0$ kcal/mol.

5.2.2.3 Stacking Energy Models

The single base pair model is not really tuned to the thermodynamic details of the structure. For example, it avoids the notion of stacking energies,

TABLE 5.1   Energy Contributions
Made by Base Pairs

$\Delta G(CG)$:	−3 kcal/mol
$\Delta G(AU)$:	−2 kcal/mol
$\Delta G(GU)$:	−1 kcal/mol

loop strains, and bulges. To consider these issues, researchers have used free energy formulas such as the following:

$$\Delta G = \Delta G_{stack} + \Delta G_{bulge} + \Delta G_{hairpin} + \Delta G_{internal} + \Delta G_{multibranch}. \quad (5.3)$$

Only ΔG_{stack} increases the stability of the structure; all other terms serve to destabilize it. Each term in this sum is calculated on the basis of a careful assessment of the manner in which the constituent nucleotides are involved in the specified secondary structure feature. For example, it may be necessary to specify that a pair of nucleotides is in a loop *and* it is adjacent to a pair of nucleotides at the top of the stem of that loop.

In research done by Freier et al. [Fĸ86], the free energy computation for a stem assumes that stacking energy contributions are made by consecutive base pairs specified by four nucleotides. For example, the G(1)::C(34) base pair below the G(2)::U(33) base pair in Figure 5.1 would make a contribution of −1.5 kcal/mol as specified at the intersection of the G::C row and the G::U column of the next table. Notice the destabilizing effect that can occur when a wobble pair is near another wobble pair.

We can also consult other tables that define the energy gain due to bulges and various loops. These energy gains are dependent on the size (number of nucleotides) involved in the structural feature. For example, based on data given in [Fĸ86] the free energy increments for loops of size 4 is +1.7, +6.7, and +5.9 depending on whether the loop is internal, a bulge, or a hairpin. As a cautionary note, the authors have stated that these values were based on untested assumptions and are particularly unreliable.

More recent estimates of these parameters are +1.7, +3.6, and +5.9. These values are from a Web page titled: "Version 3.0 free energy parameters for RNA folding at 37°." The URL for this Web page is* http://frontend. bioinfo.rpi.edu/zukerm/cgi-bin/efiles-3.0.cgi.

This page presents the parameters that make up a free energy model involving observations that go well beyond the simple stacking energies described in Table 5.2. Most significantly, there are hundreds of parameters that specify the amount of *destabilization* introduced by various types

---

* If future changes to the Web site make this page inaccessible, the reader can try to access it via Michael Zuker's homepage at: http://frontend.bioinfo.rpi.edu/zukerm/home.html. Then follow the link entitled "Doug Turner's Free Energy and Enthalpy Tables for RNA folding." This should get you to the page: http://frontend.bioinfo.rpi.edu/zukerm/rna/energy/, which contains a link to "Free Energies at 37°."

TABLE 5.2   Free Energy Contributions Made by Stacked Pairs

	A::U	C::G	G::C	U::A	G::U	U::G
A::U	−0.9	−2.2	−2.1	−1.1	−0.6	−1.4
C::G	−2.1	−3.3	−2.4	−2.1	−1.4	−2.1
G::C	−2.4	−3.4	−3.3	−2.2	−1.5	−2.5
U::A	−1.3	−2.4	−2.1	−0.9	−1.0	−1.3
G::U	−1.3	−2.5	−2.1	−1.4	−0.5	+1.3
U::G	−1.0	−1.5	−1.4	−0.6	+0.3	−0.5

of loops. In the list to follow, we describe the types of energy calculations that are included in the tables.

- **Stacking energies**

  These tables provide the ΔG values for stacked pairs. To find the needed value, you must provide the four nucleotides that are involved in the stack.

- **Terminal mismatch stacking energies (hairpin loops)**

  The base pair just below a hairpin loop and the two nucleotides in the loop, but adjacent to the base pair, do not form a stack, but there is still a ΔG value that can be assigned to them. The value is dependent on all four nucleotides involved.

- **Terminal mismatch stacking energies (interior loops)**

  This is very similar to the previous situation. Again, four nucleotides are specified in order to locate the appropriate value.

- **1 × 1 internal loop energies**

  This situation is for a pair of nucleotides that are opposite one another in a stem; however, they are not complementary and hence do not form a base pair. To get the ΔG value, you must specify six nucleotides: the two nucleotides that are responsible for this small bulge and the base pairs on either side of it.

- **1 × 2 internal loop energies**

  This corresponds to the occurrence of a nucleotide that has no opposite nucleotide on the other strand. It is a small asymmetric bulge and usually produces a bend in the stem. It is specified by five nucleotides.

- **2 × 2 internal loop energies**

  This is similar to the 1 × 1 internal loop except that there are two adjacent nucleotides which do not form complementary pairs with the two nucleotides on the opposite strand. The tables for the ΔG value are quite extensive as you must specify eight nucleotides to get the required location in the tables.

- **Single base stacking energies**

  This stacking energy arises from the last pair. There is usually a small chain that continues past this last pair and the nucleotide closest to the pair in this chain gives a small stacking energy contribution. An example of this is seen in Figure 5.4. The last base pair at the bottom of the stem is a CG pair. Just below the CG pair there is a G nucleotide in the chain that continues as an unstructured single strand.

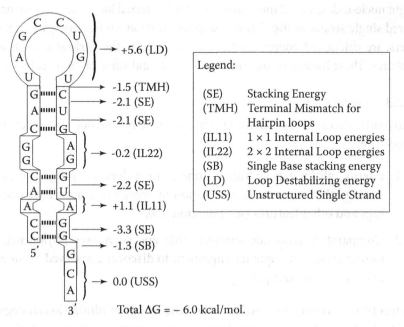

FIGURE 5.4   Example of a free energy calculation.

- **Loop destabilizing energies**

  Loops will usually raise the total ΔG value, so they have a destabilizing effect on the RNA structure. This is true for almost all loops: internal, bulge, or hairpin. In general, the amount of destabilization is roughly proportional to the loop size specified as the number of nucleotides that are in the loop.

- **Tetra-loops**

  It has been discovered that some hairpins containing four nucleotides may actually lower the free energy. The Web page presents a list of these tetra-loops. To use the table you must have the correct four nucleotides of the tetra-loop along with the base pair at the top of the stem.

- **Miscellaneous energies**

  This last category is a mixed collection of various special patterns and the change in ΔG that they provide. For example, a GGG hairpin produces a bonus of −2.2.

In Figure 5.4 we present a small example of a stem containing a loop with eight nucleotides, a 2×2 internal loop, a 1×1 internal loop, and an unstructured single strand at the 3′ end. Using the Version 3.0 free energy parameters, we calculated energy contributions made by the various structural features. These have been summed up to get a total value of −6.0 kcal/mol.

5.2.3 Computation

Currently, there are two main approaches to RNA secondary structure prediction:

1. Energy minimization: This is done using a dynamic programming strategy that relies on an estimation of energy terms for base pairings and other features (see Equation 5.3).

2. Comparative sequence analysis: This approach uses phylogenetic information and sequence alignment to discover conserved residues and covarying base pairs.

In this text, we concentrate exclusively on the energy minimization strategy.

Predicting hydrogen bonds due to tertiary structure is computationally expensive. See [RE99] for an algorithm that handles RNA secondary structure prediction with derivation of pseudoknots (running time is $O(n^6)$). Therefore, we employ simplifying assumptions for prediction of the secondary structure:

- RNA is assumed to fold into a single minimum free energy conformation.

- We do *not* try to model pseudoknots.

- The energy contribution made by a particular base pair in a stem region is independent of the neighbors of the pair.

We have already seen diagrams that show secondary structure. These are convenient for visualization but not quite suitable for a mathematical analysis. We start with some alternate representations.

5.2.3.1 Display of Secondary Structure

If we have a program that will predict secondary structure, what should it provide as output? Figures 2.36 (Chapter 2), 5.1, and 5.4 show the presence of base pairs, but they have two limitations: they are complicated to draw and it is awkward to show the occurrence of pseudoknot pairings. A simpler display technique that can capture pseudoknot information is offered by the *circular representation* of RNA secondary structure. Figure 5.5 presents an example of the circular representation of tRNA[Phe]. The consecutive nucleotides appear as small circles that are evenly distributed on a large circle. Arcs are drawn from nucleotide to nucleotide to indicate base pairing. In Figure 5.5, solid arcs are used to represent base pairings in stem structures, whereas dashed arcs are used to represent base pairings associated with pseudoknot formation. By comparing Figure 5.5 with Figure 2.36 (Chapter 2), one can see how the groups of solid arcs correspond to the stem structures in Figure 2.36. Although the pseudoknot information is a bit messy, it is easily drawn by software that can display simple graphics.

Another display strategy involves the use of an array with cells that are colored to indicate the presence or absence of base pairings. The array will have n rows and n columns for a nucleotide string with n bases. If the base at position i is paired with the base at position j, the cell at row i and column j is colored black. Otherwise, it is left as white. As this strategy produces a symmetric array, it is usually expedient to remove the redundant information and just present the upper triangle of the array.

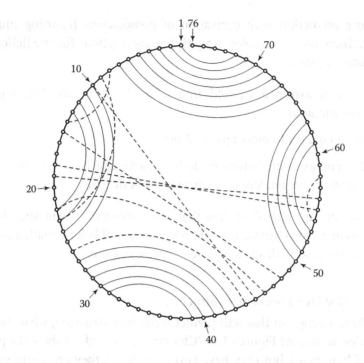

FIGURE 5.5   Circular representation of tRNA^{Phe} secondary structure.

As before, a careful inspection of the dot plot, comparing it with the structure illustrated in Figure 2.36 will reveal how the stem structures appear on the plot. In the list below, we present the various stem structures seen in tRNA^{Phe} and the corresponding cells that have been colored black to indicate base pairing in the dot plot.

Acceptor stem:	[1::72] to [7::66].
D stem:	[10::25] to [13::22].
Anticodon stem:	[26::44] to [31::39].
T stem:	[49::65] to [53::61].

In addition to these diagonal plots, there are various isolated black cells that correspond to the pseudoknots. As stated earlier, we will not attempt to predict the location of cells corresponding to pseudoknots, but it is interesting to see where they reside on a dot plot.

As it is simple and records all base pair interactions, the dot plot has the advantages of the circular plot. However, it goes one step further: it can act as a starting point for an analysis that gives us a mathematical environment for a dynamic programming algorithm that computes secondary structure.

We start by considering how the typical secondary structures appear on a dot plot. For example, Figure 5.7 shows the colored cells that specify a stem formation.

Because we are working with an array, row indices increase as we go from top to bottom whereas column indices increase as we go from left to right. Consider the position of a cell at array location (i, j). If j and i are equal, that is, $j - i = 0$, then the cell will be on the main diagonal of the plot. If $j - i = 1$, the cell is on the next diagonal just above the main diagonal. In general, the cell is on the kth diagonal if $j - i = k$. The cell in the upper right corner is distinguished by having the maximum difference between i and j. Consider the acceptor stem seen in the dot plot of Figure 5.6. The stem starts near the upper right corner with the base pair between nucleotides at positions 1 and 72 indicated by the black cell at location (1, 72). As we move along the stem, we encounter base pairings [2::71], [3::70], [4::69], etc.

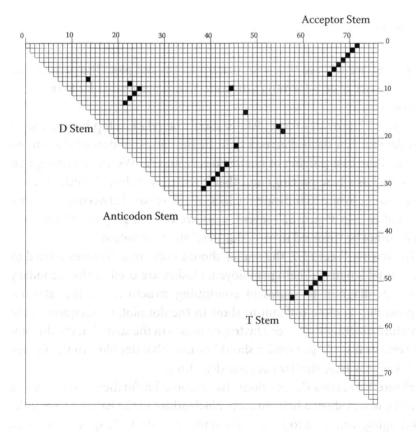

FIGURE 5.6 Dot plot for tRNA[Phe].

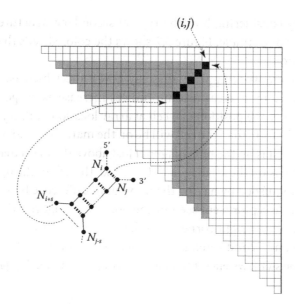

FIGURE 5.7 Dot plot for a stem.

In general, we will see this stem pattern in any dot plot. The pattern will be a sequence of cells forming a diagonal that is perpendicular to the main diagonal.

Figure 5.7 provides more details about the relationship between stem and dot plot. The dashed arrows indicate the correspondence between cell (i, j) and the base pairing of nucleotides N_i and N_j. We are assuming that the stem has s base pairings and that it ends just before the cell at $(i + s, j - s)$ colored white to indicate the lack of a base pair between nucleotides N_{i+s} and N_{j-s}. In Figure 5.7, several cells are colored gray to indicate that they cannot be involved in any additional stem formation.

In viewing Figure 5.7, the reader should note some features related to the "visual language" being employed. Dashes are used in the secondary structure diagram to represent continuing structure, but this attempt at generality is more difficult to show in the dot plot. Consequently, the structure diagram indicates s hydrogen bonds in the stem, but the dot plot is drawn as if $s = 5$. The reader should assume that dot plots in the figures to follow may have similar concrete depictions.

Figure 5.8 carries the previous discussion a bit further to show how a dot plot would show a hairpin loop. Nucleotides in the hairpin have indices ranging from $i + s$ to $j - s$. Cells in the triangle, with apex $(i + s, j - s)$, have been colored white with the array squares removed to indicate that

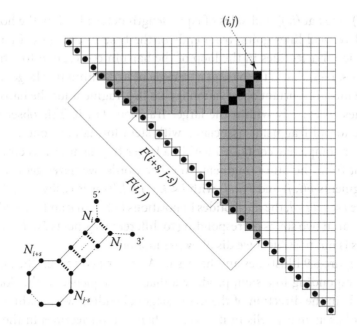

FIGURE 5.8 Dot plot for a hairpin.

no base pairings are allowed for nucleotides in the hairpin loop (recall that we do not consider pseudoknots).

Figure 5.8 introduces another idea related to the dot plot. As the cells on the main diagonal will never be used for indicating base pairs, we put them to a different use. We associate these cells with nucleotides in the primary sequence. Cell (1, 1) corresponds to nucleotide 1 and, in general, cell (i, i) corresponds to the ith nucleotide. This gives us an alternate way to locate cell (i, j). It is at the intersection of the row going right from nucleotide i on the main diagonal and the column going up from nucleotide j on the main diagonal. If you have used the triangular table on some road maps to determine the distance between two cities, you will be familiar with this indexing mechanism. It is convenient for us to imagine that the nucleotides of the primary sequence are arranged along the main diagonal because it brings in the notion of an RNA fragment and the relationship of this fragment to the dot patterns that describe its secondary structure.

We will refer to a sequence of consecutive nucleotides as a *fragment*, and we specify the fragment starting with base i and ending with base j using the notation $F(i, j)$. Note that the hairpin structure formed by the fragment is described in the dot plot by all the cells in an isosceles triangle. This triangle has a base comprising diagonal cells in the range from (i, i) to

(j, j), apex at (i, j) and sides of equal length determined by the horizontal and vertical lines going from ends of the base to the apex. In a similar fashion, nucleotides in the loop of the hairpin correspond to a fragment $F(i + s, j - s)$ that is the base of a triangle nested within this larger triangle. The most important observation is that the fragment for the hairpin loop is nested entirely within the larger fragment $F(i, j)$. This observation is true for all fragments associated with stem formations: one fragment is either completely contained within a larger fragment or it is entirely disjoint from the other fragment. In other words, we never get *overlapping* fragments $F(i, j)$ and $F(k, l)$ with $i < k < j < l$. This is easily seen. If we did have $i < k < j < l$, the nucleotides in positions between k and j would have to appear in two stems corresponding to different fragments $F(i, j)$ and $F(k, l)$. This is not possible if we disallow pseudoknots.

Figure 5.9 illustrates another issue. As described earlier, the black cells corresponding to a stem produce a diagonal sequence in the dot plot. It heads in the direction of the main diagonal ending with a white triangle that has as many cells in its base as there are nucleotides in the hairpin loop. Now recall the comment made with respect to Figure 5.2: a hairpin loop contains at least three nucleotides. This means that black cells in a stem sequence cannot contain cells from the main diagonal or from any

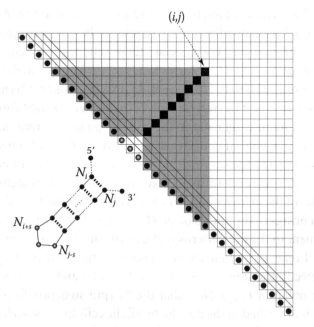

FIGURE 5.9 Dot plot for a size 3 hairpin.

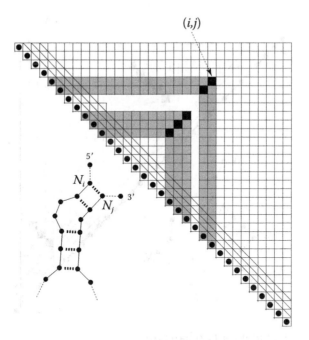

FIGURE 5.10   Dot plot for an internal loop.

of the three diagonals just above the main diagonal. The four diagonals constitute a *forbidden zone*. They are represented in the dot plot by cells with a strike-through diagonal line going from upper left to lower right. They are characterized by the inequality $j - i < 4$.

Figure 5.10 shows the dot plot of a stem containing an internal loop. One can observe the nesting of the fragments along the main diagonal and the corresponding nesting of triangles. Finally, note that the asymmetry in the number of unpaired nucleotides in the internal loop has produced a bend in the stem and a corresponding jog in the sequence of black cells that describe the two stem structures.

Figure 5.11 shows the dot plot representation of a longer fragment that has folded to make up two hairpin structures. The important observation for Figure 5.11 is that the long fragment $F(i, j)$ now includes two non-overlapped, nonnested subsidiary fragments each with its own triangle describing the structure of a hairpin.

5.2.4 Restating the Problem

Having seen these various examples, the problem of predicting secondary structure (without pseudoknots) can now be seen as the derivation of a set of nonoverlapped fragments each specifying a stem formation.

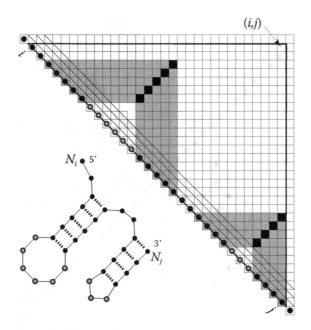

FIGURE 5.11   Dot plot for two hairpins.

To see the problem from yet another perspective, we can consider a *Tinoco* dot plot. This is an array representation of *all* possible GC, AU, and GU pairings for an RNA sequence. Any diagonal runs that are perpendicular to the main diagonal of the dot plot array will be candidates for inclusion in a solution of our problem. Selection of the "best" stems will be based on an energy minimization criterion. The primary sequence used for Figure 5.12 has been extracted from the hairpin loop in Figure 5.2, and the secondary structure for this sequence has been reproduced and displayed in the lower left corner of Figure 5.12. A dashed white line has been placed over the cells in the Tinoco dot plot that correspond to the given secondary structure. It should be clear that without this visual aid it is not immediately obvious where the best stem is situated among the helter-skelter collection of other stem candidates.

5.3 THE NUSSINOV DYNAMIC PROGRAMMING ALGORITHM

The RNA dot plot is very reminiscent of the 2-D array that we used to solve the longest common subsequence problem in Chapter 4. In fact, prediction of RNA secondary structure was done by Nussinov et al. using an algorithm that employed a 2-D score array (see [NP78]).

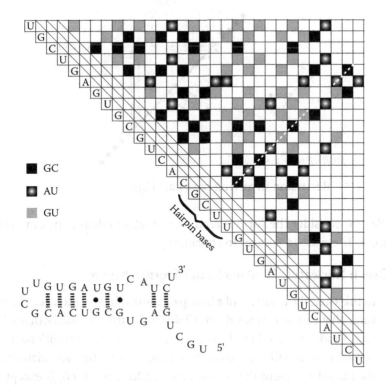

FIGURE 5.12 Tinoco dot plot for an RNA sequence.

The dynamic programming approach to the problem can be derived using the steps discussed earlier. In Step 1 we are required to identify subproblems and provide supporting analysis for the evaluation of the scoring function. On the basis of the previous discussion, it is very reasonable to consider prediction of secondary structure for a fragment as the subproblem. Considering the energy minimization issues covered in the SCIENCE section, it is desirable that a fragment assume a secondary structure that will minimize an energy score. At the outset, we start with a very simple strategy. Instead of dealing with stacking energies and loop destabilization parameters, the score for a fragment will simply be the number of bond pairings weighted by the values presented in Table 5.1. Recall that each fragment $F(i, j)$ was associated with a unique triangle in the dot plot. Each such triangle had a particular apex cell in row i and column j. The score for $F(i, j)$ will be denoted by $E(i, j)$, and we will store it in location (i, j).

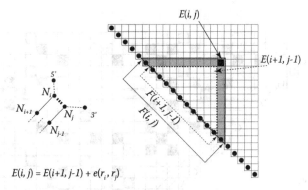

$$E(i,j) = E(i+1, j-1) + e(r_i, r_j)$$

FIGURE 5.13 Recursion involving base pair [i::j].

We now consider the various recursions that will define the evaluation of $E(i, j)$. There are four cases to consider:

Case 1: Nucleotide *i* and nucleotide *j* form a base pair

In this case, the presence of a base pair causes the score to decrease by a value that can be extracted from Table 5.1. Furthermore, both nucleotides *i* and *j* are "used up" by the base pairing, and any additional score considerations will depend on the secondary structure associated with the nested fragment that includes all nucleotides in $F(i, j)$ except for nucleotides *i* and *j*. Consequently, the needed recursion is given by

$$E(i,j) = E(i+1, j-1) + e(r_i, r_j).$$

Figure 5.13 shows a dot plot for Case 1 along with a drawing of the stem structure that shows the formation of a single base pairing of nucleotides *i* and *j*. The dashes represent the unknown structure involving the fragment $F(i + 1, j - 1)$ nested within fragment $F(i, j)$.

Case 2: Nucleotide *i* is unpaired

In this case, there is no base pair, and hence, there is no contribution to the score. As we know that nucleotide *i* is unpaired, the score for fragment $F(i, j)$ will be the same as that for the nested fragment which includes all nucleotides in $F(i, j)$ except for nucleotide *i*. Therefore, the recursion for this case is given by

$$E(i,j) = E(i+1, j).$$

The situation is described in Figure 5.14.

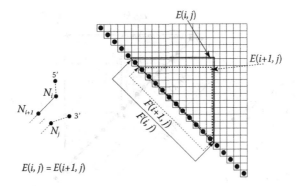

$$E(i, j) = E(i+1, j)$$

FIGURE 5.14   Recursion for unpaired nucleotide at start of fragment.

Case 3: Nucleotide j is unpaired

The analysis of this case is very similar to the previous case. The recursion is given by

$$E(i, j) = E(i, j-1).$$

and Figure 5.15 is similar to Figure 5.14.

Case 4: Bifurcation

The previous three cases are consistent with the type of dot plots we have seen in Figures 5.7, 5.8, 5.9, and 5.10. This last case arises

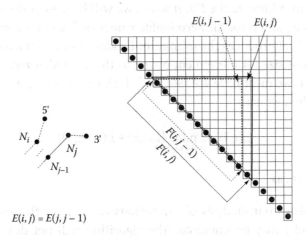

$$E(i, j) = E(j, j-1)$$

FIGURE 5.15   Recursion for unpaired nucleotide at end of fragment.

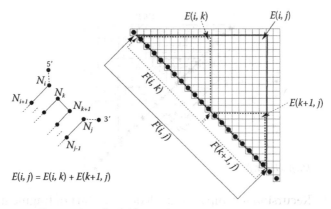

FIGURE 5.16 Recursion for bifurcation.

from a situation such as that seen in Figure 5.11. The fragment $F(i, j)$ has a secondary structure that comprises two subsidiary nonoverlapping non-nested fragments $F(i, k)$ and $F(k + 1, j)$. The score for $F(i, j)$ is calculated as the sum of the scores for $F(i, k)$ and $F(k + 1, j)$. This gives us:

$$E(i,j) = E(i,k) + E(k+1,j).$$

See Figure 5.16, which illustrates the fragments and their corresponding triangles. The computation of score $E(i, j)$ requires the computation of two other scores. This split is called a *bifurcation* because it is necessary to follow the computations associated with two separate subsidiary fragments. We have yet to discuss the value of k. As we wish to minimize the $E(i, j)$ score, we will have to evaluate the previous equation for all permissible values of k and then select the k value that produces the minimum E value. As there is a limit on the smallest size for a fragment owing to the forbidden zone described earlier, we must have $k \geq i+4$ and $j \geq k+5$. Bringing all these facts together, we get:

$$E(i,j) = \min_{i+4 \leq k \leq j-5} \{E(i,k) + E(k+1,j)\}$$

for this case.

This completes our analysis of the various cases. Note that *both* nucleotides i and j may be unpaired. The algorithm will not deal with this possibility as a distinct case because it can be handled by application of Case 2 followed by Case 3.

We now define the recursion for $E(i, j)$ by insisting that we apply the case that produces the minimum score for $E(i, j)$. This gives us the following final recursion for $E(i, j)$:

$$E(i,j) = \min \begin{cases} E(i+1,j-1)+e(r_i,r_j), \\[6pt] E(i+1,j), \\[6pt] E(i,j-1), \\[6pt] \min_{i+4 \le k \le j-5} \{E(i,k)+E(k+1,j)\} \end{cases} \tag{5.4}$$

This completes Step 1 of our dynamic programming recipe. In Step 2, we provide details that stipulate how the score values are evaluated. As described in Step 1, the score value for fragment $F(i, j)$ is designated as $E(i, j)$ and stored in location (i, j) of the array. If the RNA primary sequence contains n nucleotides, the array will have n rows and n columns. By reviewing the four cases described earlier, we see that $E(i, j)$ depends on previously evaluated $E(.,.)$ values stored in cells that are contained inside the triangle with apex at (i, j) and base comprising the fragment entries (i, i) to (j, j) along the main diagonal. The usual dynamic programming strategy is to get the scores of the smallest subproblems and use these to calculate the scores of the larger subproblems. For this algorithm, the scores for the smallest subproblems will be stored in cells that form a diagonal sequence running parallel to the main diagonal and just above the forbidden zone. This implies that evaluation of scores will consist of filling successive diagonals. We first evaluate all cells on the diagonal specified by $j = i + 4$, then the diagonal specified by $j = i + 5$, etc. The final optimal score will be the score for the entire primary sequence designated as $F(1, n)$. The corresponding $E(1, n)$ value will be stored in location $(1, n)$.

In Step 3, we define the values for the base cases. These initial values appear on the diagonal with equation $j = i + 4$. In other words, we are evaluating scores for fragments $F(i, i + 4)$, $i = 1, 2, 3, \ldots, n - 4$. As these cells are just above the forbidden zone, no score contributions are made by subsidiary fragments nested within any $F(i, i + 4)$. The values of $E(i, i + 4)$ can be extracted directly from Table 5.1. In fact, $E(i, i + 4) = e(r_i, r_{i+4})$. Note that nucleotide pairings such as AC that do not appear in Table 5.1 have score contributions that are zero.

Step 4 deals with recovery of the solution configuration. This is more of a challenge than the simple trace-back employed in the LCS problem. This is due to the bifurcation issue, which requires us to follow two different paths when Case 4 asserts itself. The trace-back algorithm is actually traversing a binary tree while it reports the occurrences of base pairings. This can be done using a *depth first search* (DFS) traversal and can be implemented using a stack to keep track of edges in the tree that must be investigated. However, it is conceptually easier to do the DFS by means of a recursive routine. A stack is still being used, but it is not explicit. It is "hidden" from the user because it is contained in the calling stack frame that is built at execution time when the recursive routine calls itself.

Pseudocode for the evaluation of $E(i, j)$ follows next. Here are some points for clarification of the pseudocode:

- Score values for fragment $F(i, j)$ are stored in E[i,j].

- The function $e(.,.)$ is designed to return the values from Table 5.1 when it is given the names of two nucleotides.

- The array entry r[i] is the name of the nucleotide at position i in the primary sequence.

- The first index of an array is 1, not 0.

- Trace-back information is stored in an array case[1..n, 1..n] of structures. The entry at array location (i, j) contains two values: case[i,j].ID identifies which of the four cases produced the minimum score for E[i,j] and case[i,j].kval contains the k value for a Case 4 situation.

- The constant highValue is a large positive value that is an initial value for the code that is computing a minimum value within a set of numbers.

```
// Initialize top diagonal of forbidden zone.
for i := 1 to n-3 do E[i,i+3] := 0;

// Initialize diagonal above forbidden zone.
for i := 1 to n-4 do E[i,i+4] := e(r[i],r[i+4]);

// There are n-5 diagonals left to fill.
for d := 5 to n-1 do
        for i := 1 to n-d do
            j := i+d;
```

```
case1E := E[i+1,j-1] + e(r[i],r[j]);
case2E := E[i+1,j];
case3E := E[i,j-1];
case4E := highValue;
// Case 4 starts at diagonal j = i + 9.
if (j > i+9) then
        for k := i+4 to j-5 do
                temp := E[i,k] + E[k+1,j];
                if (temp < case4E) then
                        case4E := temp;
                        case[i,j].kval := k;
min_E := highValue;
if (case1E < min_E AND e(r[i],r[j]) < 0) then
        min_E := case1E;
        case[i,j].ID := "case1";
if (case2E < min_E) then
        min_E := case2E;
        case[i,j].ID := "case2";
if (case3E < min_E) then
        min_E := case3E;
        case[i,j].ID := "case3";
if (case4E < min_E) then
        min_E := case4E;
        case[i,j].ID := "case4";
E[i,j] := min_E;
```

The trace-back pseudocode uses a recursive routine that is called in the mainline as:

```
get_SS(1, n).
```

```
function get_SS(i,j)
        if (case[i,j].ID == "case1") then
            output i, j;
            get_SS(i+1, j-1);

        if (case[i,j].ID == "case2") then
            get_SS(i+1, j);

        if (case[i,j].ID == "case3") then
            get_SS(i, j-1);
```

```
    if (case[i,j].ID == "case4") then
        k = case[i,j].kval;
        get_SS(i, k);
        get_SS(k+1, j];

    return;
```

Figures 5.17 and 5.18 present a small example of the dynamic programming algorithm applied to the primary sequence: GCUCAAAGUCCAG GAGG. Figure 5.18 shows how the $E(i, j)$ values have been computed with arrows indicating the cell dependencies. Recall that Case 1 involves $e(r_i, r_j)$ values in the computation of $E(i, j)$. The $e(r_i, r_j)$ values for each (i, j) cell have been put into an array and shown in Figure 5.17. Almost all the trace-back arrows are simple "up," "right," or "up-right" arrows with the very last entry in location (1, 17) being computed using Case 4. This bifurcation case involves two dependencies, namely, cell (1, 9) and (10, 17). Note that the k value is 9. During trace-back processing, two stems would be generated corresponding to the cells that are at the ends of the "up-right" arrows: (1, 9), (2, 8), (3, 7) and (17, 10), (16, 18). The $E(i, j)$ values for these cells has been rendered in a bold font. Seen as nucleotide pairings,

G				0	0	0	0	1	3	3	0	0	0	0	0	0
	C			0	0	3	0	0	0	0	3	3	0	3	3
		U		2	1	0	0	0	2	1	1	2	1	1
			C	3	0	0	0	0	3	3	0	3	3
				A	2	0	0	0	0	0	0	0	0
					A	0	0	0	0	0	0	0	0
						A	0	0	0	0	0	0	0
Minus signs have been							G	0	0	0	0	0	0
omitted to save space.							U	1	1	2	1	1
								C	3	0	3	3
									C	0	3	3
										A	0	0
											G	0
												G
													A
														G
															G

FIGURE 5.17 Dynamic programming example: $e(r_i, r_j)$ values.

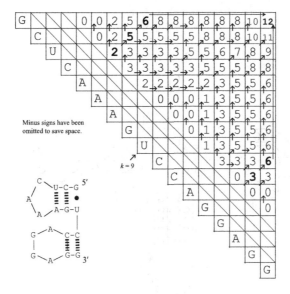

FIGURE 5.18 $E(i, j)$ score values and backtracking.

these cells give us GU, CG, UA, and CG, CG respectively. A diagram of the derived secondary structure is also shown in Figure 5.18.

5.3.1 Execution Time

Running time analysis of the pseudocode for filling of the score array reveals that the running time is $O(n^3)$ because of the three nested loops. This can be burdensome for very lengthy primary sequences. As noted earlier, the algorithm does not handle pseudoknots, and most algorithms that do predict pseudoknots have execution times with an upper bound that is significantly larger than $O(n^3)$ (see [RE99]).

5.4 THE MFOLD ALGORITHM: TERMINOLOGY

If one set out to improve the Nussinov algorithm, the introduction of strategies to handle pseudoknots would probably *not* have the highest priority. The computational model that we used in the previous section incorporated some very rough approximations in the formulation of the energy minimization principles underlying the secondary structure. The next level of sophistication would be to incorporate the stacking energies that were discussed in the SCIENCE section. This approach in combination with an account of the destabilization effects of various loops gives a more

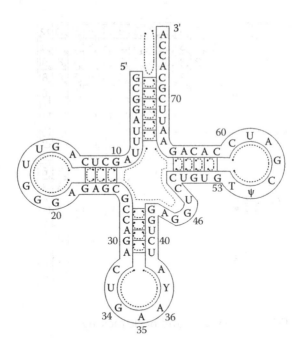

FIGURE 5.19 K-loop decomposition of tRNA[phe].

accurate representation of the energy interactions that take place during the formation of secondary structure.

The MFOLD algorithm by Zuker et al. is a dynamic program that handles loops and stacking energies. Further readings on this topic can be found in [Ms99], [Ws86], [Zm99], [Zs84] and [Zs81].

If we are given an RNA secondary structure (without pseudoknots), we can generate a unique set of *K-loops*. Figure 5.19 shows tRNA[phe] and the set of *K*-loops each represented using dotted lines. Researchers have found that rules can be formulated that allow us to assign energy values to each loop. The sum of these energy values, across all loops, will give the ΔG for the molecule. It has been demonstrated from empirical evidence that this strategy for the computation of ΔG is much more accurate than simply counting hydrogen bonds, an approach that does not account for the destabilization effects of the larger loops. Moreover, our previous calculations utilized a function such as $e(r_i, r_j)$ that recognized the contribution made by the nucleotides r_i and r_j in a base pair, but did not exploit the empirical observation that the energy contribution made by a base pair is dependent on its surrounding base pairs. This locality issue is handled by the smallest

K-loops that occur between consecutive base pairings. This accounts for the stacking energy that was described earlier.

Our computational objectives can be restated as follows: Given the primary sequence of an RNA molecule, find the set of base pairings that produces the lowest ΔG when that ΔG is calculated by summing energy contributions made by the *K*-loops.

This can be done by means of a dynamic program that is based on a formal definition of the *K*-loops. When the dynamic program is being executed, it does not have the complete secondary structure with a full set of *K*-loops as illustrated in Figure 5.19. However, as the dynamic program progresses, there will be "loop completion events" and the opportunity to bring into play the energy change made by these loops. For example, you may recall the array filling done in the last dynamic program. The algorithm evaluated $E(i, j)$ along diagonals that ran parallel to the forbidden zone, first alongside the forbidden zone, then farther away. In the calculation of $E(i, j)$ we evaluated $e(r_i, r_j) + E(i + 1, j - 1)$, and this value was then compared with the other energy terms in Equation 5.4. The $e(r_i, r_j) + E(i + 1, j - 1)$ sum corresponded to a base pairing of the nucleotides i and j. We now interpret this as a loop completion event, and the idea is to replace $e(r_i, r_j)$ with an energy evaluation that takes into account the significance of this loop completion. To apply the appropriate energy rules, the program must determine the type of loop being completed. Is it a hairpin, a bulge, a multiple loop, or part of a stacked pair? The first three types cause the free energy to increase; only the last type will cause a decrease.

Now that we have an informal description of the algorithm, we introduce some mathematical notation and make the ideas more rigorous. In the preceding discussion, the energy score was denoted by $E(i, j)$. Henceforth, we will designate the minimum folding energy as $W(i, j)$. This will distinguish the loop-dependent $W(i, j)$ from the base-pair-dependent $E(i, j)$. Also, the $W(i, j)$ notation is often used in the explanation of Zuker's algorithm when it is described in research papers. We start with the formal characterization of a *K*-loop.

Definitions

Let S denote the set of all base pairs for a given secondary structure. If $[i::j]$ is a base pair in S and $i < k < j$, then the nucleotide at position k is *accessible* from $[i::j]$ if there is *no* base pair $(i', j') \in S$ such that $i < i' < k < j' < j$. When nucleotide k is accessible from $[i::j]$, we call $[i::j]$ the *parent base pair*.

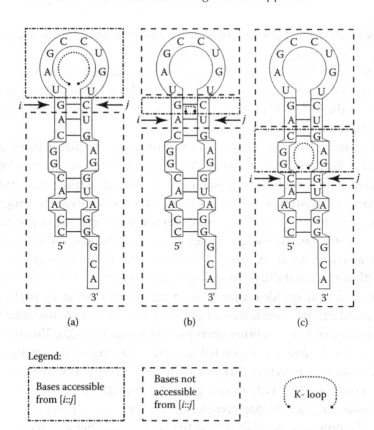

FIGURE 5.20 Accessibility of nucleotides.

Definition

The base pair [k::l] is said to be accessible from [i::j] if both nucleotides k and l are accessible from [i::j].

Figure 5.20 presents some examples. The arrows indicate the [i::j] base pair in each example. Nucleotides or base pairs enclosed by a dot-dash outline are accessible from [i::j], whereas any nucleotide inside a dashed line box is not accessible from [i::j]. A K-loop is formed by the accessible nucleotides and accessible base pairs. To show the correspondence with K-loops in Figure 5.19, it is also represented as a dotted loop.

Some notes for Figure 5.20:

- In diagram (a), we see that all nucleotides in the hairpin (U, A, G, C, C, U, G, and U) are accessible from [i::j]. These nucleotides form a hairpin loop.

- In diagram (b), G and C, at positions $i + 1$ and $j - 1$ are the only two nucleotides accessible from $[i::j]$. As they form a base pair, we have a loop that designates the stacked pair G-C over A-U.

- In diagram (c), the nucleotides (G, C, C, G, A, and G) in the dotted box are accessible from $[i::j]$. In this case, we have various unpaired nucleotides and a base pair accessible from $[i::j]$. These nucleotides form an internal loop.

- Note that the base pair $[1::28]$ at the bottom of the stem is not accessible from any existing base pair. The same can be said about the GCA "dangling" nucleotides at the 3' end. All of these nucleotides form a "null" K-loop designated as L_0.

Finally, referring to Figure 5.19, we see that the loop at the center of the clover leaf is a multiloop. Its parent is the base pair $[7::66]$, and it has three other base pairs—$[10::25]$, $[27::43]$, and $[49::65]$—that are accessible from $[7::66]$. Taking $[7::66]$ into account, this multiloop has four base pairs each at the bottom of some "branch," and so we call this multiloop a 4-loop. Note the null K-loop in the uppermost part of the figure. Now that we have seen various examples of K-loops, we can present a definition that explains the significance of the K value:

Definition

For a given K-loop, the value of K is the number of base pairs in the loop plus 1. Thus, the count includes the parent base pair $[i::j]$ and the base pairs that are accessible from this parent.

To see how K-loops relate to previously defined terms, we note that:

1. A 1-loop is a hairpin loop.

2. A 2-loop with parent $[i::j]$ corresponds to:

   a. A stacked pair if $i'- i = 1$ and $j'- j = 1$

   b. A bulge loop if either $i'- i > 1$ or $j'- j > 1$ (but not both)

   c. An interior loop if both $i'- i > 1$ and $j'- j > 1$

   where $[i'::j']$ is the base pair accessible from $[i::j]$.

3. A K-loop with $K \geq 3$ is a multiloop.

From this discussion, it can be seen that once the set of base pairs S is specified, a unique set of loops is defined. As each base pair marks a loop completion event, there are as many non-null loops as there are base pairs in S. Furthermore, each nucleotide will belong to a single loop. If R is the collection of nucleotides in the primary sequence, then we have

$$R = L_0 \cup \left(\bigcup_{[i::j] \in S} L_{[i::j]} \right) \tag{5.5}$$

where $L_{[i::j]}$ is the loop with loop completing parent $[i::j]$. As contributions to free energy are computed with respect to loops, we can write:

$$E(S) = e(L_0) + \sum_{[i::j] \in S} e(L_{[i::j]}). \tag{5.6}$$

This equation simply states that the energy $E(S)$ associated with a particular secondary structure specified by S is dependent on the energy contributions of the various K-loops. However, as stated earlier, we do not have the K-loops when the dynamic programming MFOLD algorithm begins. Consequently, just as we did for the Nussinov algorithm, we define subproblems that are related to fragments of the nucleotide sequence. The strategy is to calculate the energy contributions made by K-loops when the fragments are folded into secondary structures.

5.4.1 The MFOLD Algorithm: Recursion

We start with a description of the MFOLD algorithm that incorporates the main contributions made by the K-loops. Various refinements to this algorithm will be noted at the end of the chapter.

For $i < j$, let $W(i, j)$ be the minimum folding energy across all possible folds of the primary sequence fragment $r_i, r_{i+1}, ..., r_j$. As we are interested in loop completion events, we also define the quantity $V(i, j)$ to be the minimum folding energy across all possible folds of the primary sequence fragment $r_i, r_{i+1}, ..., r_j$ such that r_i and r_j form a base pair. Because the set of folds for $V(i, j)$ is a subset of the set of folds for $W(i, j)$, we will have:

$$W(i, j) \leq V(i, j).$$

Once again, we have a forbidden zone with cells specified by the constraint $j - i < 4$. Cells in the forbidden zone have boundary values given by $W(i, j) = V(i, j) = +\infty$. This value is consistent with the observation that one cannot have a loop completion event in the forbidden zone.

Using the same reasoning that we employed to derive Equation 5.4, we can show that:

$$W(i,j) = \min \begin{cases} V(i,j), \\ W(i+1,j), \\ W(i,j-1), \\ \min_{i \leq k \leq j-1} \{W(i,k) + W(k+1,j)\} \end{cases}. \tag{5.7}$$

This is very similar to Equation 5.4, the only difference being that previously the base pairing of nucleotides r_i and r_j led to an energy contribution of $e(r_i, r_j)$ to $E(i, j)$ whereas it now indicates a loop completion event that brings in an energy contribution of $V(i,j)$ to $W(i,j)$.

The calculation of the energy contribution $V(i,j)$ is somewhat more complicated than that for $e(r_i, r_j)$ and is dependent on the type of K-loop that is formed by the loop completion event. As we want to find the loop completion events that minimize the energy, we will demand that $V(i,j)$ be evaluated by extracting the minimum value from the set of possible configurations:

$$V(i,j) = \min \begin{cases} eh(i,j), \\ V(i+1,j-1) + es(r_i, r_j, r_{i+1}, r_{j-1}), \\ \min_{\substack{i<i'<j'<j \\ 2<i'-i+j-j'}} \{V(i',j') + ebi(i,j,i',j')\}, \\ \min_{i+1 \leq k \leq j-2} \{W(i+1,k) + W(k+1,j-1)\} + a \end{cases}. \tag{5.8}$$

Each entry in the large brackets involves a particular energy function that defines the energy contribution of some particular type of K-loop. We describe each in turn:

- The function $eh(i,j)$ is the destabilizing energy of the hairpin loop with parent [i::j]. In a simple model, this function depends on the number of residues in the hairpin. This *size* of the hairpin is easily calculated as $j - i - 1$. The value of $eh(.,.)$ for various hairpin sizes can be found in the loop destabilizing energies table provided by the Turner Group.*

---

* http://frontend.bioinfo.rpi.edu/zukerm/cgi-bin/efiles-3.0.cgi.

- The function $es(r_i, r_j, r_{i+1}, r_{j-1})$ is the stabilizing energy of the stacked pair involving base pairs $[i :: j]$ and $[i + 1 :: j - 1]$. The function value can be obtained for any possible combination of $r_i, r_j, r_{i+1}, r_{j-1}$ by consulting the stacking energies table provided by the Turner Group.

- $ebi(i, j, i', j')$ is the destabilizing energy of the bulge or internal loop with base pair $[i' :: j']$ accessible from parent $[i::j]$. For a bulge with $size \geq 1$ or an internal loop with $size \geq 4$, the loop destabilizing energies table referenced in the first point can be used to provide the value. For smaller internal loops, there are tables that are dependent on the unpaired residues within the loop. In such cases, the function is more appropriately specified as $ebi(r_i, r_j, r_{i'}, r_{j'})$. Again, the Turner group has extensive tables to handle these 1×1, 1×2, and 2×2 internal loop energy cases.

- The destabilizing energy of a multiloop will be assumed have a constant value given by a (see the Turner Group tables).

Once the W and V arrays are filled, the minimum energy can be found in $W(1, n)$.

5.4.2 MFOLD Extensions

The dynamic program just described can be enhanced further by bringing in more details that arise from various empirical studies. These have been mentioned earlier: single base stacking energies, tetra-loops, and GGG hairpins. Over the last few years, MFOLD has been repeatedly updated to capitalize on these new experimental results.

5.4.3 MFOLD Execution Time

As a final note, we mention some issues dealing with the execution times for the MFOLD algorithm. The number of i, j pairs that obey the constraint $1 \leq i < j \leq n$ is $O(n^2)$. Combining this with the observation that k ranges from $i + 1$ to $j - 2$ inclusive, we see that the computation of

$$\min_{i+1 \leq k \leq j-2} \{W(i+1, k) + W(k+1, j-1)\}$$

will involve $O(n^3)$ operations. A similar analysis should convince the reader that the computation of

$$\min_{\substack{i < i' < j' < j \\ 2 < i' - i + j - j'}} \{V(i', j') + ebi(i, j, i', j')\}$$

will require $O(n^4)$ operations. Zuker notes that this upper bound can be reduced to $O(n^3)$ operations by imposing a limit on the maximum size of the bulge or internal loop. This upper limit is usually set to 30.

5.5 EXERCISES

1. In a manner similar to that seen in Figure 5.4, calculate the total ΔG for the structural fragment shown in the following diagram:

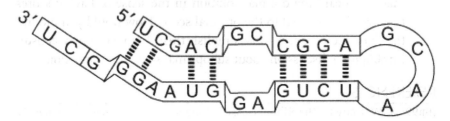

2. Consider the circular representation of RNA secondary structure (see Figure 5.5 for an example). Nucleotides are represented by points on a circle and the chords drawn between these points represent hydrogen bonds between nucleotides. Show that no two chords intersect if and only if the secondary structure is free of pseudoknots.

3. Write a program to predict secondary structure of RNA using dynamic programming. Apply your program to the RNA primary sequence: GGCCUUAGGAAACAGUUCGCUGUGCCGAAAGGUC. Compare your results with those available from the Vienna RNA Secondary Structure Prediction server.* The given primary sequence was taken from the H/ACA snoRNP illustrated in Figure 5.1. Note that the large internal loop forming the binding pocket contains nucleotides that do match as possible binding partners: U(6) – A(29), G(8) – C(26), and G(9) – C(25). What does your program do with these pairs? How does the Vienna server handle them? Do some investigating that attempts to determine whether these pairings really exist within the H/ACA snoRNP molecule before it accepts its RNA substrate. This is an interesting issue because if the pairs do exist, it would mean that the H/ACA snoRNP molecule would undergo significant changes (including base pair dissolution) to form the anti-sense binding pocket that we see illustrated in Figure 5.1.

---

* http://rna.tbi.univie.ac.at/cgi-bin/RNAfold.cgi.

4. In many applications, it is important that the secondary structure prediction achieve the minimal free energy. The dynamic programming algorithm does give an optimal solution, but it is only optimal in the sense that it selects the optimal configuration from all the configurations that are scored on the basis of the assumptions that were made. Our assumptions were simple, but introduced approximations. Now suppose that we wish to discover suboptimal solutions that are "near" the optimal solution in the sense of having scores that are almost equal to the optimal score. How would you modify the dynamic program to accomplish this? The reader may consult [MA06] for a discussion about suboptimal structure prediction.

REFERENCES

[BR99] J. W. BROWN. The Ribonuclease P Database. *Nucleic Acids Research*, **27** (1999), 314.

[ED01] S. EDDY. Non-coding RNA genes and the modern RNA world. *Nucleic Acids Research*, **27** (2001), 919–929.

[FB83] S. M. FREIER, B. J. BURGER, D. ALKEMA, T. NEILSON, AND D. H. TURNER. Effects of 3′ dangling end stacking on the stability of GGCC and CCGG double helices. *Biochemistry*, **22** (1983), 6198–6206.

[FK86] S. M. FREIER, R. KIERZEK, J. A. JAEGER, N. SUGIMOTO, M. H. CARUTHERS, T. NEILSON, AND D. H. TURNER. Improved free-energy parameters for predictions of RNA duplex stability. *Proceedings of the National Academy of Sciences of the United States of America*, **83** (1986), 9373–9377.

[HP91] H. A. HEUS AND A. PARDI. Structural features that give rise to the unusual stability of RNA hairpins containing GNRA loops. *Science*, **253** (1991), 191–194.

[HM06] C. HSIAO, S. MOHAN, E. HERSHKOVITZ, A. TANNENBAUM, AND L. D. WILLIAMS. Single nucleotide RNA choreography. *Nucleic Acids Research*, **34** (2006), 1481–1491.

[MA06] D. H. MATHEWS. Revolutions in RNA secondary structure prediction. *Journal of Molecular Biology*, **359** (2006), 526–532.

[MS99] D. H. MATHEWS, J. SABINA, M. ZUKER, AND D. H. TURNER. Expanded sequence dependence of thermodynamic parameters improves prediction of RNA secondary structure. *Journal of Molecular Biology*, **288** (1999), 911–940.

[MT06] D. H. MATHEWS AND D. H. TURNER. Prediction of RNA secondary structure by free energy minimization. *Current Opinion in Structural Biology*, **16** (2006), 270–278.

[MR04] T. F. MCCUTCHAN, D. RATHORE, AND J. LI. Compensatory evolution in the human malaria parasite Plasmodium ovale. *Genetics*, **166** (2004), 637–640.

[Np78] R. Nussinov, G. Pieczenik, J. R. Griggs, and D. J. Kleitman. Algorithms for loop matchings. *SIAM Journal of Applied Mathematics*, **35** (1978), 68–82.

[Re99] E. Rivas and S. R. Eddy. A dynamic programming algorithm for RNA structure prediction including pseudoknots. *Journal of Molecular Biology*, **285** (1999), 2053–2068.

[St02] G. Storz. An expanding universe of noncoding RNAs. *Science*, **296** (2002), 1260–1263.

[Wa78] M. S. Waterman. Secondary structure of single-stranded nucleic acids. *Studies in Foundations and Combinatorics, Advances in Mathematics Supplementary Studies*, **1** (1978), 167–212.

[Ws78] M. S. Waterman and T. F. Smith. RNA secondary structure: a complete mathematical analysis. *Mathematical Biosciences*, **42** (1978), 257–266.

[Ws86] M. S. Waterman and T. F. Smith. Rapid dynamic programming algorithms for RNA secondary structure. *Advances in Applied Mathematics*, **7** (1986), 455–464.

[Ww90] C. R. Woese, S. Winker, and R. R. Gutell. Architecture of ribosomal RNA: Constraints on the sequence of "tetra-loops." *Proceedings of the National Academy of Sciences of the United States of America*, **87** (1990), 8467–8471.

[Wf07] H. Wu and J. Feigon. H/ACA small nucleolar RNA pseudouridylation pockets bind substrate RNA to form three-way junctions that position the target U for modification. *Proceedings of the National Academy of Sciences of the United States of America*, **104** (2007), 6655–6660.

[Zm99] M. Zuker, D. H. Mathews, and D. H. Turner. Algorithms and thermodynamics for RNA secondary structure prediction: a practical guide. In: *RNA Biochemistry and Biotechnology*, J. Barciszewski and B. F. C. Clark (Eds.), Nato Science Series, High Technology, Kluwer Academic Publishers, **70** (1999), 11–43.

[Zs84] M. Zuker and D. Sankoff. RNA secondary structures and their prediction. *Bulletin of Mathematical Biology*, **46** (1984), 591–621.

[Zs81] M. Zuker and P. Stiegler. Optimal computer folding of large RNA sequences using thermodynamics and auxiliary information. *Nucleic Acids Research*, **9** (1981), 133–148.

[Zw99] C. Zwieb, I. Wower, and J. Wower. Comparative sequence analysis of tmRNA. *Nucleic Acids Research*, **27** (1999), 2063–2071.

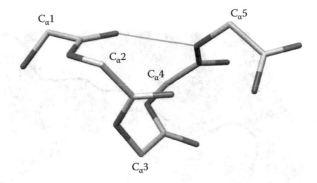

COLOR FIGURE 2.8   Formation of a hydrogen bond in an alpha helix.

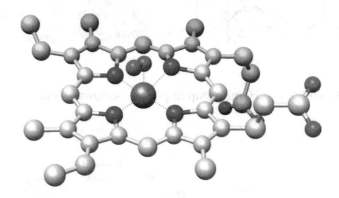

COLOR FIGURE 2.22   Heme group from 1MBO.

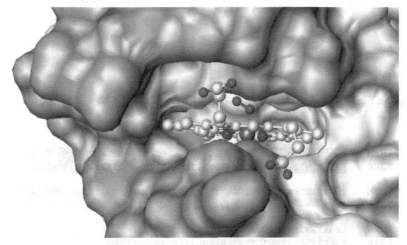

COLOR FIGURE 2.23   Heme group in its binding pocket.

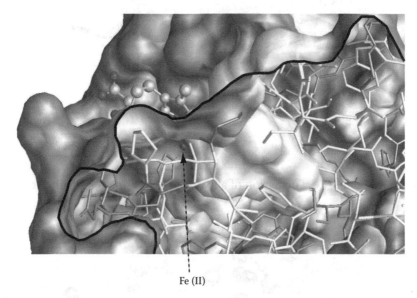

Fe (II)

COLOR FIGURE 2.24   Heme group in pocket—cutaway view.

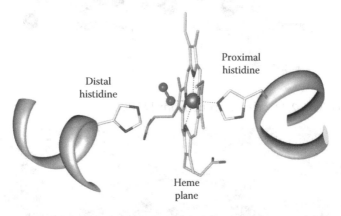

Proximal
histidine

Distal
histidine

Heme
plane

COLOR FIGURE 2.25   Heme group and histidine residues.

```
1MBN   VLSEGEWQLVLHVWAKVEADVAGHGQDILIRLFKSHPETLEKFDRFKHLKTEA
1JEB   SLTKTERTIIVSMWAKISTQADTIGTETLERLFLSHPQTKTYFPHFDLHPGSA

1MBN   EMKASEDLKKHGVTVLTALGAILKKKGHHEAELKPLAQSHATKHKIPIKYLEFI
1JEB   QLRA------HGSKVVAAVGDAVKSIDDIGGALSKLSELHAYILRVDPVNFKLL

1MBN   SEAIIHVLHSRHPGDFGADAQGAMNKALELFRKDIAAKYKELGYQG
1JEB   SHCLLVTLAARFPADFTAEAHAAWDKFLSVVSSVLTEKYR------
```

COLOR FIGURE 2.27   Sequence comparison of myoglobin and a single
chain of hemoglobin.

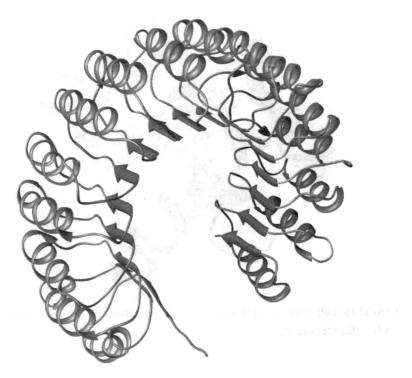

COLOR FIGURE 2.28   A tertiary structure involving repeated secondary structures (chain D of 1A4Y).

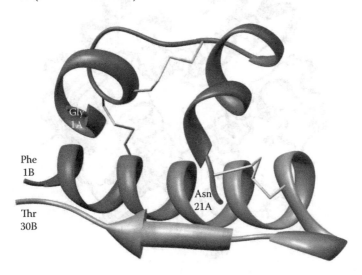

COLOR FIGURE 2.29   The insulin monomer has two chains connected by disulfide bridges.

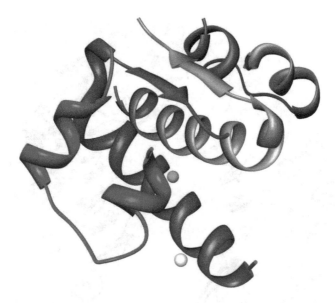

COLOR FIGURE 2.31   One of the three dimers making up the hexamer of an insulin structure.

COLOR FIGURE 2.32   Hexamer structure of insulin (PDB code 1ZNJ).

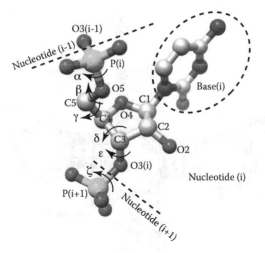

COLOR FIGURE 2.34   A nucleotide and its flexible backbone.

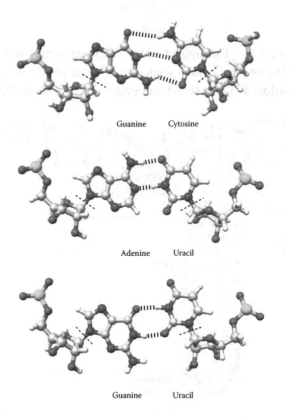

COLOR FIGURE 2.35   Hydrogen bonding between complementary purine–pyrimidine pairs.

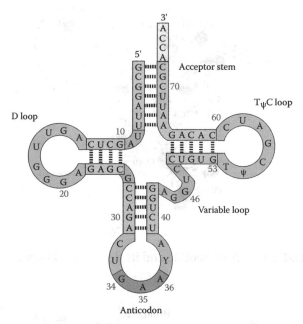

COLOR FIGURE 2.36   Example of RNA secondary structure (From H. Shi and P. B. Moore. The crystal structure of yeast phenylalanine tRNA at 1.93 A resolution: a classic structure revisited. *RNA*, **6** (2000), 1091–1105).

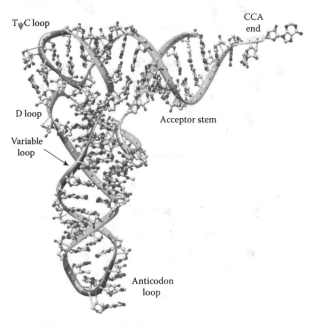

COLOR FIGURE 2.37   An example of RNA tertiary structure.

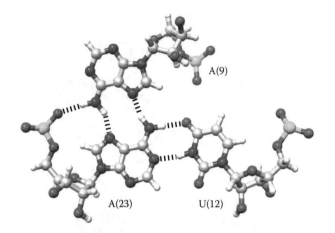

COLOR FIGURE 2.38 Pseudoknots complicate base pairings.

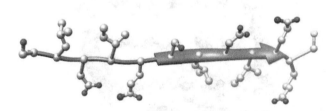

COLOR FIGURE 2.39 Peptide sequence for Exercise 1.

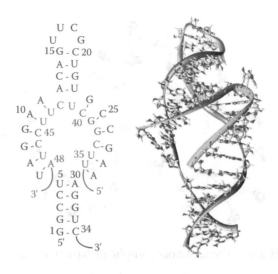

COLOR FIGURE 5.1   RNA–RNA interaction for pseudouridylation.

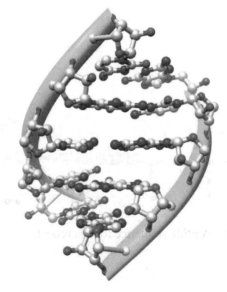

COLOR FIGURE 5.3   An example of base stacking in tRNA[Phe].

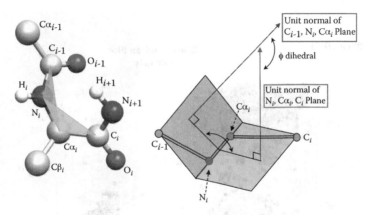

COLOR FIGURE 7.2   Phi dihedral angle.

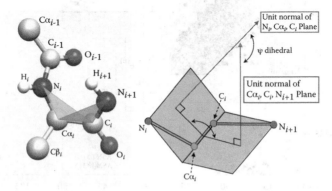

COLOR FIGURE 7.3   Psi dihedral angle.

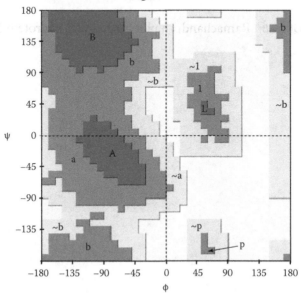

COLOR FIGURE 7.7   Ramachandran plot for 121,870 residues (463 protein structures).

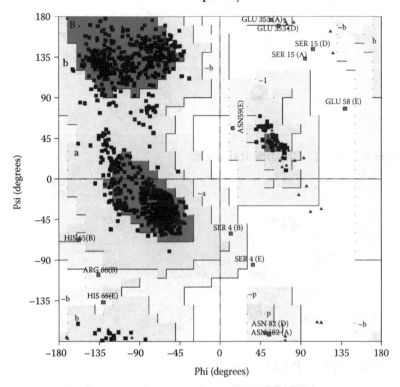

COLOR FIGURE 7.8   Ramachandran plot for residues of protein 1A4Y.

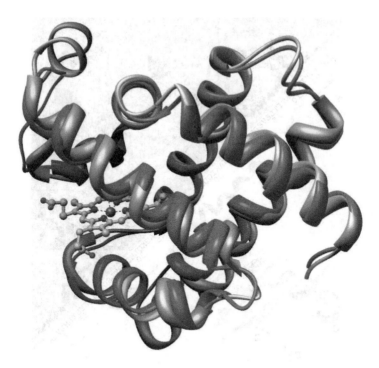

COLOR FIGURE 9.1   Structure alignment of 1MBN and 1JEB. (Courtesy of Shuo (Alex) Xiang, a graduate student at the University of Waterloo.)

COLOR FIGURE 9.3   Distance map for 2FIF.

CHAPTER 6

Protein Sequence Alignment

> The natural order of organisms is a divergent inclusive hierarchy
> and that hierarchy is recognized by taxic homology.
>
> ALEC L. PANCHEN
> (CLASSIFICATION, *EVOLUTION AND THE NATURE OF*
> *BIOLOGY,* CAMBRIDGE UNIVERSITY PRESS, 1992, P. 194)

6.1 PROTEIN HOMOLOGY

We consider a question that is fundamental for bioinformatics: Given two
protein sequences P and Q, what is the likelihood that they are homologous? This is a complex question, but it can be redefined as a simpler problem that can then be solved using a dynamic programming strategy.

We say that two proteins are homologous if they have evolved from
a common ancestor protein. The problem is important because homologous proteins usually show a significant amount of similarity in their 3-D
structure and, with that similarity, a concomitant similarity in function.*
The challenge posed by the problem arises from the fact that homologous proteins may have amino acid sequences that are substantially different, with perhaps only 30% of the amino acids matching identically.
As observed by Chothia and Lesk [CL86], the 3-D structures of proteins

---

* Strictly speaking, the similarity in function is seen for *orthologous* proteins. This is discussed
  in more detail in Chapter 6, Section 6.1.1.

diverge much more slowly in an evolutionary descent than does the amino acid sequence identity between two proteins.

Now that the question has been posed, let us pause to describe the problem in terms of the NATURE, SCIENCE, and COMPUTATION blocks introduced in Chapter 1.

6.1.1 NATURE

We start by making various statements about homology, saving problem-related assumptions for the modeling effort that is to be done for the SCIENCE section. First, note that homology may be seen in proteins that are either orthologous or paralogous [TR05,p.4]. Orthologs are homologous amino acid sequences derived from DNA sequences that have descended from a common ancestral gene. They are in different species. In other words, the genes for the orthologous proteins have diverged by speciation. Paralogs are also homologous amino acid sequences derived from DNA sequences, but the evolutionary event is a bit more complicated. In this case, the ancestral gene is duplicated, and there is an evolutionary divergence *within the same species.* More precisely: Let us suppose that gene g_A is duplicated so that the genome of the organism then has two identical copies g_A and $g_{A'}$. After the duplication event, both g_A and $g_{A'}$ will undergo an evolutionary descent *in parallel.* The vital observation is that one of these genes, say g_A, will continue to express a protein with its original or very similar biological functionality while $g_{A'}$ will not have the same pressure to conserve functionality and is free to evolve along an independent pathway leading to the expression of a protein that may have a very different functionality. Naturally, one might expect that subsequent speciation may cause an orthologous descent for both g_A and $g_{A'}$. In summary, the homology *may* indicate a common function of proteins P and Q, and that similarity of function is much more likely if the descent is strictly orthologous (we think of the "ortho" prefix as indicating "exact").

With the notions of orthologous and paralogous homology in mind, let us consider the content of the NATURE box. In mathematical models of a scientific process, we strive to predict what a natural process will do. The $h(x)$ function of the SCIENCE box acts as an approximation to $f(x, x')$, which is the measurable result of some experiment that is dependent on measurable inputs represented by the input arrow of the NATURE box in Figure 1.1.

However, our problem statement does not ask us to determine what *will* happen, but rather what *has* happened. In fact, the given proteins of the problem, P and Q, are not inputs to a natural process—they are the *results* of an evolutionary process. Consequently, there is no "homology function" $f(x, x')$ that we can initialize and then observe as a process moving forward in time. The key point of the problem is that we want to assess what happened in the past. To be consistent with this requirement, we can think of the natural process described in the NATURE box as running in *reverse* time. This sounds strange, but it is really what the problem asks of us. Our NATURE box then has the amino acid sequences P and Q as inputs and the output of this black box is the function $f(P, Q)$ that reports whether or not P and Q have a common ancestor in the past.

$$f(P,Q) = \begin{cases} 1 & \text{if } P \text{ and } Q \text{ are homologous} \\ 0 & \text{if } P \text{ and } Q \text{ are not homologous.} \end{cases} \tag{6.1}$$

This last equation clearly reflects the binary value of the answer to the question about whether P and Q are homologous—they are homologous or they are not homologous—there is no middle ground.

What is in this NATURE black box that would help us with the given problem? It would have to contain the portion of the entire evolutionary tree that describes the evolutionary descent of both P and Q. We would see P and Q as leaves with evolutionary paths leading back in time from P and Q. If P and Q are homologous, the paths intersect at the ancestor. If they are not homologous, the paths never intersect.

What observations can we extract from the NATURE black box? There will be few observations because the fossil record will typically have little, if anything, to say about a lengthy evolutionary descent.

It is reasonable to question the reverse time aspect of this description. After all, we depend on forward time descriptions to clearly describe cause-and-effect relationships, and typically scientific laws determine the unique outcome of such a relationship. If we are attempting to determine the cause of some effect, what happens if there are two or more causes that can produce the same effect? This question naturally arises when we consider the evolutionary descent from a putative ancestor protein (call it R) of one of the given proteins, say P. The amino acid sequences for R and P may differ in several places. In the evolutionary descent from R to P, it is likely that there are several intermediate sequences and we will never know their historical order because it is in DNA that vanished in the remote past. The many different pathways between R and P are essentially the many possible causes that could have produced the same final result, namely, the protein P. In fact, there may be several reasonable candidates for the ancestor protein R as well as multiple pathways leading to P. Because of these complexities, we will never be able to fully understand the binary-valued function f that we are trying to predict. It is worthwhile to draw out these issues so that we can appreciate how much more difficult the problem is when compared to a problem grounded in simple physical laws such as the Newtonian mechanics discussed in Chapter 1. Consequently, the analysis leading to a prediction of f will have to be probabilistic and, in this sense, an approximation. An exact solution of the problem is unattainable because the problem has an intrinsically daunting complexity at its core. In fact, a mathematician might consider the problem to be *ill-posed* because it does not have a unique solution.

6.1.2 SCIENCE

Despite the previous discussion, we need to design a mathematical model for the SCIENCE box, a mathematical model that attempts to formulate a function that approximates f. As f is essentially unknowable, how do we generate this approximation? The slightly cynical answer is: "We don't." We simply avoid the approximation problem by inventing a new problem that is solvable, and we argue that its solution will help us to assess the likelihood that $f = 1$. The new problem will have a solution that assesses the *similarity* of the two proteins P and Q. Moreover, as we cannot construct a verifiable exact ancestor of P and Q, it is reasonable to leave it out of the analysis, and we will not attempt to derive such an ancestor in a definitive effort to show homology.

As usual, the mathematical model is based on various assumptions and observations taken from the NATURE box. The easiest assumption is that descendants of a common ancestor would have some similarity in their amino acid sequences. This notion is consistent with the assumed evolutionary mechanism of duplication and modification described by Doolittle [Do90]:

> The vast majority of extant proteins are the result of a continuous series of genetic duplications and subsequent modifications. As a result, redundancy is a built-in characteristic of protein sequences, and we should not be surprised that so many new sequences resemble already-known sequences.

As an example of similarity, consider the two amino acid sequences:

$$P = \text{WATERMANMIKE} \quad \text{and} \quad Q = \text{MERMANETHEL}$$

To show the similarity of these two sequences, we can write them as two rows with vertical bars between matching letters:

```
P =    W A T E R M A N M I K E -
       | | | | |         : |
Q =    - - M E R M A N E T H E L
```

These two rows represent a *global alignment* of the two sequences.

Definition

An optimal global alignment of two sequences P and Q is obtained by writing the two sequences in such a way as to maximize the occurrence of matching pairs of letters, one in P and the other in Q. Gaps (written as dashes) may be introduced into either sequence or at either end of a sequence in such a way as to have both strings of equal length and to facilitate the largest numbers of matches.

Our strategy is to formulate two mathematical models with two different objectives:

1. Show similarity by generating a sequence alignment with the maximal number of matching amino acids.

2. After the alignment has been generated, evaluate their similarity to determine how likely it is that P and Q are homologous.

Note that similarity is determined by the output of the mathematical model and is intended to be an approximation of the binary-valued

$$f(P,Q).$$

function that specifies whether or not P and Q are homologous. This similarity result can have a range of values, whereas f does not; so, it is to be expected that we would never claim that P and Q have a certain "percentage of homology." This is an inappropriate use of terminology and is inconsistent with the NATURE and SCIENCE discussion up to this point.

For now, we will refer to the number of exact matches in an alignment as its score. So, the first objective is to generate an optimal alignment having the maximum score. In the next section, we discuss more complicated scores that account for *partial* matches.

6.1.2.1 Partial Matches

Another consideration in the generation of an alignment is that amino acids may be accepted as a pair even though they do not match identically. This approach takes into account the biochemical properties of the amino acids. For example, D (aspartate) and E (glutamate) are both acidic amino acids, and so they are good candidates for a partial match. Other partial matches would be biologically justified: hydrophobic amino acids (W, F, Y, L, I, V, M, A) and basic amino acids (K, R, H).

For long sequences it may be possible to generate many different optimal alignments, each having the maximal number of exact matches. When partial matches contribute to the score of an optimal alignment, the analysis becomes more complicated because we must decide how much they contribute. For example, should the score increase if the alignment brings in three partial matches at the expense of foregoing one exact match? To answer such questions, researchers have constructed *substitution matrices* that essentially define the score contributions of both exact and partial matches. Figure 6.1, adapted from http://www.ncbi.nlm.nih.gov/Class/ FieldGuide/BLOSUM62.txt, gives the BLOSUM62 (see [HH92]) version of such a substitution matrix. There are other BLOSUM matrices with suffixes such as 45, 50, 80, and 90. The BLOSUM62 matrix is often used as the standard matrix when doing ungapped alignments, whereas BLOSUM50 is often used for gapped alignments. In general, BLOSUM80 works best for sequences that show less divergence, and BLOSUM45, at the other end of the application scale, is used for sequences that show more evolutionary divergence.

```
      A  R  N  D  C  Q  E  G  H  I  L  K  M  F  P  S  T  W  Y  V
A     4 -1 -2 -2  0 -1 -1  0 -2 -1 -1 -1 -1 -2 -1  1  0 -3 -2  0
R    -1  5  0 -2 -3  1  0 -2  0 -3 -2  2 -1 -3 -2 -1 -1 -3 -2 -3
N    -2  0  6  1 -3  0  0  0  1 -3 -3  0 -2 -3 -2  1  0 -4 -2 -3
D    -2 -2  1  6 -3  0  2 -1 -1 -3 -4 -1 -3 -3 -1  0 -1 -4 -3 -3
C     0 -3 -3 -3  9 -3 -4 -3 -3 -1 -1 -3 -1 -2 -3 -1 -1 -2 -2 -1
Q    -1  1  0  0 -3  5  2 -2  0 -3 -2  1  0 -3 -1  0 -1 -2 -1 -2
E    -1  0  0  2 -4  2  5 -2  0 -3 -3  1 -2 -3 -1  0 -1 -3 -2 -2
G     0 -2  0 -1 -3 -2 -2  6 -2 -4 -4 -2 -3 -3 -2  0 -2 -2 -3 -3
H    -2  0  1 -1 -3  0  0 -2  8 -3 -3 -1 -2 -1 -2 -1 -2 -2  2 -3
I    -1 -3 -3 -3 -1 -3 -3 -4 -3  4  2 -3  1  0 -3 -2 -1 -3 -1  3
L    -1 -2 -3 -4 -1 -2 -3 -4 -3  2  4 -2  2  0 -3 -2 -1 -2 -1  1
K    -1  2  0 -1 -3  1  1 -2 -1 -3 -2  5 -1 -3 -1  0 -1 -3 -2 -2
M    -1 -1 -2 -3 -1  0 -2 -3 -2  1  2 -1  5  0 -2 -1 -1 -1 -1  1
F    -2 -3 -3 -3 -2 -3 -3 -3 -1  0  0 -3  0  6 -4 -2 -2  1  3 -1
P    -1 -2 -2 -1 -3 -1 -1 -2 -2 -3 -3 -1 -2 -4  7 -1 -1 -4 -3 -2
S     1 -1  1  0 -1  0  0  0 -1 -2 -2  0 -1 -2 -1  4  1 -3 -2 -2
T     0 -1  0 -1 -1 -1 -1 -2 -2 -1 -1 -1 -1 -2 -1  1  5 -2 -2  0
W    -3 -3 -4 -4 -2 -2 -3 -2 -2 -3 -2 -3 -1  1 -4 -3 -2 11  2 -3
Y    -2 -2 -2 -3 -2 -1 -2 -3  2 -1 -1 -2 -1  3 -3 -2 -2  2  7 -1
V     0 -3 -3 -3 -1 -2 -2 -3 -3  3  1 -2  1 -1 -2 -2  0 -3 -1  4
```

FIGURE 6.1 The BLOSUM62 substitution matrix.

The score contribution of an exact match can be found on the main diagonal. These entries have integer values ranging from 4 to 11 and take into account the relative frequency of occurrence of an amino acid. For example, the score contribution made by an exact match for W (tryptophan) is 11, the highest value, because tryptophan has the lowest frequency of occurrence. In general, the entries of the substitution matrix are based on issues other than simple frequency of occurrence. More precisely, it will describe the frequency of observed changes of characters in related protein segments or *blocks* that have been conserved in an evolutionary descent. The next section describes this in more detail.

6.1.2.2 Building a BLOSUM Matrix

The Henikoffs [Hн92] derived the BLOSUM (BLOck SUbstitution Matrix) matrices using ungapped multiple alignments of evolutionarily divergent proteins. They gathered frequency counts from blocks of conserved sequences found within these alignments.

To make these statements more clear, let us go through some steps that describe the calculations. First, we need a set of proteins that are functionally related as determined by a multiple alignment. We then extract a set of blocks from this multiple alignment. These are ungapped regions that show conserved residues. For example, the four strings

1. TWYVTHYLDKDP
2. KWYVTHFLDKDP
3. VWYVTDYLDLEP
4. TWYETHYLDMDP

represent a block that is 12 residues in width. This is the first of six blocks that can be seen in a multiple alignment of four nitrophorin proteins with Swiss-Prot accession numbers: Q94733, Q26241, Q94734, and Q6PQK2. This block represents the 12 amino acids in a beta strand that is highly conserved in this family of nitrophorins. You can inspect amino acids 21 to 32 inclusive in 1PM1 to see an example of this strand. The idea is see how often an amino acid might change in such a block, seen as having an obvious conservation of amino acids. To calculate the entries in the final substitution matrix, we will need thousands of conserved blocks, but for now we consider this single block.

Let us start by inspecting the fourth column: VVVE. We now count the number of character pairs that are matches and the number of pairs that are mismatches in that column. There are three V entries, and so there are (3 choose 2) = 3 possible VV pairs. There is a single E, so that gives us 3 VE or EV pairs. There are no EE pairs. Considering a slightly more complicated situation, we see that column one contributes a single match pair (TT). Counts for the unmatched pairs are: 2 TK, 2 TV, and 1 KV. In general, a character c_i that appears $n_i^{(k)}$ times in a column k will produce

$$f_{ii}^{(k)} = \binom{n_i^{(k)}}{2}$$
(6.2)

matched pairs, whereas mismatches involving amino acid a_i appearing $n_i^{(k)}$ times with amino acid a_j appearing $n_j^{(k)}$ times will produce

$$f_{ij}^{(k)} = n_i^{(k)} n_j^{(k)}$$
(6.3)

pairs. A character appearing only once in a column does not contribute a matched pair.

Note that the sum of all possible pairs in a column will be $s(s-1)/2$, where s is the number of sequences in a block. Therefore, we must have

$$\sum_{i=1}^{20} \sum_{j=1}^{i} f_{ij}^{(k)} = \frac{s(s-1)}{2}$$
(6.4)

for all k.

Summing the contributions across all the columns, we can define the counts for the entire block as

$$f_{ij} = \sum_{k=1}^{L} f_{ij}^{(k)}$$
(6.5)

where L is the number of columns. The observed *normalized* frequency of occurrence for a block is

$$q_{ij} = \frac{f_{ij}}{\sum_{i=1}^{20} \sum_{j=1}^{i} f_{ij}} = \frac{f_{ij}}{T}$$
(6.6)

where the total number of pairs in the block is

$$T = L \frac{s(s-1)}{2}.$$
(6.7)

This observed normalized frequency of occurrence will be compared with the expected probability of occurrence for each i, j pair. This calculation is based on the pair frequencies *observed in the blocks*. In fact, the probability of occurrence of the ith character in an i, j pair is

$$p_i = q_{ii} + \frac{1}{2} \sum_{j \neq i} q_{ij}.$$

(6.8)

From this we can compute the expected probability of occurrence e_{ij} for each pair i, j as

$$e_{ij} = \begin{cases} p_i^2 & i = j \\ 2p_i p_j & i \neq j. \end{cases}$$

(6.9)

An *odds ratio* matrix can now be calculated where each entry is q_{ij}/e_{ij}. This could act as a substitution matrix, but its use in the evaluation of the similarity across two sequences would require multiplication of entries taken from the matrix. To avoid extensive multiplication, we instead define matrix entries as logarithms of these ratios. Then, the multiplications are achieved via simple addition. With this as our strategy, the log odds ratio is

$$s_{ij} = \log_2 \left(\frac{q_{ij}}{e_{ij}} \right).$$

(6.10)

The value in the BLOSUM matrix is actually $2s_{ij}$ rounded to the nearest integer. Therefore, they have "half-bit" units.

Note that if the observed frequencies are as expected, then $s_{ij} = 0$. If the observed frequencies are less than expected, then $s_{ij} < 0$, and if they are more than expected, then $s_{ij} > 0$.

As noted earlier, this analysis applies to a single block. In practice we would use thousands of blocks. The counts of all possible pairs in each column of each block in the entire collection of blocks would be evaluated.

There is still a final issue to consider. It may be that blocks contain sequences that are very similar. In such a situation, these sequences will be overrepresented in the calculations and they will skew the results. In fact, the sequences of the nitrophorin example introduced earlier are really only 4 of 8 sequences in Block IPB002351A available from the Blocks Database.*

```
TWYVTHYLDKDP
TWYVTHYLDKDP
KWYVTHFLDKDP
VWYVTDYLDLEP
VWYVTDYLDLEP
TWYVTHYLDKDP
TWYETHYLDMDP
VWYVTDYLDLEP
```

The sequences are from proteins with different accession numbers. In some cases, the sequences in the block are exact copies or very similar. To deal with this type of sequence redundancy, Henikoff and Henikoff resorted to a clustering strategy. For example, in generating the BLOSUM80 matrix, two sequences were put into the same cluster if they had a sequence identity in their aligned columns that was greater than or equal to 80%. A sequence is placed into a cluster if it has a sequence identity that is greater than or equal to 80% with respect to *any* other member of the cluster.

Inspection of the four sequences given at the beginning of this section gives us six different sequence pairs and computing their percentages of sequence identity we obtain the following:

```
1,2          10/12 = 83.33%
1,3          8/12 = 66.66%
1,4          8/12 = 66.66%
2,3          7/12 = 58.33%
2,4          8/12 = 66.66%
3,4          7/12 = 58.33%
```

Consequently, we get three clusters for the 80% threshold, namely, {1, 2}, {3}, and {4}. If we bring in the extra sequences that are part of Block IPB002351A, we get the following three clusters:

```
TWYVTHYLDKDP
TWYVTHYLDKDP
TWYVTHYLDKDP
KWYVTHFLDKDP
------------
```

---

* Blocks WWW Server: http://blocks.fhcrc.org.

```
VWYVTDYLDLEP
VWYVTDYLDLEP
VWYVTDYLDLEP
------------
TWYETHYLDMDP
```

If we had chosen 62% as the threshold, then all the sequences would have collapsed into a single cluster.

When we processed a block before the notion of clusters was introduced, each sequence was essentially making up a single cluster. The calculation of T in Equation 6.7 was done in terms of s the number of sequences in a block. The motivation for clustering is that we are going to disallow the extraction of pairs with entries from the same cluster. If pairs are formed from two amino acids, each in a *different* cluster, then the definition of T changes:

$$T = L\frac{c(c-1)}{2} \tag{6.11}$$

where c is the number of clusters. This will require that Equation 6.4 change and each column k is subject to the requirement that

$$\sum_{i=1}^{20}\sum_{j=1}^{i} f_{ij}^{(k)} = \frac{c(c-1)}{2}. \tag{6.12}$$

This can be arranged if the characters in a column are given *weights* that are used in the computations of $f_{ii}^{(k)}$ and $f_{ij}^{(k)}$. We replace Equations 6.2 and 6.3 with

$$f_{ii}^{(k)} = \sum_{m=1}^{c-1}\sum_{n=m+1}^{c} \frac{R(i,m)}{r_m}\frac{R(i,n)}{r_n} \tag{6.13}$$

and

$$f_{ij}^{(k)} = \sum_{m=1}^{c-1}\sum_{n=m+1}^{c}\left[\frac{R(i,m)}{r_m}\frac{R(j,n)}{r_n} + \frac{R(j,m)}{r_m}\frac{R(i,n)}{r_n}\right] \tag{6.14}$$

where $R(i, l)$ represents the number of times that amino acid a_i appears in cluster l, and r_l is the number of sequences in cluster l. Note that a column character in a cluster has a weight that varies directly with the number of times that the character appears in the cluster and inversely with the number of sequences in the cluster.

For example, the pair TK had a count of 2 in the four sequence case. In the six sequence case, the K in column one cannot be paired with a T from its own cluster and therefore it can only contribute to the TK pair count by being paired with the T in the last cluster. As K is part of a cluster with four entries, its significance is deflated by a factor of 4 and so $f_{TK}^{(1)}$ is ¼. Here is a more complicated example. Consider a column with three clusters:

Cluster 1	P
	M
	M
	M

Cluster 2	P

Cluster 3	P
	M

We compute

$$\frac{R(P,1)}{r_1}=\frac{1}{4} \qquad \frac{R(P,2)}{r_2}=\frac{1}{1} \qquad \frac{R(P,3)}{r_3}=\frac{1}{2}$$

$$\frac{R(M,1)}{r_1}=\frac{3}{4} \qquad \frac{R(M,2)}{r_2}=\frac{0}{1} \qquad \frac{R(M,3)}{r_3}=\frac{1}{2}.$$

Using Equations 6.13 and 6.14, we get

$$f_{PM}=\left(\frac{1}{4}\times\frac{0}{1}\right)+\left(\frac{3}{4}\times\frac{1}{1}\right)+\left(\frac{1}{4}\times\frac{1}{2}\right)+\left(\frac{3}{4}\times\frac{1}{2}\right)+\left(\frac{1}{1}\times\frac{1}{2}\right)+\left(\frac{0}{1}\times\frac{1}{2}\right)$$

$$=\frac{3}{4}+\frac{1}{8}+\frac{3}{8}+\frac{1}{2}=\frac{14}{8}$$

$$f_{PP}=\left(\frac{1}{4}\times\frac{1}{1}\right)+\left(\frac{1}{4}\times\frac{1}{2}\right)+\left(\frac{1}{1}\times\frac{1}{2}\right)=\frac{1}{4}+\frac{1}{8}+\frac{1}{2}=\frac{7}{8}$$

$$f_{MM}=\left(\frac{3}{4}\times\frac{0}{1}\right)+\left(\frac{3}{4}\times\frac{1}{2}\right)+\left(\frac{0}{1}\times\frac{1}{2}\right)=\frac{3}{8}.$$

Note that

$$f_{PM}+f_{PP}+f_{MM}=\frac{14}{8}+\frac{7}{8}+\frac{3}{8}=\frac{24}{8}=3=\frac{c(c-1)}{2}$$

as expected from Equation 6.12.

Now that we know how to calculate $f_{ii}^{(k)}$ and $f_{ij}^{(k)}$, we can complete the computation of the BLOSUM matrix as done earlier when no clustering was used.

In our description of the construction of BLOSUM matrices, it was stated that clustering of sequences was done using a threshold specification. For example, in calculations for the BLOSUM62 matrix, two sequences were in the same cluster if they had a percent sequence identity that was 62 or higher. As stated earlier, researchers have created several BLOSUM matrices. Those in common use include BLOSUM45, BLOSUM50, BLOSUM62, BLOSUM80, and BLOSUM90. It should now be clear why a matrix such as BLOSUM50 is suitable for studies of more divergent sequences. If the sequences used in the construction of the matrix are very similar, they end up in the same cluster and there is no attempt to incorporate the effects of the few changes that may have occurred in their evolutionary descent.

6.1.2.3 Gaps

Consider the following two alignments:

Alignment #1:

```
M  I  C  H A E L  S  W A T E R M A N
   |  |     | |    :        | | | | |
-  E  T  H - E L  M  - - - E R M A N
0 -3 -1  8 0 5 4 -1  0 0 0 5 5 5 4 6
```

Alignment #2:

```
M  I  C H A E L S W  A T E R M A N
   |  |   | |        :    | | | | |
-  E  T H - E L - -  M - E R M A N
0 -3 -1 8 0 5 4 0 0 -1 0 5 5 5 4 6
```

The numbers in each column of the alignment are the score contributions given by the BLOSUM62 matrix. The score for the entire alignment is simply the sum of these contributed scores. This means that Alignment #1 has a score of 37, and Alignment #2 also has a score of 37. However, most alignment programs will offer Alignment #1 as the better alignment because Alignment #2 initiates a gap four times, while the first does so only three times. To support this choice, it is possible for the alignment score to incorporate a term that leads to an extra penalty whenever a new run of gap symbols is started.

Let us consider why a gap appears. A gap is a sequence of one or more spaces drawn as dashes in the alignment display. Each space can represent an insertion or a deletion. For example, the space below the first A represents one of the following:

1. The insertion of an A into the first string, which would have been MICHELSWATERMAN prior to the insertion

2. The deletion of some letter that was originally between H and E of ETHELMERMAN

Very often, we do not have enough evidence to determine which scenario occurred, and so either possibility is acceptable and the space is referred to as an indel (*in*sertion or *del*etion). The key fact is that indels are assumed to be much less likely than substitutions from a biological perspective. Consequently, their initiation and extension lead to the addition of negative contributions (penalties) in the score computation. These penalties have been assessed using empirical observations gathered from several examples of homologous proteins that have been previously aligned.

6.1.2.4 Summary

We are now in a position to summarize the main points of the mathematical model for global sequence alignment.

Given two strings representing the residues in two proteins P and Q, an optimal global alignment seeks to demonstrate maximal similarity by writing one string beneath the other so that residues in columns either identically or partially match. The objective of attaining maximal similarity must address the following issues:

- Gaps (written as dashes) may be introduced into either sequence or at either end of a sequence in such a way as to have both strings of equal length and to facilitate the largest numbers of matches.

- Although gaps may be introduced, we attempt to do this in a parsimonious fashion, and so, in the formulation of the problem, we reduce the score of an alignment by introducing a *gap penalty*.

6.1.3 COMPUTATION

The optimal global alignment problem can be solved using dynamic programming. This solution of the problem was done in 1970 by Needleman and Wunsch (see [Nw70]).

In the design of a dynamic programming algorithm, the most important issue is the specification of a subproblem. We know that a subproblem should be smaller in some sense and capable of contributing to the solution of a larger problem. The idea of aligning substrings of the given strings should immediately come to mind.

It is possible to define subproblems that have too many degrees of freedom in their specification. For example, we might state that a subproblem should be the global alignment of two substrings extracted from P and Q, that is, $P[i..j]$ and $Q[k..m]$, where the notation $[i..j]$ represents the string of contiguous characters between and including character positions i and j. This subproblem would be parameterized by four integers, i, j, k, and m. This approach would involve evaluation of all cells in a 4-D array and this would be an extremely costly evaluation for lengthy strings.

In fact, we need not go to such an extreme when specifying substrings in the definition of our subproblem. It will be shown in Lemma 6.1 that the selected substrings can be *prefixes* of the given strings.

Definition

The ith prefix of a string X denoted by X_i is the substring of X consisting of the first i characters of X, that is, $X_i = x_1, x_2, \ldots, x_{i-1}, x_i = X_{i-1}, x_i$

Note the use of comma-separated lowercase letters to represent consecutive characters within the prefix.

With this agreement about notation, we proceed with the subproblem specification described in the following subsections.

6.1.3.1 Subproblem Specification

The dynamic programming subproblem will be parameterized by two arguments, i and j, and its solution will be the optimal global alignment of the two prefixes P_i and Q_j.

6.1.3.2 Scoring Alignments

We score an alignment by adding up the scores derived from each column. Suppose that after adding dashes, we get an alignment that involves m_{ij} columns when we do a global alignment of P_i and Q_j. Then the score for this alignment will be given by

$$NW(i,j) = \sum_{k=1}^{m_{ij}} s_c(k) \tag{6.15}$$

where $s_c(k)$ is the score for the kth column in the alignment. We have used the notation $NW(i, j)$ to emphasize that this score is for the global alignment algorithm due to Needleman and Wunsch. On the basis of the earlier discussion concerning gaps and partial matches, we can define the column score using

$$s_c(k) = \begin{cases} B(p'_k, q'_k) & \text{in a column with a match} \\ -d & \text{in a column with a dash.} \end{cases} \tag{6.16}$$

In this equation, the quantity d represents the penalty that is assigned for the occurrence of a gap. The gap penalty d is a positive quantity, so the presence of any gap in the alignment will cause this penalty to be subtracted from the total alignment score. We use the notation p'_k and q'_k to indicate the residues of P_i and Q_j that appear in the kth column when the global alignment is written out, complete with dashes. The function $B(row\_residue, column\_residue)$ is used to extract an entry from a row and column of the substitution matrix that is being used for the alignment. In a previous section we used the BLOSUM62 matrix as an example of such a substitution matrix.

6.1.3.3 Suitability of the Subproblem

There are undoubtedly many ways to define a "subproblem," many of them ill-fitted to the needs of a dynamic programming algorithm. However, our specification of a subproblem *is* appropriate for the dynamic programming algorithm. This will be demonstrated by proving that there is a relationship between the subproblem specified as the optimal global alignment of prefixes P_i and Q_j and other optimal global alignment subproblems that work on shorter prefix strings. This gives us a recursive evaluation leading to the computation of the full solution: The full problem depends on smaller subproblems, each of which depends on even smaller subproblems, etc., until eventually we have subproblems so small that they are trivial to solve. We need the following lemma.

Lemma 6.1

Let P_i and Q_j be prefixes of the strings P and Q, respectively. If we consider the optimal global alignment of P_i and Q_j, there are three possible cases when describing the appearance of the last column in the alignment:

a. Character p_i matches character q_j in the optimal global alignment, which can then be drawn as

$$\begin{aligned} P_{i-1}, p_i \\ Q_{j-1}, q_j. \end{aligned} \tag{6.17}$$

Claim: In this case, the matching of characters seen prior to p_i and q_j is, in fact, the optimal global alignment of P_{i-1} and Q_{j-1}.

b. Character p_i does not participate in any match; that is, the optimal global alignment can be drawn as

$$P_{i-1}, p_i$$
$$Q_j, -.$$
(6.18)

Claim: In this case, the matching of characters seen prior to p_i is the optimal global alignment of P_{i-1} and Q_j.

c. Character q_j does not participate in any match; that is, the optimal global alignment can be drawn as

$$P_i, \quad -$$
$$Q_{j-1}, q_j.$$
(6.19)

Claim: In this case, the matching of characters seen prior to q_j is the optimal global alignment of P_i and Q_{j-1}.

Proof of claim for Case (a):
Assume the matching of P_{i-1} and Q_{j-1} as seen in Case (a) is *not* the optimal global alignment of P_{i-1} and Q_{j-1}. Then, suppose we replaced this alignment with the optimal global alignment of P_{i-1} and Q_{j-1}. This will raise the score associated with the columns prior to the last column. Then simply appending p_i to P_{i-1} and q_j to Q_{j-1}, we would get an alignment with a higher score than that seen in Case (a). This is a contradiction because we have already assumed that Case (a) represents an optimal global alignment.

Proof of claim for Case (b):
Reasoning along similar lines: Suppose the matching of P_{i-1} and Q_j as seen in Case (b) is *not* the optimal global alignment of P_{i-1} and Q_j. Then, suppose we replaced this alignment with the optimal global alignment of P_{i-1} and Q_j. This will raise the score associated with the columns prior to the last column. Then simply appending p_i to P_{i-1} and writing a dash below it, we would get an alignment that is characterized as being Case (b) but with a higher score. This is a contradiction because we have already assumed that Case (b) represents an optimal global alignment.

Proof of claim for Case (c):
The proof is a simple variation of the proof for Case (b).

Computation of NW(i, j)

Lemma 6.1 is a formal discussion that brings out the relationships among subproblems and serves to justify the formulation of our subproblem as the optimal global alignment of P_i and Q_j. More important for us at the moment is the implications these cases provide for the dependencies among the scores associated with the subproblems. Each case in Lemma 6.1 defines a possible relationship between $NW(i, j)$ and the score derived from the solution of some smaller alignment problem.

Case (a):

By matching the last characters in each string of residues, the score will be incremented by an amount that is accessed from the substitution matrix:

$$NW(i,j) = NW(i-1, j-1) + B(p_i, q_j). \tag{6.20}$$

Case (b):

By placing a dash in the bottom row of the alignment, we decrease the score in going from $NW(i-1, j)$ to $NW(i, j)$:

$$NW(i,j) = NW(i-1, j) - d. \tag{6.21}$$

Case (c):

By placing a dash in the top row of the alignment, we decrease the score in going from $NW(i, j-1)$ to $NW(i, j)$:

$$NW(i,j) = NW(i, j-1) - d. \tag{6.22}$$

Considering these possibilities, we see that there are three ways to define $NW(i, j)$. As we wish to *maximize* the alignment score, it makes sense to always choose the option that will maximize $NW(i, j)$. Consequently, we use the following equation for the computation of $NW(i, j)$:

$$NW(i,j) = \max \begin{cases} NW(i-1, j-1) + B(p_i, q_j), \\ NW(i-1, j) - d, \\ NW(i, j-1) - d \end{cases}. \tag{6.23}$$

We now have enough information about the mathematical model to go through the dynamic programming design steps suggested in Section 4.4 of Chapter 4. These steps are summarized as follows:

1. Identification of subproblems:

   The subproblem will be parameterized by two arguments, i and j, and its solution will be the optimal global alignment of the two prefixes P_i and Q_j.

2. Specify evaluation of score values:

   We will use the notation $|P|$ and $|Q|$ to represent the number of residues in proteins P and Q, respectively. The $NW(i, j)$ score values will be retained in an array with $|P| + 1$ rows and $|Q| + 1$ columns. There will be an extra row "0" and an extra column "0" to store initial values required by the recursion. Considering the dependencies of $NW(i, j)$ on other score values in the array, we see that the score stored in cell (i, j) depends on scores stored in $(i-1, j-1)$, $(i-1, j)$, and $(i, j-1)$. Consequently, it will be sufficient to evaluate the array entries in a row-by-row fashion. We will need to know the score of the full global alignment: This will be found in bottom right corner of the array (the intersection of the last row and last column).

3. Set values for the base case:

   The entries in row 0 and column 0 will hold scores that correspond to penalties that are introduced because of the insertion of initial dashes that might appear in the alignment before the characters of either the P string or Q string. As we evaluate scores along row 0, the penalties increase so that $NW(0, j) = -d^*j$. Similarly, for column 0, we have $NW(i, 0) = -d^*i$.

4. Recover the solution configuration:

   The trace-back operation involves keeping track of the cell that produced the maximum value when $NW(i, j)$ was calculated using Equation 6.23. This can be done using a separate trace-back array having $|P|$ rows and $|Q|$ columns. The entry stored in cell (i, j) of this array would be a code indicating whether the maximum value was taken from $(i-1, j-1)$, $(i-1, j)$, or $(i, j-1)$. When describing a trace-back operation, one can superimpose the trace-back array over the score array with trace-back dependencies illustrated by means of arrows.

		E	T	H	E	L	M	E	R	M	A	N
	0	← -1	← -2	← -3	← -4	← -5	← -6	← -7	← -8	← -9	← -10	← -11
M	↑ -1	← -2	↖ -2	← -3	← -4	↖ -2	↖ 0	← -1	← -2	← -3	← -4	← -5
I	↑ -2	← -3	↑ -3	← -4	← -5	↖ -2	↑ -1	← -2	← -3	↖ -1	← -2	← -3
C	↑ -3	← -4	↑ -4	← -5	← -6	↑ -3	↑ -2	← -3	← -4	↑ -2	↖ -1	← -2
H	↑ -4	↖ -3	← -4	↖ 4	← 3	← 2	↑ 1	← 0	← -1	← -2	↑ -2	↖ 0
A	↑ -5	↑ -4	↖ -3	↑ 3	↖ 3	← 2	← 1	← 0	← -1	← -2	↖ 2	← 1
E	↑ -6	↖ 0	← -1	↑ 2	↖ 8	← 7	← 6	↖ 6	← 5	← 4	← 3	← 2
L	↑ -7	↑ -1	↖ -1	↑ 1	↑ 7	↖ 12	← 11	← 10	← 9	← 8	← 7	← 6
S	↑ -8	↑ -2	↖ 0	↑ 0	↑ 6	↑ 11	↖ 11	↖ 11	← 10	← 9	↖ 9	← 8
W	↑ -9	↑ -3	↑ -1	↑ -1	↑ 5	↑ 10	↑ 10	↑ 10	← 9	↖ 9	← 8	← 7
A	↑ -10	↑ -4	↑ -2	↑ -2	↑ 4	↑ 9	↑ 9	↑ 9	↖ 9	← 8	↖ 13	← 12
T	↑ -11	↑ -5	↖ 1	← 0	↑ 3	↑ 8	↖ 8	↑ 8	↑ 8	↖ 8	↑ 12	↖ 13
E	↑ -12	↑ -6	↑ 0	↖ 1	↖ 5	↑ 7	↑ 7	↖ 13	← 12	← 11	↑ 11	↑ 12
R	↑ -13	↑ -7	↑ -1	↑ 0	↑ 4	↑ 6	↑ 6	↑ 12	↖ 18	← 17	← 16	← 15
M	↑ -14	↑ -8	↑ -2	↑ -1	↑ 3	↖ 6	↖ 11	↑ 11	↑ 17	↖ 23	← 22	← 21
A	↑ -15	↑ -9	↑ -3	↑ -2	↑ 2	↑ 5	↑ 10	↑ 10	↑ 16	↑ 22	↖ 27	← 26
N	↑ -16	↑ -10	↑ -4	↖ -2	↑ 1	↑ 4	↑ 9	↖ 10	↑ 15	↑ 21	↑ 26	↖ 33

FIGURE 6.2 Global alignment example.

6.1.3.4 A Global Alignment Example

To provide an example, we use the same strings seen earlier in Section 6.1.2.2. This time, we will assume a gap penalty of $d = -1$. The results are displayed in Figure 6.2.

Working with the trace-back, one gets the following alignment:

```
 -  M  I  C  H  A  E  L  S  W  A  T  E  R  M  A  N
    |        |     |  |           |  |  |  |  |  |
 E  T  -  -  H  -  E  L  -  -  -  M  E  R  M  A  N
-1 -1 -1 -1  8 -1  5  4 -1 -1 -1 -1  5  5  5  4  6
```

The pseudocode for this problem is reasonably similar to the longest common subsequence problem of Chapter 4 and is left as an exercise for the reader.

6.2 VARIATIONS IN THE GLOBAL ALIGNMENT ALGORITHM

For many problems solved by dynamic programming, the computed solutions are provably optimal. For example, in Bentley's problem (see Exercise 2 of Chapter 4) we have a very simple optimization objective, and the algorithm will always extract the subsequence of numbers with the maximum sum. The pairwise alignment of two protein sequences is a much more complicated problem because of the many parameters that must be properly set before the dynamic program can proceed. This complexity is related to the mathematical model for the treatment of gaps and the handling of partial matches.

As described earlier, substitution matrices such as BLOSUM62 are used to specify the score contribution of a partial match. Here are some typical approaches for the computation of penalties arising from gaps:

- Linear gap model

  The presence of a dash introduces a penalty that is seen in the recursion as a decrease of the score by the value d_L. In the simplest model, the dependence is linear and the presence of a gap with g dashes leads to a decrease of gd_L.

- Affine gap model

  The affine gap model uses a penalty formula given by $d_A + (g - 1)e$, where d_A is the gap-open penalty and e is the gap-extension penalty. Typically, $e < d_A$ and this means that the insertion of a long sequence of dashes into the alignment is penalized less than it would be if a linear model was used.

Typical values are $d_L = 8$ for the linear case and $d_A = 12$, $e = 2$ for the affine case. Although these values get good results in practice, there is no rigorous theory or rule stipulating the optimal values for these parameters. Furthermore, the choice of substitution matrix may not be obvious. One can find guidelines for their use (see [ZB07], p. 84), but the choice is often decided by empirical observation and experience rather than a mathematically rigorous decision process.

6.3 THE SIGNIFICANCE OF A GLOBAL ALIGNMENT

Given the amount of choice in the construction of an alignment algorithm, how does one objectively choose the best alignment for a given pair of sequences? Ideally, we want some quantitative assessment that indicates the "goodness" of an alignment. The score calculation, which is part of the global alignment algorithm, should give us a reasonable assessment of this sequence similarity. However, we must be careful about the biological meaning associated with this best score. The algorithm is simply giving us the optimal alignment determined by the best score across all possible alignments (considering the gap model) *for that particular pair of sequences*. The global alignment algorithm would carry out the same type of computations, eventually reporting the alignment with the best score even for random sequences that were completely unrelated.

In other words, the alignment algorithm will always produce a "best" alignment for any pair of sequences, related or not. Furthermore, the score produced by the alignment procedure does not have an intrinsic biological significance. Evolutionary processes are extremely complicated, and it is impossible to build a scoring model that reports on the biological significance of a particular alignment without the necessity of comparing this score with the scores of any other alignments. To give the alignment score a significance that is hopefully more relevant in the biological sense, we invoke an extra assumption. This assumption states that two aligned sequences will have a significant similarity because of an evolutionary process if the ordering of residues in the given pair of sequences contributes to a high score. To clarify this point, suppose we went through the steps of the following experiment:

1. Given the sequences P and Q, we derive the best alignment using the global alignment algorithm and we designate the score value for this alignment as $\text{Score}(P, Q) = NW(|P|, |Q|)$.

2. We scramble (randomly permute) the letters in P to get P_S and then in Q to get Q_S.

3. We use the same global alignment algorithm to align the scrambled sequences and we designate the computed score as $\text{Score}(P_S, Q_S)$.

If the given P and Q were unrelated, then there would be very few instances of a sequence pattern appearing in both P and Q. We would expect Score (P, Q) to be approximately the same as $\text{Score}(P_S, Q_S)$. However, if P and Q did have some sequence similarity, we would expect a significant difference with $\text{Score}(P, Q)$ being more than $\text{Score}(P_S, Q_S)$.

The previous discussion gives us an intuitive grasp of the situation, but our experiment is not useful until we quantify the procedure and thus eliminate vague phrases such as "approximately the same as" and "significant difference."

6.3.1 Computer-Assisted Comparison

We can modify the previous procedure so that it becomes a well-defined decision process. We follow the computer-assisted comparison methodology described by Doolittle (see [Do81], p. 152). As before, we perform Step 1 aligning the given protein sequences P and Q to obtain Score(P, Q).

However, this time we repeat Steps 2 and 3 r times to get several alignments of scrambled sequences. This will generate r alignment scores:

$$\{\text{Score}\,(P_{S_i},Q_{S_i})\,|\,i=1,2,\ldots,r\} \tag{6.24}$$

which are averaged and a standard deviation is computed. Specifically,

$$\mu = \sum_{i=1}^{r} \text{Score}(P_{S_i},Q_{S_i}) \tag{6.25}$$

and the unbiased estimate of σ:

$$s = \sqrt{\frac{1}{r-1}\sum_{i=1}^{r}[\text{Score}(P_{S_i},Q_{S_i})-\mu]^2}\,. \tag{6.26}$$

The value of Score(P, Q) is then compared with the mean of the scores from the scrambled pairs and the number of standard deviations above or below the mean is computed:

$$\frac{\text{Score}\,(P,Q)-\mu}{s}. \tag{6.27}$$

When Score(P, Q) is 3 or more standard deviations above the mean, the alignment is considered to represent a biologically meaningful similarity for P and Q. If Score(P, Q) falls between 0 and 3 standard deviations above the mean value, we consider this comparison to be indecisive. It is still possible that P and Q have a common ancestor, but the alignment algorithm is unable to demonstrate enough similarity to show this.

6.3.2 Percentage Identity Comparison

In addition to the computer-assisted comparison described in Section 6.3.1, there are several other approaches for the evaluation of an alignment score. The simplest approach is to count the number of identical matches in an alignment. Then, *percent identity* can be derived by dividing this count by the length of the aligned region and multiplying by 100. It should be noted that each column in the alignment of two random proteins would have a 1 out of 20, or 5%, chance of producing an identical match. This calculation implicitly assumes that all amino acids occur with the same probability. It is well known that this is not the case. The most frequent amino acids, glycine, alanine, and leucine, account for roughly one quarter of all residues and any one of them typically appears about four times more often than any one of the least frequent amino acids, tryptophan, methionine, histidine, and cysteine. Doolittle [Do81] notes that when these observed frequencies are taken into account, the percent identity for random sequences goes up from 5 to 6%. Consequently, we would expect the alignment of homologous proteins to have a percent identity that is well above 6%, but then we must ask: "What is the threshold of percent identity above which we will accept the alignment as producing evidence of homology?"

In 1999 Burkhard Rost reported the results of an experiment that provides this threshold (see [Ro99]). He analyzed more than a million sequence alignments between protein pairs with known structures. As the structures were known, he was able to determine whether a pair of proteins had similar or nonsimilar structure. This structure comparison was used to provide a definitive test of homology for the protein pairs. The following points describe some of the main results:

- When the percent identity is high, over 40% for long alignments, the sequence alignment will unambiguously distinguish between similar and nonsimilar structure.

- For alignments showing more than 30% sequence identity, about 90% of the pairs are homologous.

- For alignments showing less than 25% sequence identity, less than 10% of the pairs were homologous.

The region between 35 and 20% identity was designated as the "twilight zone." Here, pair homology may exist but cannot be reliably ensured without extra evidence. More than 95% of all pairs in the twilight zone had nonsimilar structure. Sequence percent identity below 20% has been called the "midnight zone" and typically requires an assessment of structure as the alignment is not reliable.*

6.4 LOCAL ALIGNMENT

It is possible that two proteins have evolved from a distant ancestor and the evidence of this is seen as a conserved domain. Suppose that for each protein, the portion of the primary sequence belonging to this domain

---

* Recall from Chapter 2, Section 2.2.3.3, that the globin family has sequence identity at around 15% but structural analysis reveals their homology.

is small in comparison to the rest of the protein that corresponds to a sequence bearing no relation to the other protein. Under these circumstances, a global alignment of the proteins may obscure the similarity of the domain because the algorithm is forced to find an alignment that deals with the entire sequence of each protein. The insertion of gaps necessary to achieve this global alignment will cause a reduction of the score and a test to evaluate the significance of the alignment score may be undeservedly low. In this situation, the discovery of homology is best handled by computing a *local* rather than a global alignment. We want the best alignment between shorter regions of the given proteins. It is best to think of global alignment as most applicable for the alignment of two proteins that have roughly the same length and that are somewhat similar over their entire lengths.

The dynamic programming algorithm that produces such an alignment is the Smith–Waterman algorithm (see [Sw81]). The dynamic program for Smith–Waterman uses the same 2-D array that we employed for the Needleman–Wunsch global alignment algorithm. However, because we only want to align shorter regions of the given pair of proteins, the trace-back path will not necessarily start in the lower right corner of the array and it will not necessarily end in the upper left corner of the array.

If the reader has done the exercises in Chapter 4, he or she will have gathered some valuable insights about the required behavior of this alignment algorithm. In the solution of Bentley's problem (Chapter 4, Exercise 2), we fill a linear array with scores that are the sums of consecutive numbers. As we wish to maximize our sum, we can choose to start with an empty purse (i.e., sum) at any time during the computation of increasing sums. In this way, a negative value in the purse is replaced with a zero value, and this excludes the possibility that a negative value in the purse brings down an otherwise high positive accumulation over subsequent cells of the array.

In summary, the trace-back path will start with the highest score and, in general, it will traverse only part of the array, ending when we encounter a cell with score value zero. Moreover, we will not assign penalties when the given strings P and Q do not produce alignments at the ends of the strings. In other words, dashes at either end are inserted with no penalty. This means that even though the alignment algorithm may use a linear or affine gap model, the initial scores set up in the cells of the first row and column will be zero and never negative. Furthermore, during the evaluation of scores in the array, we can change any negative score (computed in the same manner as Needleman–Wunsch) by setting such a score to 0. Consequently, the recursion becomes

$$SW(i,j) = \max \begin{cases} SW(i-1,j-1)+B(p_i,q_j), \\ SW(i-1,j)-d, \\ SW(i,j-1)-d, \\ 0 \end{cases}. \tag{6.28}$$

We have used the notation $SW(i, j)$ to emphasize that this score is for the local alignment algorithm due to Smith and Waterman. If you compare this with Equation 6.23, you will see that they are the same except that Equation 6.28 has a zero in the max function argument list so that the cell entry never becomes negative.

If you recall Bentley's problem, there was an essential stipulation that made the problem more interesting. The given numbers were both positive and negative. If all numbers are of the same sign, the solution is always trivial: All positive means that we take *all* the numbers to form the sum and all negative means we take none or the least negative, depending on how the problem is specified for this degenerate case. In a similar fashion, it is expected that values for partial matches provided by the substitution matrix will have a mix of both positive and negative entries. The idea is that we want to derive *local* alignments, and the score for aligning unrelated sequence regions should be negative. We do not want a local alignment to be arbitrarily extended in either direction by partial matches that always contribute a positive increment to the score.

To provide a small example, we show the results of a local alignment between the strings "MNAMRETAW" and "SNAMELDEEN." This time, we have assumed a gap penalty of $d = -8$. The results are displayed in Figure 6.3.

Working with the trace-back, one gets the following alignment:

```
N   A   M   R
|   |   |   |
N   A   M   E
```

Note that this corresponds to finding the maximum value of 15 and then taking the diagonal path up and to the left until a zero entry is encountered.

		S	N	A	M	E	L	D	E	E	N
	0	0	0	0	0	0	0	0	0	0	0
M	0	0	0	0	5	0	2	0	0	0	0
N	0	1	6	0	0	5	0	3	0	0	6
A	0	1	0	10	2	0	4	0	2	0	0
M	0	0	0	2	15	7	2	1	0	0	0
R	0	0	0	0	7	15	7	0	1	0	0
E	0	0	0	0	0	12	12	9	5	6	0
T	0	1	0	0	0	4	11	11	8	4	6
A	0	1	0	4	0	0	3	9	10	7	2
W	0	0	0	0	3	0	0	1	6	7	3

FIGURE 6.3 Local alignment example.

6.5 EXERCISES

1. Start with the definitions of $f_{ii}^{(k)}$ and $f_{ij}^{(k)}$ (see Equations 6.13 and 6.14) and prove that Equation 6.12 is true. Hint: You will need to realize that, for any cluster l we have

$$\sum_{i=1}^{20} R(i,l) = r_l.$$

2. Is the myoglobin of pig more similar to that of horse or human? Obtain the following protein sequences:

    a. 1PMB: Pig myoglobin (use chain A)

    b. 1WLA: Horse myoglobin

    c. 2MM1: Human myoglobin

    Perform a pairwise sequence global alignment for proteins 1PMB and 1WLA recording the percent identities. Do the same for proteins 1PMB and 2MM1. What do you observe?

3. Obtain amino acid sequences for the following two proteins:

    a. Swiss-Prot entry Q13873: Human bone morphogenetic protein receptor type-2 (precursor), length: 1038 AA.

    b. Swiss-Prot entry P36897: TGF-beta receptor type-1 (precursor), length: 503 AA.

    Perform a pairwise sequence *global* alignment for these proteins. Is there any evidence of significant sequence similarity? Report the pairwise sequence alignment, but this time do a local alignment. Is there any evidence of significant similarity for any regions within these proteins?

4. Obtain amino acid sequences for the following two proteins:

   a. Chain B of PDB code 1VRU: HIV-1 reverse transcriptase

   b. Chain B of PDB code 1QAI: Moloney murine leukemia virus reverse transcriptase.

   Each protein functions as a reverse transcriptase, but they are from different retroviruses. Perform a pairwise sequence local alignment for these proteins. Is there any evidence of significant sequence similarity? What is the percent identity for the longest over-lap? Is it in the twilight zone as defined by Rost? Use two concur-rently running Chimera sessions to isolate the parts of each protein that are involved in the overlap, displaying each region in its own Chimera window. Try to rotate and position the displays in such a way as to give a visual verification of structural similarity.

5. Chain A of the human placental ribonuclease inhibitor–angiogenin complex (PDB code 1A4Y) contains several loop–helix–loop–strand subsequences. In Figure 2.28 it is clear that these subsequences show a remarkable similarity in structure. Work with the UCSF Chimera application to fetch the protein and use the Tools → Structure_ Analysis → Sequence menu to pull out the primary sequences cor-responding to these substructures. For example, L(62) is the start of the loop-helix-loop-strand subsequence LRSNELGDVGVHCV LQGLQTPSCKIQKLS, and this is followed by LQNCCLTGAGCGV LSSTLRTLPTLQELH, with several others after that. There are 16 helices in all. Arbitrarily select 10 pairs of such subsequences, and calculate the sequence global alignment for each pair. Do you find significant sequence similarity? Repeat these evaluations working with local alignments.

6. Read the research paper by Dundas et al. (see [DB07]). The paper dis-cusses several examples of structurally similar proteins with different chain topologies. For example, two proteins may have domains with very similar hydrophobic cores, but the helices and strands making up this core have different connectivity. The different connectivity includes both circular permutations and noncyclic permutations of

various chain segments in the primary sequence. For one or more examples of such protein pairs, use Chimera to illustrate the similarity of structure for chain segments that overlap in 3-D space. Use a sequence alignment tool (global or local) to investigate the percent identity for chain segments that have a similar tertiary structure.

REFERENCES

[CL86] C. CHOTHIA AND A. M. LESK. The relation between the divergence of sequence and structure in proteins. *Embo Journal*, **5** (1986), 823–826.

[Do81] R. F. DOOLITTLE. Similar amino acid sequences: Chance or common ancestry? *Science*, **214** (1981), 149–159.

[Do90] R. F. DOOLITTLE. Searching through sequence databases. In *Methods in Enzymology*, Vol. 183, *Molecular Evolution: Computer Analysis of Protein and Nucleic Acid Sequences*, Doolittle, R. F. Ed., Academic Press, New York, 1990, pp. 99–110.

[DB07] J. DUNDAS, T. A. BINKOWSKI, B. DASGUPTA, AND J. LIANG. Topology independent protein structural alignment. *BMC Bioinformatics*, **8** (2007), 388.

[HH92] S. HENIKOFF AND J. HENIKOFF. Amino acid substitution matrices from protein blocks. *Proceedings of the National Academy of Sciences of the United States of America*, **89** (1992), 10915–10919.

[Nw70] S. B. NEEDLEMAN AND C. D. WUNSCH. A general method applicable to the search for similarities in the amino acid sequence of two proteins. *Journal of Molecular Biology*, **48** (1970), 443–453.

[Ro99] B. ROST. Twilight zone of protein sequence alignments. *Protein Engineering*, **12** (1999), 85–94.

[Sw81] T. F. SMITH AND M. S. WATERMAN. Identification of common molecular subsequences. *Journal of Molecular Biology*, **147** (1981), 195–197.

[Tr05] A. TRAMONTANO. *The Ten Most Wanted Solutions in Protein Bioinformatics*. Chapman & Hall/CRC, 2005.

[ZB07] M. ZVELEBIL AND J. BAUM. *Understanding Bioinformatics*. Garland Science, 2007.

CHAPTER 7

Protein Geometry

Arithmetic! Algebra! Geometry! Grandiose trinity! Luminous triangle! Whoever has not known you is without sense!

ISIDORE LUCIEN DUCASSE
(Comte de Lautreamont; 1846–1870)

Algebra exists only for the elucidation of geometry.

WILLIAM EDGE (1904–1997)

Geometry is the science of correct reasoning on incorrect figures.

GEORGE POLYA (1887–1985)

The human heart likes a little disorder in its geometry.

LOUIS DE BERNIERES (1954–)

7.1 MOTIVATION

The previous chapters have stressed the link between protein functionality and its dependence on protein structure. In light of this, one of the goals of structural bioinformatics is to aid the biochemist in modeling molecular functionality on a computer. This modeling is the final step in the change of experimental setting: in vivo → in vitro → in silico. Although reaction mechanisms are essentially the final statement in characterizing molecular interactions, there is often a need to track conformational changes and other geometric aspects of the molecules. In order to carry out these operations, we need to perform calculations that measure the various positional properties of atoms in a macromolecule. We start with the more elementary geometric calculations and work toward the more challenging computations.

7.2 INTRODUCTION

The geometry of molecules deals with computations related to bond length and interatomic distances in general, bond angles, and dihedral angles. More complicated calculations deal with the construction of molecular surfaces and quantities such as charge densities. These are calculations related to a static molecule. In more dynamic settings we may attempt to evaluate these quantities as they change with time owing to flexibility of the molecule. For now, we assume that the model is static and that we know the positions of atoms in 3-D space. In particular, we will be dealing with proteins that have been analyzed to the extent that the coordinates of all atoms are known with fairly reasonable accuracy. Typically, we will get these coordinates from a file that is downloaded from the Protein Data Bank (PDB).

7.3 CALCULATIONS RELATED TO PROTEIN GEOMETRY

7.3.1 Interatomic Distance

If the positions of atoms in 3-D space are given by (x, y, z) coordinates, we can use the standard Pythagorean calculation of distance. If atom A has coordinates $a = (a_x, a_y, a_z)^T$ and atom B has coordinates $b = (b_x, b_y, b_z)^T$, the distance between A and B is given by

$$d(A,B) = \sqrt{(a_x - b_x)^2 + (a_y - b_y)^2 + (a_z - b_z)^2}. \tag{7.1}$$

This is the same as the norm calculation:

$$\|a - b\| = \sqrt{(a-b)^T(a-b)}. \tag{7.2}$$

7.3.2 Bond Angle

Recall from linear algebra that an inner product of normalized vectors u and v can be viewed as the cosine of the angle between these vectors. This is expressed using the following formula:

$$\cos\theta = \langle u,v \rangle / (\|u\| \|v\|). \tag{7.3}$$

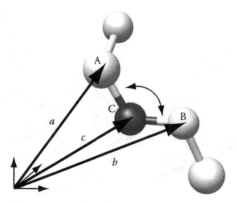

FIGURE 7.1   Bond angle defined via vectors.

Consider two atoms A and B (with coordinate vectors a and b), both bonded to a third atom C with coordinates given by vector c as in Figure 7.1.

If we set $u = a - c$ $v = b - c$, the cosine of the bond angle will be given by

$$\cos\theta = \frac{(a-c)^{\mathrm{T}}(b-c)}{\|a-c\|\|b-c\|}. \tag{7.4}$$

Once this quantity is calculated, the required bond angle can be easily derived by taking the inverse cosine of the quantity on right-hand side of the last equation.

7.3.3 Dihedral Angles

7.3.3.1 Defining Dihedral Angles

Consider atoms in the protein backbone: N_1 $C\alpha_1$ C_1 N_2 $C\alpha_2$ C_2 N_3 $C\alpha_3$ C_3 N_4 ... . As we move from residue to residue, the corresponding bond lengths do not change much.

For example, $N_i - C\alpha_i$ has approximately the same bond length as $N_j - C\alpha_j$ for arbitrary i and j. Similarly, corresponding bond angles tend to be the same. This type of observation cannot be made for dihedral angles. Dihedral angles are due to a "swivel" action around a single bond and are primarily responsible for defining the position of the backbone atoms in 3-D space. Be sure to understand that changing a dihedral angle will not affect bond angles between successive atoms in the backbone. In fact, we could keep all bond angles constant while moving N_i relative to N_{i+1} by using a swiveling rotation around the $C_i - C\alpha_i$ bond.

A bond angle is determined by a sequence of three bonded atoms, but a dihedral angle is defined by a sequence of four consecutive bonded atoms.

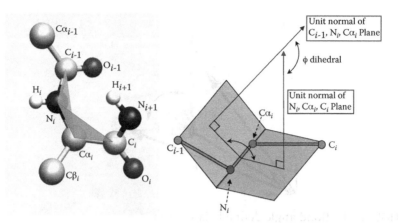

FIGURE 7.2 (See color insert following page 72) Phi dihedral angle.

In general, the first three atoms and the last three atoms of this sequence define two planes that intersect in a line that is coincident with the bond between the middle two atoms. For example, Figure 7.2 shows how the four consecutive atoms C_{i-1} N_i $C\alpha_i$ C_i of the backbone define two planes that intersect in the line that is coincident with the $N_i - C\alpha_i$ bond. The angle between these two planes is referred to as the ϕ (phi) angle associated with residue i. In a similar fashion, Figure 7.3 shows how the four consecutive atoms N_i $C\alpha_i$ C_i N_{i+1} of the backbone define two planes that intersect in the line that is coincident with the $C\alpha_i - C_i$ bond. The angle between these two planes is referred to as the ψ (psi) angle associated with residue i. Because the backbone atoms go through consecutive repetitions of the atoms N, Cα and C, we have discussed only two of the three possible starting

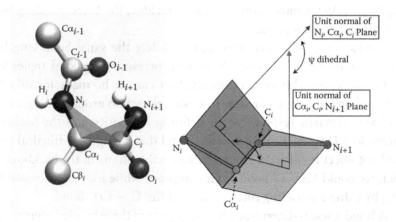

FIGURE 7.3 (See color insert following page 72) Psi dihedral angle.

positions for a sequence of four atoms that define a dihedral angle. In fact, there is another dihedral angle to be considered, namely, the dihedral angle that is defined by the four atom sequence $C\alpha_i\ C_i\ N_{i+1}\ C\alpha_{i+1}$. This is called the ω (omega) angle, and its middle two atoms are the carbon and nitrogen of the peptide bond. The peptide bond has a partial double-bond character owing to resonance effects. Because of this and the steric interactions between the two side chains attached to the alpha carbons atoms at either end of the sequence, the ω dihedral angle typically shows a trans configuration for this four-atom sequence. In other words, $\omega = 180°$. As a consequence, $C\alpha_i, C_i, N_{i+1}, C\alpha_{i+1}$, the O atom attached to the carbonyl C_i atom, and the H atom on the nitrogen are all very close to being coplanar. This planar nature of the peptide bond usually puts the side chains in positions that are as far apart as possible.

Our definition of a dihedral angle as the angle between two planes has to be made more precise. As an angle is determined by the intersection of two lines, we will need to derive two such lines from the planes defined by the sequence of four atoms. For example, we could have each line lie within a plane and have both lines perpendicular to the line of intersection of the two planes. The analytic derivation required by this approach is somewhat lengthy and is less preferable to a strategy that generates the needed lines by computing the normals of the two planes. These normals are illustrated in Figures 7.2 and 7.3. A short deduction using Euclidean geometry will show that the dihedral angle between the two planes is equal to the angle between the two normals.

As the normal to a plane defined by two intersecting lines (in our case, two consecutive bonds) is easily computed using a cross product formula, we can use this formula as a convenient "algebraic subroutine."

7.3.3.2 Computation of a Normal

Two lines that intersect in 3-D space define a plane. To construct a normal to this plane, we calculate a *cross product*. Suppose we have two vectors u and v. The cross product is a vector that is perpendicular to both u and v, and it has a magnitude that is equal to the area of the parallelogram spanned by the vectors. It can be shown that this is

$$u \times v = \det \begin{bmatrix} i & j & k \\ u_1 & u_2 & u_3 \\ v_1 & v_2 & v_3 \end{bmatrix} = (u_2 v_3 - u_3 v_2)i + (u_3 v_1 - u_1 v_3)j + (u_1 v_2 - u_2 v_1)k.$$

$$(7.5)$$

The result of this computation is a vector that is expressed as a linear combination of the three basis vectors i, j, and k. Alternatively, we could express the right-hand side of the previous equation using a column vector notation:

$$u \times v = \begin{bmatrix} u_2 v_3 - u_3 v_2 \\ u_3 v_1 - u_1 v_3 \\ u_1 v_2 - u_2 v_1 \end{bmatrix}. \tag{7.6}$$

To see why these formulas are correct, consider the following argument: If the magnitude of the normal is the area of the parallelogram defined by u and v, its magnitude is given by the base times height formula $\|u\|h = \|u\|\|v\|\sin\theta$, where h is the height of the parallelogram and θ is the angle between u and v. So, we can write

$$\begin{aligned} \|u \times v\|^2 &= \|u\|^2 \|v\|^2 \sin^2\theta = \|u\|^2 \|v\|^2 (1 - \cos^2\theta) \\ &= \|u\|^2 \|v\|^2 - (\|u\|\|v\|\cos\theta)^2 \\ &= \|u\|^2 \|v\|^2 - (u^T v)^2. \end{aligned} \tag{7.7}$$

Continuing

$$\|u \times v\|^2 = \|u\|^2 \|v\|^2 - (u^T v)^2 = \left(\sum_{k=1}^{3} u_k^2\right)\left(\sum_{k=1}^{3} v_k^2\right) - \left(\sum_{k=1}^{3} u_k v_k\right)^2. \tag{7.8}$$

We can now apply Lagrange's Identity to see that the last expression is actually

$$\sum_{i=1}^{2}\sum_{j=i+1}^{3} (u_i v_j - u_j v_i)^2. \tag{7.9}$$

To prove this, you can simply expand both expressions (a somewhat tedious procedure).*

Now note that

$$\left\| \det \begin{bmatrix} i & j & k \\ u_1 & u_2 & u_3 \\ v_1 & v_2 & v_3 \end{bmatrix} \right\|^2 = \|(u_2 v_3 - u_3 v_2)i + (u_3 v_1 - u_1 v_3)j + (u_1 v_2 - u_2 v_1)k\|^2 \tag{7.10}$$

and we see that this is the same as the double sum just given.

---

* A proof of the general case can be found at: en.wikipedia.org/wiki/Lagrange's_identity.

So far we have shown that by working with this "area of parallelogram" definition, we get a vector that has a magnitude given by

$$\left| \det \begin{bmatrix} i & j & k \\ u_1 & u_2 & u_3 \\ v_1 & v_2 & v_3 \end{bmatrix} \right|. \tag{7.11}$$

Is the vector defined by this determinant perpendicular to both u and v? Yes. If you calculate the inner product of this vector with u or v, you get 0. For example,

$$\begin{bmatrix} u_2 v_3 - u_3 v_2 \\ u_3 v_1 - u_1 v_3 \\ u_1 v_2 - u_2 v_1 \end{bmatrix}^{\mathrm{T}} \begin{bmatrix} u_1 \\ u_2 \\ u_3 \end{bmatrix} \equiv 0. \tag{7.12}$$

So, we have shown that the determinant-based calculation for the cross product has given the correct magnitude and we have the "perpendicular to both u and v" requirement. The resulting vector could be illustrated by either of the two diagrams contained in Figure 7.4. We now have to decide which of these is correct, as they both show a vector perpendicular to both u and v. For $u \times v$ we use the "right-hand rule" that is also seen in the usual 3-D Euclidean i, j, k coordinate system: If the fingers of the right-hand curl around the k unit normal with fingers going in the direction from i to j then the thumb points in the positive direction assigned to k. Similarly, if the fingers of the right-hand curl around the $u \times v$ normal with fingers going in the direction from u to v, then the thumb points in the positive direction assigned to $u \times v$. So, the first diagram is for $u \times v$, and the next diagram is for $v \times u$ (or for $-(u \times v)$).

One last vector algebra issue: The cross product evaluated using the determinant does not necessarily have a unit length. Recall that our

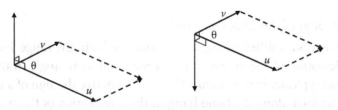

FIGURE 7.4 Right-hand rule.

dihedral angle calculation requires vectors that are unit normals. In such a situation, we will have to normalize the cross product vectors. Given $u \times v$ we calculate

$$\frac{u \times v}{\|u \times v\|}. \tag{7.13}$$

7.3.3.3 Calculating the Phi Dihedral Angle

Suppose we want to calculate the ϕ dihedral angle as shown in Figure 7.2. For simplicity, we will just use the atom names to label vectors that contain the coordinates of these atoms. We let $u = N_i - C_{i-1}$ and $v = C\alpha_i - N_i$. Then,

$$u \times v = (u_2 v_3 - u_3 v_2)i + (u_3 v_1 - u_1 v_3)j + (u_1 v_2 - u_2 v_1)k. \tag{7.14}$$

Dividing this by its norm will give the unit normal (call it $n(C_{i-1}, N_i, C\alpha_i)$).

Similarly, to calculate the other required normal, we let $v = C\alpha_i - N_i$ and $w = C_i - C\alpha_i$. Then calculate

$$v \times w = \det \begin{bmatrix} i & j & k \\ v_1 & v_2 & v_3 \\ w_1 & w_2 & w_3 \end{bmatrix}$$

$$= (v_2 w_3 - v_3 w_2)i + (v_3 w_1 - v_1 w_3)j + (v_1 w_2 - v_2 w_1)k.$$

Dividing this by its norm will give the second unit normal (call it $n(N_i, C\alpha_i, C_i)$).

Finally, with both unit normals computed, we can get the value of the ϕ dihedral for this alpha carbon using the inverse cosine function:

$$\phi = \tau(C_{i-1}, N_i, C\alpha_i, C_i) = \arccos(n(C_{i-1}, N_i, C\alpha_i) \bullet n(N_i, C\alpha_i, C_i)). \tag{7.15}$$

7.3.3.4 Sign of the Dihedral Angle

By convention, a dihedral angle is assumed to be in the range $[-\pi, \pi]$. As the calculation of the arccos function may lead to an angle in the range $[0, \pi]$, we typically have to adjust the sign. To derive the sign of a dihedral angle, we look along the bond lying in the intersection of the two planes (be sure to look in the increasing i direction as shown in Figure 7.5).

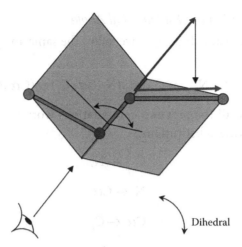

FIGURE 7.5 Sign of the dihedral angle.

Consider the unit normal of the plane defined by the first three atoms. Compute its inner product with the vector going from the third atom to the last atom. If this inner product is positive (as in the diagram), then the sign of the dihedral angle is positive, otherwise it is negative. In the case of ϕ this computation is

$$n(C_{i-1}, N_i, C\alpha_i) \bullet (C_i - C\alpha_i). \tag{7.16}$$

For positive dihedral angles the projection of the unit normal is in the same direction as the last bond. For negative dihedral angles the projection of the unit normal is in the direction opposite to the last bond. In both cases, we are looking down along the central bond in the increasing i direction. The views for positive and negative dihedral angles are given in Figure 7.6.

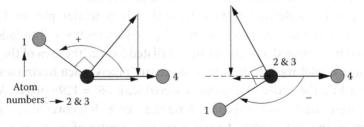

FIGURE 7.6 Using a projection to determine the sign of the dihedral angle.

7.3.3.5 Calculating the Psi Dihedral Angle

To calculate the ψ dihedral angle, we follow the same analysis but this time compute

$$\psi = \tau(N_i, C\alpha_i, C_i, N_{i+1}) = \arccos(n(N_i, C\alpha_i, C_i) \bullet n(C\alpha_i, C_i, N_{i+1})). \quad (7.17)$$

In other words, all the previous computations work to get ψ if you perform the following substitutions:

$$C_{i-1} \leftarrow N_i$$

$$N_i \leftarrow C\alpha_i$$

$$C\alpha_i \leftarrow C_i \qquad (7.18)$$

$$C_i \leftarrow N_{i+1}.$$

7.4 RAMACHANDRAN PLOTS

Although the dihedral angles in a polypeptide can change with the conformation of the chain, they cannot freely adopt just any arbitrary values. It has been known since the early 1960s that steric collisions will prohibit certain combinations of the phi and psi angles on either side of a given alpha carbon atom. An observed combination of the phi and psi angles is often represented by a single point in a 2-D plot that has a horizontal axis for phi ranging over values $[-\pi, \pi]$ (i.e., −180° to +180°) and a vertical axis for psi ranging over the same values. A collection of many such points can be used to record the observed phi–psi combinations corresponding to a set of alpha carbon atoms. This type of scatter plot is called a Ramachandran plot because Ramachandran et al. (see [RR63]) were the first people to study the allowable phi–psi angles and the dependency of their values on the type of residue associated with the alpha carbon.

Figure 7.7 is derived from a Ramachandran scatter plot of 121,870 phi–psi combinations taken from 463 protein structures. The following procedure was used to generate this pixilated figure: The area of the original scatter plot was subdivided into square "pixels" each having a side of length 10°. The entire plot is thus covered with $36^2 = 1296$ pixels. A pixel was given a dark color and characterized as "core" if it covered at least 100 points in the scatter plot. A pixel was given a color of medium intensity and characterized as "allowed" if it was not a core pixel but covered at least 8 points in the scatter plot. Pixels in the "generous" region are obtained by extending an allowed region with a border area of width 20°. The generous

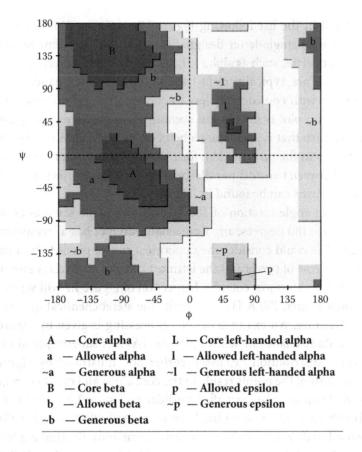

A — Core alpha   L — Core left-handed alpha
a — Allowed alpha   l — Allowed left-handed alpha
~a — Generous alpha   ~l — Generous left-handed alpha
B — Core beta   p — Allowed epsilon
b — Allowed beta   ~p — Generous epsilon
~b — Generous beta

FIGURE 7.7 (See color insert following page 72) Ramachandran plot for 121,870 residues (463 protein structures).

region is given a color that is lighter than an allowed region. Any pixel that is not core, allowed, or generous is given the lightest color or simply left as white. So, in summary, the darker areas correspond to regions of the plot where the density of points is highest and other colors represent regions of lower density. White regions or those with the lightest color are used to denote areas of the plot where there is the least likelihood of observing a point, for example, $\phi = 90°$ and $\psi = -90°$. The plot of Figure 7.7 was created using PROCHECK*, a program that checks the stereochemical quality of protein structures (see [Lm93] and [Mm92]).

---

* http://www.biochem.ucl.ac.uk/~roman/procheck/procheck.html.

As indicated in the list following Figure 7.7, various regions of the plot are labeled with single-letter designations that indicate some secondary structure properties such as alpha helices, beta strands, and epsilon turns. Alpha helices are typically right-handed, although left-handed helices do occur but with considerably less frequency. The epsilon region of the Ramachandran plot is for phi–psi combinations that are usually associated with a turn that is just before a helix or beta strand. A good example of this occurs in the protein 1JAL for residue VAL287, which occurs on a tight turn between two anti-parallel beta strands. More information about two-residue turns can be found in [Mp94].

Because an angle rotation of $180 + n$ degrees is the same as an angle rotation of $n - 180$ degrees, any Ramachandran plot has a "wraparound" continuity. We could consider the uppermost row of pixels to be adjacent to the lowest row of pixels and the leftmost column of pixels is essentially adjacent to the rightmost column. Inspection of Figure 7.7 will verify this.

As noted earlier, PROCHECK checks the stereochemical quality of a protein structure. An example of such processing is given by Figure 7.8, where the Ramachandran scatter plot for 1A4Y is superimposed on the background distribution discussed earlier and illustrated by Figure 7.7. Note that most of the points reside in the core alpha and core beta regions. This is what you would expect after considering Figure 2.28. After the protein's backbone conformation has been computed, PROCHECK software can be used to detect the existence of any anomalous dihedral angles. For example, Figure 7.8 indicates that residues SER 15 (chains A and D), GLU 58 (chain E), and SER 4 (chain E) all have phi–psi angles that produce points in the white region. A point in the white region may indicate an error in the computation of the conformation or it may simply be a valid entry that rarely occurs.

To some extent, Ramachandran plots can also show how the dihedral angles on either side of an alpha carbon atom are affected by the type of residue associated with that alpha carbon. For example, the glycine residue essentially lacks a side chain, and so it does not have the steric constraints of the other amino acids. Figure 7.9 shows the scatter plot for phi–psi angles that are limited to glycine residues. One can see that the distribution is somewhat different from the general case illustrated in Figure 7.7. Although there are fewer constraints on dihedral angles for glycine, there are still some distribution patterns that can be observed, and Figure 7.9 shows five clusters within the scatter plot. An in-depth analysis of this plot can be found in [Hb05].

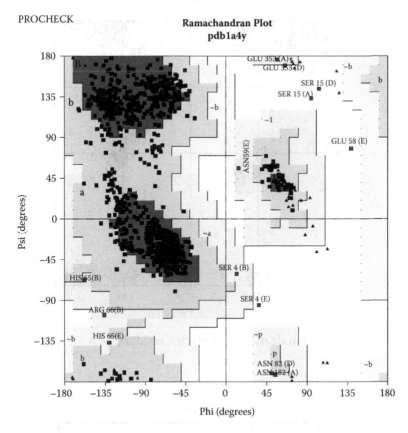

FIGURE 7.8 (See color insert following page 72) Ramachandran plot for residues of protein 1A4Y.

Although glycine provides an example of dihedral angles that are subject to few steric constraints, we find that the phi–psi angles for proline are severely constrained. The PROCHECK version of the Ramachandran plot for the 42 proline residues of 1A4Y is given in Figure 7.10. We see that phi is constrained to values that are within a narrow band typically ranging from $-60°$ to $-90°$, depending on the secondary structure of its local environment. Inspection of Figure 7.11 shows why this is expected. The phi dihedral angle just before the proline alpha carbon is severely restricted because the chain arising from the alpha carbon loops back to bond with the backbone nitrogen atom bonded to that alpha carbon. The psi angle after the alpha carbon has much less restriction, as can be seen in the plot of Figure 7.10.

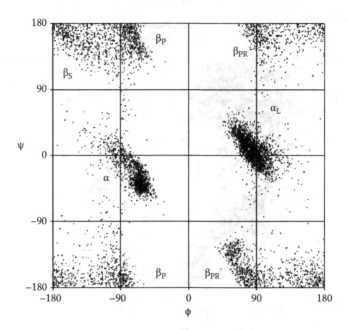

FIGURE 7.9   Ramachandran plot for glycine residues.

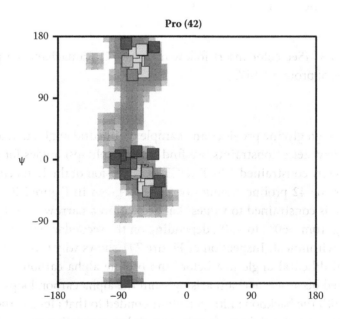

FIGURE 7.10   Ramachandran plot for proline residues.

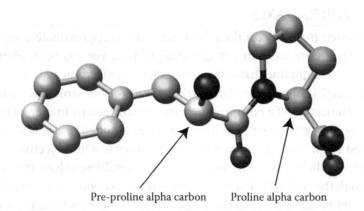

Pre-proline alpha carbon   Proline alpha carbon

FIGURE 7.11   Phe 70 a pre-proline residue in 1AG9.

Another interesting example of a constraint is seen in "pre-proline" residues. In this case, the side chain of the residue prior to proline bends away from the proline loop in order to avoid a steric clash. This is clearly illustrated in Figure 7.11, where the phenylalanine ring of residue 70 in 1AG9 veers away from the proline loop in the next residue. Figure 7.12 (taken from [HB05]) gives the Ramachandran plots and associated analysis for pre-proline residues.

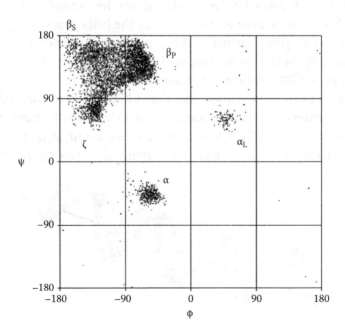

FIGURE 7.12   Ramachandran plot for pre-proline residues.

7.5 INERTIAL AXES

Very often in a structural analysis, we want to approximate a secondary structural element with a single straight line. For example, Figure 7.13 shows a straight line that acts as the longitudinal axis of a helix.

Figure 7.14 shows the alpha carbon atoms of the same helix and gives us an indication of a reasonable strategy for the determination of the axis: We should have the straight line positioned among the atoms so that it is closest to all these atoms in a least-squares sense. Our objective is to find this "best" helix axis. A reasonable strategy would be to have the axis pass through the centroid of all the atoms with a direction chosen to minimize the sum of the squares of the perpendicular distances from the atoms to the helix axis.

More precisely, if $\|d_i\|$ represents the perpendicular distance between atom $\mathbf{a}^{(i)}$ and the helix axis, then we calculate the axis direction so that it minimizes the sum:

$$S = \sum_{i=1}^{N} \|d_i\|^2. \tag{7.19}$$

Depending on the needs of the application, it is possible to define the position of this axis using a different set of atoms, for example, all the atoms in the backbone or even all the atoms in the helix (backbone plus side chains). Before going further in this discussion, it should be stated that such an axis could also be computed to go among the atoms of a beta strand, in fact, any arbitrary set of atoms.

We now provide an analysis that shows how the axis is dependent on the coordinates of the chosen atoms. To make the analysis simpler, we first consider one atom, call it $\mathbf{a}$, with position vector $a = (a_x, a_y, a_z)^{\mathrm{T}}$. As noted earlier, this vector is using a frame of reference that is an x, y, z coordinate

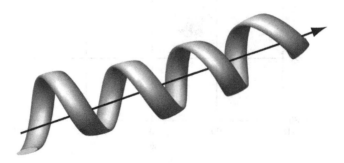

FIGURE 7.13   Longitudinal axis for a helix.

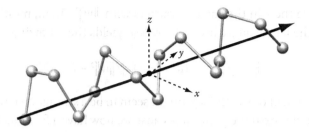

FIGURE 7.14 Longitudinal axis defined relative to the alpha carbon atoms of the helix.

system with origin positioned at the centroid of the atoms under consideration (see Figure 7.14). The perpendicular distance from the atom to the axis is illustrated in Figure 7.15. We are trying to determine the values of w_x, w_y, and w_z that define the directions of the axis. These are set up as components of a unit vector w and so $w = (w_x, w_y, w_z)^T$ with $w^T w = 1$. The required axis is essentially a scalar multiple of w, and d represents the perpendicular vector going from atom **a** to this axis.

Some trigonometry defines the square of the norm of d as

$$\|d\|^2 = \|a\|^2 \sin^2 \theta = \|a\|^2 [1 - \cos^2 \theta] = \|a\|^2 \left[1 - \frac{(a^T w)^2}{\|a\|^2 \|w\|^2} \right]. \qquad (7.20)$$

We have replaced $\cos\theta$ with its equivalent formulation in terms of a dot product. As $w^T w = \|w\|^2 = 1$, we can remove $\|w\|^2$ from the denominator

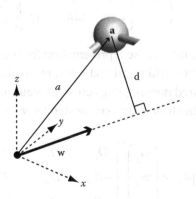

FIGURE 7.15 The relationship between the axis and one atom.

and replace the 1 in the square brackets with $\|w\|^2$. Then, multiplying the terms in the large square brackets by $\|a\|^2$ yields the following:

$$\|d\|^2 = \|a\|^2 \bullet 1 - (a^T w)^2 = \|a\|^2 \|w\|^2 - (a^T w)^2. \qquad (7.21)$$

The replacement of 1 with $\|w\|^2$ might seem to be a strange step in this derivation, but it is extremely useful because we now have a formula for $\|d\|^2$ in which w appears in each term as a quadratic. We will see that this leads to an elegant matrix formulation for $\|d\|^2$ and lays the groundwork for a subsequent strategy to solve our minimization problem. Deriving the matrix formulation for $\|d\|^2$ is straightforward but somewhat daunting owing to the lengthy expressions that are involved. We continue as follows:

$$\|d\|^2 = \left(a_x^2 + a_y^2 + a_z^2 \right)\left(w_x^2 + w_y^2 + w_z^2 \right) - (a_x w_x + a_y w_y + a_z w_z)^2$$
$$= A w_x^2 + B w_y^2 + C w_z^2 - 2[F w_y w_z + G w_z w_x + H w_x w_y] \qquad (7.22)$$

where

$$A = a_y^2 + a_z^2 \qquad F = a_y a_z$$
$$B = a_z^2 + a_x^2 \qquad G = a_z a_x \qquad (7.23)$$
$$C = a_x^2 + a_y^2 \qquad H = a_x a_y.$$

Now the quadratic formulation pays off and we can write:

$$\|d\|^2 = \begin{pmatrix} w_x \\ w_y \\ w_z \end{pmatrix}^T \begin{bmatrix} A & -H & -G \\ -H & B & -F \\ -G & -F & C \end{bmatrix} \begin{pmatrix} w_x \\ w_y \\ w_z \end{pmatrix}. \qquad (7.24)$$

This gives us an elegant concise representation for the $\|d\|^2$ value related to a single atom $\mathbf{a}$, but you will recall that we want to minimize S, which is the sum of all such squared norms going across all N atoms in the set. We let the coordinates of the ith atom $\mathbf{a}^{(i)}$ be represented by $a^{(i)} = (a_x^{(i)}, a_y^{(i)}, a_z^{(i)})^T$, then

$$S = \sum_{i=1}^{N} \|d_i\|^2 = \begin{pmatrix} w_x \\ w_y \\ w_z \end{pmatrix}^T \begin{bmatrix} D_{xx} & -E_{xy} & -E_{xz} \\ -E_{yx} & D_{yy} & -E_{yz} \\ -E_{zx} & -E_{zy} & D_{zz} \end{bmatrix} \begin{pmatrix} w_x \\ w_y \\ w_z \end{pmatrix} \qquad (7.25)$$

where

$$D_{xx} = \sum_{i=1}^{N} \left(a_y^{(i)}\right)^2 + \left(a_z^{(i)}\right)^2 \qquad E_{yz} = E_{zy} = \sum_{i=1}^{N} a_y^{(i)} a_z^{(i)}$$

$$D_{yy} = \sum_{i=1}^{N} \left(a_z^{(i)}\right)^2 + \left(a_x^{(i)}\right)^2 \qquad E_{xz} = E_{zx} = \sum_{i=1}^{N} a_z^{(i)} a_x^{(i)} \qquad (7.26)$$

$$D_{zz} = \sum_{i=1}^{N} \left(a_x^{(i)}\right)^2 + \left(a_y^{(i)}\right)^2 \qquad E_{xy} = E_{yx} = \sum_{i=1}^{N} a_x^{(i)} a_y^{(i)}.$$

This gives a concise representation for S:

$$S = w^{\mathrm{T}} T w \qquad (7.27)$$

where

$$T = \begin{bmatrix} D_{xx} & -E_{xy} & -E_{xz} \\ -E_{yx} & D_{yy} & -E_{yz} \\ -E_{zx} & -E_{zy} & D_{zz} \end{bmatrix} \qquad (7.28)$$

is dependent only on the components of the vectors holding the atomic positions of all atoms in the given set. In physics, T is called the *inertial tensor* and it often appears in discussions involving studies of rotational inertia.

We can now formulate our minimization problem as a Lagrange multiplier problem in which we minimize S subject to the constraint that $w^{\mathrm{T}}w = 1$. The Lagrangian will be

$$L(w, \lambda) = w^{\mathrm{T}} T w - \lambda(w^{\mathrm{T}} w - 1) \qquad (7.29)$$

where λ is the Lagrange multiplier. Proceeding with the usual differentiation, we get

$$\frac{\partial L}{\partial w} = 2Tw - 2\lambda w \qquad (7.30)$$

and setting this to zero gives us the following eigenvalue problem:

$$Tw = \lambda w. \qquad (7.31)$$

The 3×3 tensor matrix T is symmetric, and considering that S is always positive for any vector w, we see that T is also positive definite.

Consequently, all three eigenvalues are positive. Note that once we have an eigenvalue, eigenvector pair, we can write

$$S = w^T T w = \lambda w^T w = \lambda. \tag{7.32}$$

Therefore, we have the solution to our problem. Once T is calculated, we compute its eigenvalues and the required w (specifying the directions for our axis) will be the eigenvector corresponding to the smallest eigenvalue.

It should also be noted that there are three eigenvectors produced by this procedure and because of the symmetry of T, they are mutually orthogonal and can be used as an orthogonal basis for the set of atoms under consideration. In some applications, this *inertial frame of reference* is a useful construct and the eigenvectors are called the principle axes of inertia.

Various research studies have used the inertial axis as a simplifying geometric construct to represent secondary structural elements. This is especially important when one wishes to compare the tertiary structure of related proteins. The following references are good starting points: [AQ86], [AM88], [OB92], and [YS99].

Finally, here is an interesting note related to the principal axes of inertia computed for an entire protein (and with atomic masses taken into account): Some researchers have observed that, with high frequency, at least one principal axis penetrates the surface of a protein in a region used for ligand binding (see [FR00]).

7.6 EXERCISES

1. When considering beta sheets, are consecutive residues of a strand closer than neighboring residues on different strands? To answer this question, let us assume that the "coordinates of a residue" are given by the coordinates of the beta carbon atom within a residue. Write a program that reads a PDB file and calculates the average distance between C_β atoms of consecutive residues in a beta strand. See the dotted line in Figure 7.16 for an example of one such distance. The program should also calculate the average distance between C_β atoms of hydrogen-bonded residues that are in different strands of the beta sheet (see the dashed line in Figure 7.16). Test your program on the parallel and antiparallel beta sheets of baker's yeast topoisomerase II (PDB code 1BJT).

2. It has been observed that various types of bonds in a protein backbone tend to have very similar lengths if you observe a large number

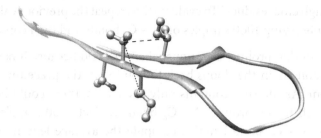

FIGURE 7.16 Distances between beta carbons.

of them. In this exercise, we gather evidence to test this assertion. In Figure 7.17, we see the peptide bond between Serine 6 and Leucine 7 in chain A of the protein 1A4Y.

Using Chimera to calculate bond lengths we get:

SER 6 C_α – SER 6 C   1.530Å

SER 6 C – LEU 7 N   1.328Å

LEU 7 N – LEU 7 C_α 1.452Å

Are these observations typical for C_α – C, C – N and N – C_α bond lengths? To answer this, do the following: Write a program that will read PDB files to extract the coordinates of C_α, C, and N atoms. For each type of bond, get 1000 observations of bond length and use this data to calculate both the average length and the standard deviation. You should also report the minimum and maximum lengths for each type of bond.

You should observe a rather narrow range of possibilities for each type of bond. Does this also apply to the C_α – C_β bond of

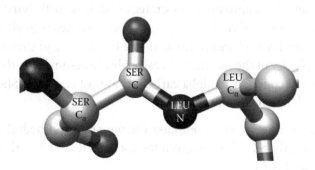

FIGURE 7.17 Bonds in the peptide backbone.

nonglycine residues? To evaluate this, repeat the previous calculations by observing 1000 samples of $C_\alpha - C_\beta$ bonds and then compare.

Recall that proline is an unusual amino acid because it has a three-carbon chain that loops back to bond with the preceding nitrogen atom. As the ring contains only five atoms, there could be a slight stretch in the proline $N - C_\alpha$ bond. Collect 1000 samples of proline $N - C_\alpha$ bond lengths, compute the average length, and compare with the average $N - C_\alpha$ bond lengths of nonproline residues. Finally, glycine is unusual because of its small size. Using the same approach as done for proline, collect 1000 samples of glycine $N - C_\alpha$ and $C_\alpha - C$ bond lengths, compute their average lengths, and compare these values with the average $N - C_\alpha$ and $C_\alpha - C$ bond lengths of nonglycine residues.

3. It has also been observed that bond angles in a protein backbone tend to have a small variance. In this exercise, we gather evidence to test this assertion. Referring again to Figure 7.17, Chimera reports the following bond angles:

SER 6 C_α – SER 6 C – LEU 7 N   116.086

SER 6 C – LEU 7 N – LEU 7 C_α 121.881

Are these observations typical for $C_\alpha - C - N$ and $C - N - C_\alpha$ bond angles? As in the previous question, get 1000 observations of each type of bond angle and use this data to calculate both the average and the standard deviation. You should also report the minimum and maximum lengths for each type of angle. Are there any significant differences in these observations if you restrict the amino acids to be within a helix?

4. Consider the approximation strategy of dealing with "virtual bonds" between successive alpha carbon atoms. The idea is to eliminate the coordinates of C and N atoms in the backbone and thus reduce the complexity of the model. Is it reasonable to assume that the distance between consecutive alpha carbons is almost constant? Show results to back up your claim.

5. Write a function that computes the phi and psi dihedral angles for alpha carbon $C_\alpha[i]$ when given the coordinates of $C[i-1]$, $N[i]$, $C_\alpha[i]$, $C[i]$, and $N[i+1]$.

As in the previous questions, get 1000 observations of each type of dihedral angle and use this data to calculate both the average and the standard deviation. You should also report the minimum and maximum lengths for each type of angle. The observed data spread should be much wider than that computed for bond lengths and bond angles.

What are the significant differences in these observations if you restrict the amino acids to be within a helix? Repeat these calculations for beta sheets and loops.

6. Use the function that you programmed for Question 5 to generate a Ramachandran plot. To do this, write a mainline program that reads a PDB file and then uses the function to compute all the phi and psi dihedral angles. If you do not have the programming skills to generate such a plot, you may pass the (phi, psi) pairs to Microsoft Excel or an application such as MATLAB to compute the needed scatter plot. Test your program on protein 1A4Y to see if it will reproduce a plot that is the same as the black points in Figure 7.8.

7. Copy and then modify the mainline program of the previous question so that you can compute the Ramachandran plots corresponding to some particular type of residue. The program should accept a list of one or more proteins as input. Test the program by doing a Ramachandran plot for glycine residues and then compare with Figure 7.9.

8. Write a program that accepts a list of coordinates of N atoms as input and generates the following output:

a. A vector representing the centroid of the set of atoms

b. A direction vector w defining the inertial axis for this set of atoms

c. The beginning and end points of this axis

The axis points required in part (c) would be the two projections of the atoms onto the axis that are the farthest apart. For example, if the given list of atoms is a set of alpha carbons $C_\alpha[i]$, $i = 1, ..., N$, then the beginning and end points for the axis will be the projections of $C_\alpha[1]$ and $C_\alpha[N]$ onto the computed axis. The axis itself is a line that goes through the centroid and has the same direction as the vector *w.* Test your program by doing computations (a), (b), and (c) on the helices of 1MBN.

Cautionary note: Does 1MBN have six helices or eight? If you only look at the secondary structure designations of the 1MBN PDB file, you will see six subsequences that indicate helix structure. However, a visual inspection of the structure provides an interesting observation: Two of these helices contain sharp turns and so there is reasonable evidence to argue that there are eight helices each with its own axis.* How can you modify your "helix axis discovery" code so that the software computations of helix axes provide the same set of axes as a visual inspection would?

9. Read the paper by Kachalova et al. ([KP99]). In Figure 2 of their paper, they illustrate the change in heme geometries that are seen in myoglobin bound with and without carbon monoxide. This amounts to a change in the position of two helices with respect to the heme group. Perform a computation in which you evaluate the angle of this change. You will need two PDB files for myoglobin (with and without CO). Consider SCOP as good place to start a search for these files. Compute the inertial axes of the helices that are involved. Compute the angle of a helix axis relative to the plane of the heme group. Show how this changes when myoglobin goes from deoxyMb to MbCO.

REFERENCES

[Am88] R. A. ABAGYAN AND V. N. MAIOROV. A simple quantitative representation of polypeptide chain folds: Comparison of protein tertiary structures. *Journal of Biomolecular Structure & Dynamics*, **5** (1988), 1267–1279.

[AQ86] J. ÅQVIST. A simple way to calculate the axes of an α-helix. *Computers & Chemistry*, **10** (1986), 97–99.

[FR00] J. FOOTE AND A. RAMAN. A relation between the principal axes of inertia and ligand binding. *Proceedings of the National Academy of Sciences of the United States of America*, **97** (2000), 978–983.

[HB05] B. K. HO AND R. BRASSEUR. The Ramachandran plots of glycine and pre-proline. *BMC Structural Biology*, **5** (2005), http://www.biomedcentral.com/1472-6807/5/14.

---

* One of these turns contains a proline residue within the helix. Recall that proline often produces a bend in the backbone structure.

[Kp99] G. S. KACHALOVA, A. N. POPOV, AND H. D. BARTUNIK. A steric mechanism for inhibition of CO binding to heme proteins. *Science*, **284** (1999), 473–476.

[Lm93] R. A. LASKOWSKI, M. W. MACARTHUR, D. S. MOSS, AND J. M. THORNTON. PROCHECK: a program to check the stereochemical quality of protein structures. *Journal of Applied Crystallography*, **26** (1993), 283–291.

[Mp94] C. MATOS, G. A. PETSKO, AND M. KARPLUS. Analysis of two-residue turns in proteins. *Journal of Molecular Biology*, **238** (1994), 733–747.

[Mm92] A. L. MORRIS, M. W. MACARTHUR, E.G. HUTCHINSON, AND J. M. THORNTON. Stereochemical quality of protein structure coordinates. *Proteins*, **12** (1992), 345–364.

[Ob92] C. A. ORENGO, N. P. BROWN, AND W. R. TAYLOR. Fast structure alignment for protein databank searching. *Proteins: Structure, Function, and Genetics*, **14** (1992), 139–167.

[Rr63] G. N. RAMACHANDRAN, C. RAMAKRISHNAN, AND V. SASISEKHARAN. Stereochemistry of polypeptide chain configurations. *Journal of Molecular Biology*, **7** (1963), 95–99.

[Ys99] M. M. YOUNG, A. G. SKILLMAN, AND I. D. KUNTZ. A rapid method for exploring the protein structure universe. *Proteins*, **34** (1999), 317–332.

CHAPTER 8

Coordinate Transformations

> Knowing reality means constructing systems of transformations
> that correspond, more or less adequately, to reality.
>
> JEAN PIAGET

8.1 MOTIVATION

If we have the 3-D coordinates of atoms in a molecule, transformations
can be applied to these coordinates in order to produce translations and
rotations of the molecule in its 3-D frame of reference. Operations such as
these are needed in various applications. We can list some examples:

1. *Structural alignment of proteins*: Translate and rotate a protein in
   3-D space so that it is superimposed, as much as possible, on a different protein (see [KA76] and [KA78]).

2. *Overlaying small molecules*: This is similar to structural alignment of
   proteins, but the objective is to overlap one ligand over another so as
   to superimpose atoms that have similar chemical traits (see [WD06]).

3. *Movement of parts of molecules*: We may wish to change the conformation of a molecule by rotating a part of it around some designated bond.

4. *Docking*: Compute the movement of a molecule in the study of docking maneuvers that investigate the dynamic positioning of two molecules leading up to a bound state.

This chapter will introduce some of the basic mathematics related to coordinate transformations, and Chapter 9 will deal with applications such as structural alignment.

8.2 INTRODUCTION

We will assume that we are given the 3-D coordinates for each atom in a set P of atoms that may represent an entire molecule or some part of it. The coordinates of any one atom will be stored in a 3×1 column vector, and because there are many of these vectors we will distinguish them by using a superscript index. If we have N atoms, then the vector of coordinates for atom i will be stored in the column vector $p^{(i)}$, $i = 1, 2, \ldots, N$. Adopting set notation, $P = \{p^{(i)} \mid i = 1, 2, \ldots, N\}$. We will assume that each $p^{(i)}$ is a set of coordinates in a 3-D Euclidean coordinate system with axes $e^{(1)}$, $e^{(2)}$, and $e^{(3)}$. We will use the following notation if we need to access the components of $p^{(i)}$:

$$p^{(i)} = \left(p_1^{(i)}, p_2^{(i)}, p_3^{(i)} \right)^{\mathrm{T}}. \tag{8.1}$$

We can also write

$$p^{(i)} = \sum_{j=1}^{3} p_j^{(i)} e^{(j)}. \tag{8.2}$$

When doing a coordinate transformation, our task is to apply the same arithmetic computation to each of the $p^{(i)}$, so that the entire set of atoms undergoes some movement in 3-D space, either a translation or rotation. Because translation is the simplest case, we will start with that.

8.3 TRANSLATION TRANSFORMATIONS

To accomplish a translation we simply add a 3-D column vector of constants to each $p^{(i)}$ vector. If vector d has components d_1, d_2, and d_3, then $p^{(i)} + d$ will represent the coordinates of the atom i after it is moved d_1 units in the $e^{(1)}$-direction, d_2 units in the $e^{(2)}$-direction, and d_3 units in the $e^{(3)}$-direction.

8.3.1 Translation to Place the Centroid at the Origin

The centroid of a set of atoms is calculated by taking the average of all the column vector coordinates in the set. If all the atoms had the same mass, then the calculation would be to compute the center of mass (or center of gravity) for the set. If the centroid is represented by the 3-D column vector $p^{(c)}$, then

$$p^{(c)} = \frac{1}{N} \sum_{i=1}^{N} p^{(i)}. \tag{8.3}$$

To move the entire set of atoms so that their centroid is at the origin of the coordinate system, we subtract $p^{(c)}$ from each of the $p^{(i)}$. This, essentially, gives us new coordinates for the set of atoms, written as follows: $\{x^{(i)} \mid i = 1, 2, \ldots, N\}$ where

$$x^{(i)} = p^{(i)} - p^{(c)}. \tag{8.4}$$

If we now use these new coordinates in the computation of the centroid we see that it is at the origin. In fact,

$$\sum_{i=1}^{N} x^{(i)} = \sum_{i=1}^{N} \left(p^{(i)} - p^{(c)} \right) = \sum_{i=1}^{N} p^{(i)} - \sum_{i=1}^{N} p^{(c)} \tag{8.5}$$

$$= Np^{(c)} - Np^{(c)} = 0.$$

In summary, when the centroid is shifted to the origin, the new coordinates of the atoms satisfy the condition

$$\sum_{i=1}^{N} x^{(i)} = 0. \tag{8.6}$$

In later chapters, we will see that this condition greatly simplifies various calculations, and so provides a significant motivation for a centroid-based new coordinate system.

8.4 ROTATION TRANSFORMATIONS

If an atom has coordinates given by $p^{(i)}$, we can regard the atom as being at the terminus of a vector that starts at the origin and stops at the point $p^{(i)}$. We can then define a rotation transformation for the atom by determining the final position of the atom when its position vector is rotated about the origin through some particular angle. However, the specification of this final position is somewhat complicated when the rotation is done in a

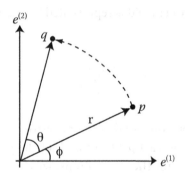

FIGURE 8.1 Rotation in the plane.

3-D space setting. However, in a 2-space setting, the situation is much easier to analyze, so we start with rotations in the plane.

8.4.1 Rotation Transformations in the Plane

In the next section, we will see that a rotation transformation applied to a point p can be accomplished by a premultiplication of p by a special matrix. Some elementary trigonometry can be utilized to derive this matrix. We assume that p lies in the $e^{(1)}$, $e^{(2)}$ plane, and we wish to rotate p through an angle of θ so that it assumes a final position given by the column vector q.[*] The situation is illustrated by Figure 8.1. We are assuming that the initial position vector p makes an angle ϕ with the $e^{(1)}$ axis.

If we assume that the distance from point p to the origin is r, then we can write

$$p_1 = r \cos \phi$$
$$p_2 = r \sin \phi \tag{8.7}$$

and similarly for the final position q,

$$q_1 = r \cos(\phi + \theta)$$
$$q_2 = r \sin(\phi + \theta). \tag{8.8}$$

---

[*] All our derivations assume that we are rotating a point within a fixed frame of reference. Various other texts involve explanations that analyze rotations by rotating the axes, but we avoid this approach in the belief that a single frame of reference provides a simpler analysis.

Expanding these last equations gives

$$q_1 = r(\cos\phi\cos\theta - \sin\phi\sin\theta)$$
$$q_2 = r(\cos\phi\sin\theta + \sin\phi\cos\theta).$$

(8.9)

Then simplification by substituting the components of p gives

$$q_1 = p_1\cos\theta - p_2\sin\theta$$
$$q_2 = p_1\sin\theta + p_2\cos\theta$$

(8.10)

which can be written as

$$\begin{bmatrix} q_1 \\ q_2 \end{bmatrix} = \begin{bmatrix} \cos\theta & -\sin\theta \\ \sin\theta & \cos\theta \end{bmatrix}\begin{bmatrix} p_1 \\ p_2 \end{bmatrix} = R_\theta \begin{bmatrix} p_1 \\ p_2 \end{bmatrix}$$

(8.11)

or more succinctly as

$$q = R_\theta p.$$

(8.12)

Note that $R_\theta^T R_\theta = I$, the identity matrix. In fact, the columns $C^{(1)}$ and $C^{(2)}$ of this rotation matrix are orthonormal:

$$C^{(i)T}C^{(i)} = \sin^2\theta + \cos^2\theta = 1 \qquad i = 1,2$$
$$C^{(i)T}C^{(j)} = -\cos\theta\sin\theta + \cos\theta\sin\theta = 0 \quad i \neq j.$$

(8.13)

The equation $R_\theta^T R_\theta = I$ is an important property of the rotation matrix. It makes the transformation *isometric*, and it is this property that we will want when we consider rotations in 3-D space.

When a real $n \times n$ matrix M has the property $M^T M = I$, it is said to be *orthogonal*. Note the equivalent description of this property: The columns (or rows) of M provide an orthonormal basis for $\mathbb{R}^n$.

8.4.2 Rotations in 3-D Space

It is possible to develop a 3-D version of the rotation matrix with entries that are determined by the angles that specify the 3-D rotation. The derivation of this matrix is more of a challenge than the derivation of the 2-D matrix because in 3-D space, a rotation matrix is defined by three angles. Think of the roll, pitch, and yaw angles that specify the angular position of

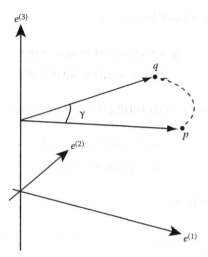

FIGURE 8.2   Rotation through angle γ around the $e^{(3)}$ axis.

an aircraft. The simplest approach is to generate the 3-D rotation matrix as a product of three elementary matrices:

R_α causes a rotation around the $e^{(1)}$ axis through an angle given by α.

R_β causes a rotation around the $e^{(2)}$ axis through an angle given by β.

R_γ causes a rotation around the $e^{(3)}$ axis through an angle given by γ.

We start with R_γ because it is very similar to the matrix that we have just seen for a 2-D rotation in the $e^{(1)}$, $e^{(2)}$ plane. The scene is illustrated in Figure 8.2.

It is important to understand that this transformation does not change the $e^{(3)}$ coordinate of p. In other words, $q_3 = p_3$. Furthermore, looking down the $e^{(3)}$ axis, you would see a situation that is essentially the same as that of Figure 8.1, except that the positive (counterclockwise) rotation is labeled γ instead of θ. Consequently, by modifying the 2-D rotation transformation developed earlier, the relationship between q and p would be given by

$$q = \begin{bmatrix} q_1 \\ q_2 \\ q_3 \end{bmatrix} = \begin{bmatrix} \cos\gamma & -\sin\gamma & 0 \\ \sin\gamma & \cos\gamma & 0 \\ 0 & 0 & 1 \end{bmatrix} \begin{bmatrix} p_1 \\ p_2 \\ p_3 \end{bmatrix} = R_\gamma p. \qquad (8.14)$$

Now consider rotation around the $e^{(2)}$ axis as illustrated in Figure 8.3.

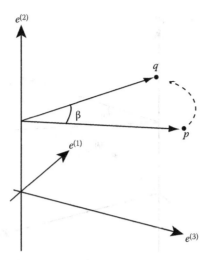

FIGURE 8.3 Rotation through angle β around the $e^{(2)}$ axis.

Note that this is the same as Figure 8.2 except that γ is replaced by β and the axes have changed: $e^{(3)}$ is now $e^{(2)}$, $e^{(2)}$ is now $e^{(1)}$, and $e^{(1)}$ is now $e^{(3)}$. Be careful to observe that Figure 8.3 has axes labeled to preserve the right-handedness of the system. We can now take the last equation and make the same replacements to get

$$\begin{bmatrix} q_3 \\ q_1 \\ q_2 \end{bmatrix} = \begin{bmatrix} \cos\beta & -\sin\beta & 0 \\ \sin\beta & \cos\beta & 0 \\ 0 & 0 & 1 \end{bmatrix} \begin{bmatrix} p_3 \\ p_1 \\ p_2 \end{bmatrix}. \tag{8.15}$$

An inspection of this transformation will verify that it keeps the $e^{(2)}$ coordinate invariant and otherwise changes p so as to move it through an angle of β around the $e^{(2)}$ axis. However, we are not done yet because the vectors in this last equation do not represent q and p. We can do a rotational "upshift" of the rows in vector with the p components if we compensate by doing a rotational left-shift of the columns in the matrix. Doing this, we get the same computation by writing

$$\begin{bmatrix} q_3 \\ q_1 \\ q_2 \end{bmatrix} = \begin{bmatrix} -\sin\beta & 0 & \cos\beta \\ \cos\beta & 0 & \sin\beta \\ 0 & 1 & 0 \end{bmatrix} \begin{bmatrix} p_1 \\ p_2 \\ p_3 \end{bmatrix}. \tag{8.16}$$

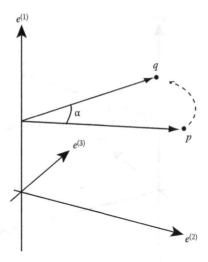

FIGURE 8.4   Rotation through angle α around the $e^{(1)}$ axis.

Then, we can get the left-hand side of this equation to have the components of q by doing a rotational upshift of this vector and then doing the same upshift of rows in the matrix (the p vector does not change). This gives

$$q = \begin{bmatrix} q_1 \\ q_2 \\ q_3 \end{bmatrix} = \begin{bmatrix} \cos\beta & 0 & \sin\beta \\ 0 & 1 & 0 \\ -\sin\beta & 0 & \cos\beta \end{bmatrix} \begin{bmatrix} p_1 \\ p_2 \\ p_3 \end{bmatrix} = R_\beta p. \qquad (8.17)$$

Finally, we derive the matrix for rotation around the $e^{(1)}$ axis as illustrated in Figure 8.4. Again, we label the axes in a way that ensures a right-handed system.

Using the same reasoning as before, we perform the appropriate replacements on the R_γ system leading to

$$\begin{bmatrix} q_2 \\ q_3 \\ q_1 \end{bmatrix} = \begin{bmatrix} \cos\alpha & -\sin\alpha & 0 \\ \sin\alpha & \cos\alpha & 0 \\ 0 & 0 & 1 \end{bmatrix} \begin{bmatrix} p_2 \\ p_3 \\ p_1 \end{bmatrix}. \qquad (8.18)$$

A rotational downshift of the vector with the p components and a compensatory rotational right-shift of the matrix give

$$\begin{bmatrix} q_2 \\ q_3 \\ q_1 \end{bmatrix} = \begin{bmatrix} 0 & \cos\alpha & -\sin\alpha \\ 0 & \sin\alpha & \cos\alpha \\ 1 & 0 & 0 \end{bmatrix} \begin{bmatrix} p_1 \\ p_2 \\ p_3 \end{bmatrix}. \qquad (8.19)$$

To finish the derivation, we do a rotational downshift of rows in the vector with the q components and the same downshift of the matrix rows:

$$q = \begin{bmatrix} q_1 \\ q_2 \\ q_3 \end{bmatrix} = \begin{bmatrix} 1 & 0 & 0 \\ 0 & \cos\alpha & -\sin\alpha \\ 0 & \sin\alpha & \cos\alpha \end{bmatrix} \begin{bmatrix} p_1 \\ p_2 \\ p_3 \end{bmatrix} = R_\alpha p. \tag{8.20}$$

In many applications, it is customary to use these rotational matrices in a particular order that is agreed upon at the outset of a discussion. A popular approach is to apply the rotation about $e^{(1)}$, then $e^{(2)}$, and finally $e^{(3)}$. This can be done using a single matrix product, as in $q = A_{\gamma\beta\alpha} p$ if we define the composite rotation matrix $A_{\gamma\beta\alpha}$, using

$$A_{\gamma\beta\alpha} = R_\gamma R_\beta R_\alpha$$

$$= \begin{bmatrix} \cos\gamma\cos\beta & -\sin\gamma\cos\alpha+\cos\gamma\sin\beta\sin\alpha & \sin\gamma\sin\alpha+\cos\gamma\sin\beta\cos\alpha \\ \sin\gamma\cos\beta & \cos\gamma\cos\alpha+\sin\gamma\sin\beta\sin\alpha & -\cos\gamma\sin\alpha+\sin\gamma\sin\beta\cos\alpha \\ -\sin\beta & \cos\beta\sin\alpha & \cos\beta\cos\alpha \end{bmatrix}.$$

$$\tag{8.21}$$

It should be stressed that the order of the operations is important. Even though each elementary rotation deals with a single angle, they cannot be computed without changing the final composite rotation. For example, $R_\gamma R_\beta R_\alpha \neq R_\alpha R_\beta R_\gamma$ in general.

Once again, with a bit of tedious calculation, we can affirm that this rotation matrix is subject to the orthogonality property:

$$A_{\gamma\beta\alpha}^{\mathrm{T}} A_{\gamma\beta\alpha} = I. \tag{8.22}$$

8.5 ISOMETRIC TRANSFORMATIONS

Now that we have the complete representation of a 3-D rotation matrix do we finally have all the groundwork necessary to discuss applications that utilize 3-D rotation matrices? No, not yet. The previous derivation is very useful for situations in which we know the explicit angles that define the 3-D rotation. For example, in an application that displays a molecule on a computer screen, a mouse drag could lead to the generation of α, β, and γ and the subsequent computation of the $A_{\gamma\beta\alpha}$ matrix needed to update the visualization. However, there are many situations in which we do not know the rotation angles. In these cases, a rotation matrix must be computed to perform some task that is determined by an application requirement that is not expressed using explicit rotation angles. For example, in Chapter 9, we want a rotation matrix that satisfies an optimization requirement. A protein P is to be rotated in such a way as to maximize its overlap with another protein Q. In these situations, the optimization algorithm will determine the entries of the rotation matrix, but this must be done while observing the additional constraint that the final matrix is a legitimate rotation matrix.

In this section, we derive the two mathematical conditions that will characterize the constraints necessary for a matrix to be a valid rotation matrix. We will show that the transformation must be a *nonreflective isometry*. We begin by discussing the notion of isometry.

8.5.1 Our Setting Is a Euclidean Vector Space

To establish a general setting we consider a linear operator T such that $T: \mathbb{R}^n \to \mathbb{R}^n$. Because the vector space $\mathbb{R}^n$ is of finite dimension, we can assume that the transformation T is defined by $T(x) = Ax$ for any vector $x \in \mathbb{R}^n$. We will need a distance function $d(x, y)$ in our discussion, and this will be the metric induced by the usual Euclidean norm: $d(x, y) = \|x - y\|$.

> DEFINITION 8.1: The linear operator T is said to be an isometry if it preserves distances. In other words, if $T(p)$ and $T(q)$ are the images of p and q under the transformation T, then the distance between $T(p)$ and $T(q)$ will be the same as the distance between p and q. That is to say, $\|T(p) - T(q)\| = \|p - q\|$.

8.5.2 Orthogonality of A Implies Isometry of T

Suppose A is an orthogonal matrix, i.e., $A^T A = I$. The orthogonality of A gives T a very valuable property: Inner products are preserved under T. This means that for any p and q in $\mathbb{R}^n$ we have

$$\langle T(p), T(q) \rangle = \langle p, q \rangle \tag{8.23}$$

or in matrix form

$$(Ap)^T (Aq) = p^T q. \tag{8.24}$$

This is true because $A^T A = I$, and so $(Ap)^T (Aq) = p^T A^T A q = p^T q$.

The preservation of inner products gives us

$$\begin{aligned}
\|T(p)-T(q)\|^2 &= \|Ap-Aq\|^2 \\
&= \|A(p-q)\|^2 \\
&= (p-q)^T A^T A(p-q) \\
&= \|p-q\|^2 .
\end{aligned}$$

(8.25)

Now, by taking the positive square root of each side, we can demonstrate that orthogonality of A leads to isometry of T.

8.5.3 Isometry of T Implies Orthogonality of A

We have just shown that orthogonality implies isometry. We now demonstrate the implication going in the opposite direction. If $\|T(p)-T(q)\|=\|p-q\|$ then, setting $q=0$ and relying on T being a linear operator, we get the invariance of the norm under T, that is,

$$\|T(p)\|=\|p\|.$$

(8.26)

This implies that

$$\|Ap\|^2=\|p\|^2 \quad \text{for all} \quad p \in \mathbb{R}^n.$$

(8.27)

Note that

$$\begin{aligned}
&\|T(p)-T(q)\| = \|p-q\| \Rightarrow \\
&\|Ap-Aq\|^2 = \|p-q\|^2 \Rightarrow \\
&(Ap-Aq)^T (Ap-Aq)=(p-q)^T(p-q) \Rightarrow \\
&\|Ap\|^2 -2(Ap)^T(Aq)+\|Aq\|^2 =\|p\|^2 - 2p^T q +\|q\|^2
\end{aligned}$$

(8.28)

and applying Equation 8.27 we get

$$(Ap)^T(Aq)= p^T q \quad \text{for all} \quad p,q \in \mathbb{R}^n$$

(8.29)

which tells us that if T is an isometry then we have the preservation of dot products under the transformation T. We can rewrite the last equation as

$$p^T A^T Aq = p^T q.$$

(8.30)

As this is true for all $p,q \in \mathbb{R}^n$ we can set p and q to be any of the standard basis vectors of $\mathbb{R}^n$; more formally, we let $p = e^{(i)}$ and $q = e^{(j)}$ with $1 \le i, j \le n$. This gives us

$$e^{(i)T} A^T A e^{(j)} = e^{(i)T} e^{(j)} = \delta_{ij} \tag{8.31}$$

where δ_{ij} is the Kronecker delta symbol that always has the value 0 unless $i = j$ in which case it is 1. Noting that $e^{(i)T} A^T$ is really the ith column of A, and $A e^{(j)}$ is the jth column of A, we see that dot products of different columns in A will be 0 and a column "dotted" with itself will produce a value of 1. In other words, A is orthogonal.

8.5.4 Preservation of Angles

As the angle θ between two vectors p and q can be defined by using an inner product,

$$\cos(\theta) = \frac{p^T q}{\|p\| \|q\|} \tag{8.32}$$

we see that a transformation using orthogonal matrix A will also preserve angles between vectors because both the numerator and the denominator of this expression are unchanged by the application of an isometric transformation. Expressed as an equation, we have

$$\text{angle}(p, q) = \text{angle}(Ap, Aq). \tag{8.33}$$

8.5.5 More Isometries

If we have a set of vectors and consider a linear transformation that does a simple translation of all these vectors, then it is easy to see that this transformation is an isometry. Similarly, a transformation that does a rotation of all the vectors with respect to some axis in the space will likewise preserve distances, and so it is an isometry. Do these two types of transformation exhaust the possibilities for an isometry? Algorithms that perform transformation on molecular coordinates would find it very convenient to assume that this is true. However, it is not true that an isometry is always equal to some combination of a translation and rotation operation because there is yet another linear transformation that preserves distances. The reflection operation is such a linear transformation. If we image all the vectors in a set reflected in a mirror that is held in some position in $\mathbb{R}^3$, then it is easy to verify that distances are preserved and so reflection is an isometry.

When you see yourself in a mirror, you are looking at a reflection that represents an isometry of yourself. That "other person" in the mirror has a right eye that is, at least superficially, a copy of your left eye. More significant to our discussion is the fact that at the molecular level your reflection has chiral centers that are all enantiomers of your own. Your proteins have L-amino acids, the virtual person in the mirror has D-amino acids. These considerations carry a cautionary message. If we utilize an algorithm that derives an orthogonal transformation matrix that attempts to rotate all the atomic coordinates, we must take care to ensure that the algorithm does not inadvertently introduce an unwanted reflection. In the next section, we look at this more closely and develop a constraint that will help us eliminate reflections.

8.5.6 Back to Rotations in the Plane

In this section we again derive the 2-D rotation matrix, but this time we show that the derivation can be accomplished by starting with the notion of an isometric transformation.

We begin with a simple linear isometry in the Euclidean plane, that is, $T : \mathbb{R}^2 \to \mathbb{R}^2$, and we look at its effect on an arbitrary point $p = (p_1, p_2)^T$ in the plane. We can write this as

$$T(p) = Ap = A \begin{pmatrix} p_1 \\ p_2 \end{pmatrix}. \tag{8.34}$$

We can also consider p to be a linear combination of standard basis vectors, that is,

$$p = p_1 e^{(1)} + p_2 e^{(2)} \tag{8.35}$$

where $e^{(1)} = (1,0)^T$ and $e^{(2)} = (0,1)^T$. Because T is a linear transformation,

$$T(p) = p_1 T(e^{(1)}) + p_2 T(e^{(2)})$$
$$= p_1 A e^{(1)} + p_2 A e^{(2)} \tag{8.36}$$
$$= p_1 u^{(1)} + p_2 u^{(2)}$$

where $u^{(1)}$ and $u^{(2)}$ are the columns of A. So the effect of T on p is essentially specified once you know what T does to the unit vectors $e^{(1)}$ and $e^{(2)}$, and this action is defined by the columns of A.

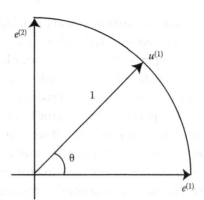

FIGURE 8.5   Rotation of a unit vector through angle θ.

We now want to derive an isometry T that works on vectors in such a way as to rotate them about the origin of the plane. As T preserves distances, we can assume that $u^{(1)}$ and $u^{(2)}$ are also points on the unit circle along with $e^{(1)}$ and $e^{(2)}$. In fact, we can specify the transformation of T on $e^{(1)}$ to be a counterclockwise rotation through an angle θ, as illustrated in Figure 8.5.

A straightforward trigonometric calculation then gives

$$u^{(1)} = \begin{pmatrix} \cos(\theta) \\ \sin(\theta) \end{pmatrix} \qquad (8.37)$$

assuming $e^{(1)}$ and $e^{(2)}$ is our frame of reference. What about $u^{(2)} = T(e^{(2)})$?
Our isometric constraints insist that $u^{(2)}$ is also on the unit circle due to preservation of distance with respect to the origin, and the preservation of angles implies that $u^{(1)} \perp u^{(2)}$ because $e^{(1)} \perp e^{(2)}$. This gives us two scenarios for $u^{(2)}$ (see Figure 8.6).

Computations for $u^{(2)}$ give

$$u^{(2)} = \begin{pmatrix} -\sin(\theta) \\ \cos(\theta) \end{pmatrix} \quad \text{and} \quad u^{(2)} = \begin{pmatrix} \sin(\theta) \\ -\cos(\theta) \end{pmatrix} \qquad (8.38)$$

respectively, and the corresponding transformation matrices are

$$A = \begin{pmatrix} \cos(\theta) & -\sin(\theta) \\ \sin(\theta) & \cos(\theta) \end{pmatrix} \quad \text{and} \quad A = \begin{pmatrix} \cos(\theta) & \sin(\theta) \\ \sin(\theta) & -\cos(\theta) \end{pmatrix}. \qquad (8.39)$$

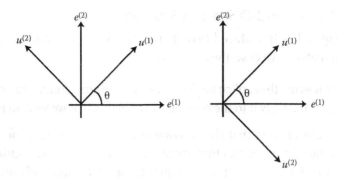

FIGURE 8.6   Two choices for the rotation.

Both matrices for A are orthogonal, and both are valid isometric transformations of the $e^{(1)}$, $e^{(2)}$ pair because in each case we have the preservation of distances and angles. If we want T to be a rotation in the usual sense, with both $e^{(1)}$ and $e^{(2)}$ carried through the same angle relative to $e^{(1)}$, then we will want the first case in Figure 8.6. The important issue is that the second transformation essentially involves a reflection in the $e^{(1)}$ axis followed by a counterclockwise rotation through an angle θ.

Now note that $\det(A) = 1$ for the case that we prefer, whereas the transformation involving the reflection has $\det(A) = -1$. This constraint can also be applied to 3×3 rotation matrices. In Chapter 9, we will insist that the determinant of a rotation matrix for molecules has a determinant equal to 1 so that no reflections are introduced in the rotation.

As a final observation, we write out the explicit transformation that operates on a vector p in such a way as to rotate it through an angle θ to produce a vector q:

$$q = T(p) = Ap = A\begin{pmatrix} p_1 \\ p_2 \end{pmatrix}$$

$$= \begin{pmatrix} \cos(\theta) & -\sin(\theta) \\ \sin(\theta) & \cos(\theta) \end{pmatrix}\begin{pmatrix} p_1 \\ p_2 \end{pmatrix}$$

(8.40)

yielding the same rotation transformation that we derived earlier:

$$q_1 = p_1 \cos(\theta) - p_2 \sin(\theta)$$

$$q_2 = p_1 \sin(\theta) + p_2 \cos(\theta).$$

(8.41)

8.5.7 Rotations in 3-D Space: A Summary

This chapter has introduced two approaches for the generation of a 3-D rotation matrix. We describe them as follows:

- By knowing the explicit angles that specify the rotation, we can compute $A_{\gamma\beta\alpha}$ and apply it to the coordinates of points that we wish to rotate.

- We can assume that the rotation matrix has a "generic" form consisting of nine entries that are to be determined by the requirements of some application (for example, an optimization problem). However, if the matrix computed in this way is to be a legitimate rotation matrix, it must also satisfy two constraints: it must be orthogonal and its determinant must be 1.

8.6 EXERCISES

1. Recall the warning following Equation 8.21: In general, $R_\gamma R_\beta R_\alpha \neq R_\alpha R_\beta R_\gamma$. Create an example to show that this is true. Hint: It may suffice to use "easy" settings of the α, β, and γ parameters (for example, particular multiples of $\pi/2$ or $\pi/4$).

2. Demonstrate which of the following matrices represent isometric transformations:

$$
a) \begin{pmatrix} \dfrac{1}{\sqrt{2}} & 0 & \dfrac{-1}{\sqrt{2}} \\[2mm] \dfrac{1}{\sqrt{6}} & \dfrac{-2}{\sqrt{6}} & \dfrac{1}{\sqrt{6}} \\[2mm] \dfrac{1}{\sqrt{3}} & \dfrac{1}{\sqrt{3}} & \dfrac{1}{\sqrt{3}} \end{pmatrix} \quad
b) \begin{pmatrix} e^{i\alpha} & 0 & 0 \\ 0 & e^{i\beta} & 0 \\ 0 & 0 & e^{i\gamma} \end{pmatrix} \quad
c) \begin{pmatrix} 1 & 0 & 0 \\[2mm] 0 & 1 & \dfrac{1}{\sqrt{2}} \\[2mm] 0 & -1 & \dfrac{1}{\sqrt{2}} \end{pmatrix}.
$$

3. Write a program that accepts the following input:

- The coordinates of a molecule written in Protein Data Bank (PDB) format

- The specification of a dihedral angle (for example, the designation of the four atoms that define this angle)

- A value θ that represents an angle

The output of the program should be the coordinates of the molecule transformed, so that the specified dihedral angle is increased by the amount θ. Test the program by altering the dihedral angle within the backbone of a protein that can act as a "hinge." For example, chain A of 6PAX has a backbone that links two quite distinct hydrophobic cores. Use GLY 69 C, GLY 70 N, GLY 70 C_α, and GLY 70 C to determine a phi dihedral angle and let θ be 20°.

4. Essay project: Many viruses, such as Simian Virus 40 (SV40), have an icosahedral coat. In the case of SV40, one can obtain the coordinates of all atoms that make up one viral pentamer (see PDB code 1SVA). One might casually assume that having the coordinates of the atoms that make up one face of a 3-D regular polytope (dodecahedron or icosahedron, for example) would easily lead to the generation of coordinates for all the other faces using rotation matrices, such as $A_{\gamma\beta\alpha}$, each specified by a particular setting of the α, β, and γ parameters. One would then have the entire set of coordinates for a virus coat and the ability to display it using protein visualization software. Although such a rotation matrix does play a role in the algorithmic assembly process, the entire 3-D reconstruction process has various computational issues. For example, one must know the lattice structure of the icosahedron to be assembled. Write an essay that discusses these points. You may consult the following papers to get a start: [Bo99], [Sb04]. The Virus Particle Explorer (VIPERdb) Web site is also recommended (http://viperdb.scripps.edu/index. php). If you go to this Web site and do a search on 1SVA, you can get a visualization of the entire virus coat with the ability to see how the 6 chains of 1SVA combine to form one of the 20 faces of an icosahedron. Incidentally, the VIPERdb Web site provides a list of rotation matrices that are suitable for these investigations.

REFERENCES

[Bo99] T. S. BAKER, N., H. OLSON, AND S. D. FULLER. Adding the third dimension to virus life cycles: Three-dimensional reconstruction of icosahedral viruses from cryo-electron micrographs. *Microbiology and Molecular Biology Reviews*, **63** (1999), 862–922.

[Ka76] W. KABSCH. A solution for the best rotation to relate two sets of vectors. *Acta Crystallographica*, **A32** (1976), 922–923.

[KA78] W. KABSCH. A discussion of the solution for the best rotation to relate two sets of vectors. *Acta Crystallographica*, **A34** (1978), 827–828.

[SB04] M. SITHARAM AND M. BONA. Counting and enumeration of self-assembly pathways for symmetric macromolecular structures. *Advances in Bioinformatics and Its Applications*, Eds.: M. He, G. Narasimhan, S. Petoukhov, (2004), 426–436.

[WD06] G. WOLBER, A., A. DORNHOFER, AND T. LANGER. Efficient overlay of small organic molecules using 3-D pharmacophores. *J. Comput. Aided Mol. Des.*, **20** (2006), 773–788.

CHAPTER 9

Structure Comparison, Alignment, and Superposition

With the increasing availability of protein structures, the generation of biologically meaningful 3D patterns from the simultaneous alignment of several protein structures is an exciting prospect: active sites could be better understood, protein functions and protein 3D structures could be predicted more accurately.

JEAN-CHRISTOPHE NEBEL

The existence of very remote protein similarities (beyond what can be detected through the comparison of sequence data alone) is known from the comparison of protein structures. Structure comparison therefore provides the criterion of "truth" towards which sequence alignment algorithms should aim.

NIGEL P. BROWN, CHRISTINE A. ORENGO,
AND WILLIAM R. TAYLOR [Bo96]

Comparison of structures is more difficult than sequence comparison because of the non-local similarity score; it is not even clear if the optimal solution of structure comparison exists.

YUZHEN YE AND ADAM GODZIK

241

All science is either physics or stamp collecting.

ERNEST RUTHERFORD
BRITISH CHEMIST AND PHYSICIST (1871–1937)

9.1 MOTIVATION

The somewhat uncharitable "stamp collecting" statement made by Rutherford followed the momentous physics experiments that were revealing the structure of matter at its subatomic level. At that time, much of biology was still concerned with morphological comparison, categorization, and phylogenetic analysis of plants and animals. In his younger days, Rutherford would have seen very little of the experimental studies that would eventually provide strong links between organic chemistry and biology. Fast forward to the 21st century, and scientists have learned a great deal about both physics and biology. In particular, both fields of study are revealing extensive realms of complexity. We can take some smug satisfaction in contemplating that Rutherford would probably be dismayed by the huge collection of atomic particles that some people now characterize as the "elementary particle zoo" [Ho97], containing about 200 members.

For biology, the stamp collecting mission still continues in some sense. The Linnaean taxonomy, introduced by Carolus Linnaeus (1707–1778), now has a counterpart at the molecular level. We seek to organize and classify a huge collection of proteins that are responsible for life processes. Concepts related to structural similarity are fundamental to this goal.

Structure comparison deals with various algorithms that analyze two or more proteins in an effort to describe structural similarities. Why should we want to do this? The primary goal is to understand functionality, but this a very complicated endeavor. In Chapter 2, we presented various behavioral aspects of a protein's structure. Reviewing these points, we see that protein structure must provide for the following "life-cycle functionality":

- *Uniqueness*: The peptide chains must fold into a particular low-energy conformation that has the desired biological utility. The protein should not have other competitive low-energy conformations that lead to biological dysfunction.

- *Stability*: Once the protein is in its folded state, it must retain that folded state even though some changes in conformation may occur due to flexibility requirements.

- *Interaction*: The final conformation must have an interaction capability, either a ligand binding site or a surface for protein–protein interactions. This capability may also include the flexibility just noted.

- *Degradation*: In many cases, the structure should be degradable so that the protein is conveniently disassembled and its constituent amino acids used in the construction of other proteins.

Unfortunately, each of these functional behaviors is generally difficult to relate to structure. In some cases, the mechanism of a binding site may be thoroughly studied in a drug design project, and the interactions of the drug with the binding site are known with high precision. The conformational changes of a protein when it interacts with a ligand are one of the motivations for structural comparisons.

However, in most situations, the connection between function and structure is much more difficult to assess, and we must rely on observation in lieu of theory. In other words, when a new protein is being studied, we typically compare its structure with that of similar proteins in an attempt to draw inferences about its behavior based on the known behavior of the other proteins in the same category.

How do we organize this categorization? Because the final structure must satisfy all of the aforementioned requirements, evolutionary processes will have produced structures that have attempted to optimize fitness criteria with multiple objectives. Although a particular protein strives to satisfy multiple objectives, it is very difficult to determine the structural properties that are *most* important to its functionality. For example, do we focus on the configuration of atom types in a binding site, or do we stress the organization of secondary structural elements that lead to the stability of a hydrophobic core? The question has no definitive answer.

As an example, consider the alignment of 1MBN and 1JEB as illustrated in Figure 2.27. If we focus on this pair of proteins, it seems that the region of highest sequence conservation corresponds to the 16 residues at the very beginning of the protein. This is a helix that does not seem to be directly related to the heme-binding site. The region does contain the sixth amino acid that leads to sickle cell anemia if it mutates to valine in hemoglobin, but this is only a single residue. One may speculate that there seems to be some additional critical functionality that is being preserved by this helix. To investigate this issue more completely, it would be necessary to consider other proteins in the globin family.

In summary, we do structural comparisons for various reasons: We may wish to study the conformational change that is seen in an interaction between a protein and another molecule. We may also study the geometry of binding sites and seek to understand the chemical mechanism of that interaction. The other issues (uniqueness, stability, and degradation) are somewhat more difficult to assess. We understand that a hydrophobic core is necessary for stability, but understanding the attainment of all three objectives and its relation to the consecutive arrangement of amino acids in a peptide chain is a major challenge. In other words, we do not have algorithms that can examine a primary sequence and then accurately predict whether the peptide would fold to a unique and stable conformation that could be enzymatically degraded at a later time. This puts us back into "stamp collecting mode." To understand or predict a protein's behavior, we must consider its membership in a family of proteins with similar structure. Therefore, we are concerned with the structural differences between a given protein and other proteins that have an evolutionary relationship with it. In particular, what has been structurally conserved?

We may also study the **structural plasticity** of a family. This topic deals with the degree of structural variability seen in a family. It seems that some families show more acceptance of such diversity than others. Algorithms that deal with multiple structure alignment are important for these investigations. Experiments such as those reported in [VB96] and [SB06] have demonstrated the effect that amino acid insertions have had on stability and the tendency to maintain local helix formation by propagating the effects of the insertion into a bend or loop. Another interesting discussion about structural plasticity and its relation to evolution may be found in [Mo07].

Finally, it should be mentioned that proteins can assume similar folds even though there is no obvious evolutionary relationship among them. These **analogs** may be the result of a limited number of configurations available when packing helices and sheets to form a hydrophobic core (see [OJ03]).

Research in structural comparison seems to reflect the wide variety of concerns and viewpoints on the topic. To get some perspectives on this, the reader may consult [Bo96] written by Brown, Orengo, and Taylor. This paper suggests a systematic nomenclature that helps organize the diverse topics contained in approximately 50 or so research papers written on protein structure comparison methodologies prior to 1996. Since then, there have been considerably more papers written with a corresponding increase in the range of comparison methods.

9.2 INTRODUCTION

9.2.1 Specifying the Problem

To understand the wide range of topics related to structural comparison, we list some of the attributes that characterize a study of the problem:

- *Level of detail*: A comparison study is done at a particular resolution or level of structural detail. Comparison algorithms may deal with the positions of various types of structure:

  - Individual atoms

  - Locations of residues (specified by coordinates of the alpha carbon atoms, beta atoms, or center of mass of the side chains)

  - Position and orientation of secondary structural elements (for example, treating helices as cylinders)

  - Similarity of folds at the tertiary structure level

- *Physicochemical properties*: The quantification of similarity may only involve the relative positions of structural entities in 3-D space, or it may also include additional descriptive attributes that are important for some application. For example, an atom may be categorized with a physicochemical label that describes it as being a hydrogen bond donor, a hydrogen bond acceptor, having a positive charge or a negative charge, or being hydrophobic. Hydrophobicity or hydrophilicity may also be assigned to entire residues. Depending on the objectives of the comparison, a property may be associated with a point in space, or it may be spread out over some portion of a surface constructed around the underlying atoms.

- *Extent of comparison*: The comparison may be global, dealing with the entire protein, or more local, being restricted in some fashion. Various types of restriction may be applied limiting the comparison to a protein domain, a contiguous subsequence of amino acids, or the subset of atoms in a binding pocket.

- *Number of proteins*: A comparison study may involve one, two, or several proteins. In the case of a single protein, the study typically compares the change in atom positions when the protein changes its conformation. Good examples of this would be the conformational change associated with an allosteric movement or the shift of side chains in binding pocket during ligand docking. When two proteins are involved, the objectives of the study usually depend on how different the proteins are in sequence composition. A simple case would be the study of conformational change due to the substitution of one or two residues. A more complicated comparison would deal with two proteins that have different lengths and perhaps very little sequence identity when a sequence alignment is performed. In the latter case, it is typical that the proteins have a homologous relationship, and the study is designed to assess the extent of structural similarity that has been conserved in an evolutionary descent. Studies involving many proteins are often designed to evaluate similarities in proteins that are closely related as established by sequence alignment. Possible objectives include an analysis of all proteins in the group with the intention of characterizing atom types in a binding site (looking for similarities) or an analysis of evolutionary divergence (this time looking for differences). Conversely, studies may establish similarities in proteins due to structural conformations only independent of sequence similarity. If the sequence similarity is then found to be very low, it is reasonable evidence that structural similarity is due to evolutionary convergence.

These points should convince the reader that there are many approaches to structural comparison.* In fact, an extended treatment of the various methodologies that are used could form the subject matter of an entire book. To set a limit on this extensive material, we will continue with the discussion of only a few comparison algorithms. First, we discuss the typical categories of the comparison algorithms. After that, we will present brief descriptions of representative algorithms.

9.3 TECHNIQUES FOR STRUCTURAL COMPARISON

Suppose we are given two proteins that we will designate as P and Q. Algorithms for structural comparison of P and Q generally fall into two broad categories:

A. **Superposition algorithms.** These algorithms try to maximize some type of overlap of two proteins in 3-D space. The overlap may deal with all the residues of the protein or some subset of the residues. If subsets of residues are being considered, then there must be some specification of atom pairs that will participate in the overlap.

---

* Further discussion can be found in [EJ00].

B. *Algorithms that compare geometric relationships within a protein.* We start by defining a set of *relationships* between elements in the protein P. For example, we might calculate, for each alpha carbon in P, a set of distances to all the other alpha carbons in P. Each distance is seen as a relationship between the two atoms. Then do the same calculations for Q. The comparison algorithm then compares the relationships found in P with those found in Q. There are many variations on this theme: Instead of atoms, use fragments or secondary structure components such as helices or strands. Instead of all such elements, concentrate on only those that are nearby or within the same domain. The strategies for comparing the relationships can also vary from one algorithm to the next.

9.4 SCORING SIMILARITIES AND OPTIMIZING SCORES

In all comparison algorithms we try to evaluate some successful demonstration of similarity. All comparison algorithms should clearly specify:

a. *Score evaluation.* Given two proteins P and Q, a comparison algorithm may reveal several different possibilities, each demonstrating similarity between P and Q. For example, there may be several ways to overlap two proteins in 3-D space. Consequently, it is necessary to have some measure of similarity. This evaluation of the similarity is usually called a *score*.

b. *Score optimization.* Typically, there will be a strategy to optimize this score. For example, the algorithm will derive a best structural alignment, or compute some type of correspondence for the atoms of P and Q, so that the score will be optimized.

When we discuss the various comparison algorithms, both of these points will be covered.

9.5 SUPERPOSITION ALGORITHMS

9.5.1 Overview

A frequent concern is whether two proteins have the same or very similar structure. An assessment of this can be done by attempting to superimpose the two proteins in 3-D space. The proteins may have the same residues, or they may be very similar (homologs, for example). There are various situations when such a comparison would be calculated:

- The proteins may have the same sequence but differ in conformation. The difference might be due to a backbone flexing of the protein or may be due to a change in conformation when a ligand is present in a binding site.

- The proteins may have almost the same sequence. For example, a mutation has caused some amino acid to change. What is the effect on conformation?

- The proteins may have more extensive differences in their sequences, but it is possible that they nonetheless show much similarity in conformation. The question is, "how similar are these conformations?"

- The proteins may have considerable differences in their sequences, but it is possible that they share similar structure in various regions. We would want to compare these similar regions. Recall that structure is more conserved than sequence.

In some situations, a transformation of coordinates is necessary to compare different conformations of the same protein because the coordinates are derived from PDB files that assume different orientations of the protein. For example, 1IZI, 1MSN, and 1EBW all deal with the same protein, namely, **HIV protease**. A casual comparison of the coordinates of the first two atoms in each file gives:

```
1IZI
ATOM    1  N   PRO  A  1  -12.600   38.218    3.719
ATOM    2  CA  PRO  A  1  -12.444   38.367    2.244

1MSN
ATOM    1  N   PRO  A  1    0.421   40.709   18.682
ATOM    2  CA  PRO  A  1   -0.422   39.511   18.905

1EBW
ATOM    1  N   PRO  A  1   29.101   40.309    5.484
ATOM    2  CA  PRO  A  1   30.105   39.343    4.986
```

So, it is clear that the PDB does not store proteins in any "standard" orientation.

9.5.2 Characterizing the Superposition Algorithm

In this algorithm we consider the structural comparison of two proteins that we will designate as P and Q. We will not consider any physico-chemical properties, just atomic coordinates. So, at this level of detail, it is expected that the input to the algorithm will be the 3-D coordinates of the atoms involved in the comparison. The algorithm will require that we specify a set called an *equivalence*. This is a set of matching atom pairs that are to be brought into 3-D alignment. Each pair will designate an atom in P and its corresponding atom in Q. The algorithm will then find the translation and rotation operations that minimizes the sum of squares of the distances between the atoms in the pairs.

In the description of the algorithm, we will assume that this equivalence is generated using some technique that is appropriate to the application that uses the structure overlap algorithm. Various approaches can be listed:

- If the proteins have the same sequence, then the equivalence can be generated simply by matching up the corresponding atoms. Then, we can extract the subset of these pairs that is appropriate to the needs of the study. For example, we might deal solely with the alpha carbons in the backbone or some part of it.

- If the proteins have a different sequence, then a sequence alignment between P and Q can identify matching pairs of residues, and then the corresponding alpha carbon atoms can be used to define the equivalence.

In some cases, the sequence similarity is so vague that it is quite difficult to generate the equivalence set by doing a sequence alignment. In these cases, a different algorithm is required, one that attempts to do a structural comparison using coordinate information with little or no help from a sequence alignment.

9.5.3 Formal Problem Description

Although different choices could be made for the atoms in an equivalence set, we will assume that the superposition algorithm deals with alpha carbon atoms. We are given two sequences of alpha carbon 3-D coordinates:

$$P = \{p^{(i)}\}_{i=1}^{|P|} \qquad Q = \{q^{(i)}\}_{i=1}^{|Q|} \tag{9.1}$$

where $|P|$ and $|Q|$ are the number of residues in protein P and protein Q, respectively.

We start with a formal description of the equivalence set followed by the notion of an alignment. Some definitions:

DEFINITION 9.1: An equivalence is a set of pairs

$$\{(p^{(\alpha_1)}, q^{(\beta_1)}), (p^{(\alpha_2)}, q^{(\beta_2)}), \ldots, (p^{(\alpha_N)}, q^{(\beta_N)})\} \qquad (9.2)$$

indicating the correspondence between the amino acids in P and Q.

DEFINITION 9.2: An alignment M for P and Q is an equivalence such that $\alpha_1 < \alpha_2 < \ldots < \alpha_N$ and $\beta_1 < \beta_2 < \ldots < \beta_N$.

We extract the alpha carbons from each list in the alignment as follows:

$$M(P) = (p^{(\alpha_1)}, p^{(\alpha_2)}, \ldots, p^{(\alpha_N)})$$
$$M(Q) = (q^{(\beta_1)}, q^{(\beta_2)}, \ldots, q^{(\beta_N)}). \qquad (9.3)$$

The structural overlap of the two proteins P and Q will involve modifying the coordinates of atoms to do the following:

1. All the atoms of each protein are moved (translated) to the origin, so that the centroids of $M(P)$ and $M(Q)$ coincide.

2. An optimal rotation of protein P is then done to get the maximal amount of overlap between $M(P)$ and $M(Q)$.

What do we mean by "maximal overlap"? Our measure of success will be to minimize the sum of squares of norms that measure the distance between matching alpha carbons. This is minimizing in the **least squares** sense

$$E(M(P_{transformed}), M(Q)) = \frac{1}{2} \sum_{i=1}^{N} \|Rp^{(\alpha_i)} + T - q^{(\beta_i)}\|^2. \qquad (9.4)$$

Our objective is to find the rotation matrix R and the translation vector T that will minimize E. Relating this to the issues discussed in Section 9.4, we consider E to be an evaluation of the similarity score, and the purpose of the algorithm is to optimize this score by calculating R and T so that E is minimized.

9.5.4 Computations to Achieve Maximal Overlap

We start by defining the centroids of the alpha carbons used in the superimposition.

Let

$$p^{(c)} = \frac{1}{N}\sum_{i=1}^{N} p^{(\alpha_i)} \qquad q^{(c)} = \frac{1}{N}\sum_{i=1}^{N} q^{(\beta_i)}. \tag{9.5}$$

Then let

$$x^{(i)} = p^{(\alpha_i)} - p^{(c)} \qquad y^{(i)} = q^{(\beta_i)} - q^{(c)}. \tag{9.6}$$

We will now consider $x^{(i)}$ and $y^{(i)}$, $i = 1, 2, \ldots, N$ to be the coordinates of the matching alpha carbons in proteins P and Q, respectively. For each case we have essentially translated the entire protein so that its centroid is at the origin.*

Recall from the previous chapter that

$$\sum_{i=1}^{N} x^{(i)} = 0 \qquad \sum_{i=1}^{N} y^{(i)} = 0. \tag{9.7}$$

So, working with our new coordinate system, we see that we want to find R and T to minimize

$$E = \frac{1}{2}\sum_{i=1}^{N} \|Rx^{(i)} + T - y^{(i)}\|^2. \tag{9.8}$$

Expanding this, we get

$$
\begin{aligned}
E &= \frac{1}{2}\sum_{i=1}^{N} (Rx^{(i)} + T - y^{(i)})^{\mathrm{T}}(Rx^{(i)} + T - y^{(i)}) \\
&= \frac{1}{2}\sum_{i=1}^{N} ((Rx^{(i)} - y^{(i)})^{\mathrm{T}} + T^{\mathrm{T}})((Rx^{(i)} - y^{(i)}) + T) \\
&= \frac{1}{2}\sum_{i=1}^{N} \left[\|Rx^{(i)} - y^{(i)}\|^2 + (Rx^{(i)} - y^{(i)})^{\mathrm{T}}T + T^{\mathrm{T}}(Rx^{(i)} - y^{(i)}) + T^{\mathrm{T}}T \right] \\
&= \frac{1}{2}\sum_{i=1}^{N} \left[\|Rx^{(i)} - y^{(i)}\|^2 + T^{\mathrm{T}}T \right].
\end{aligned}
\tag{9.9}
$$

---

* If atoms in the alignment do not have the same atomic weight, we may decide to calculate the "center of mass".

Note that in the second last line the coefficients of T sum to zero because the centroid of each protein is centered at the origin. So, we end up with

$$E = \frac{1}{2} \sum_{i=1}^{N} \left[\| Rx^{(i)} - y^{(i)} \|^2 + \| T \|^2 \right].$$ (9.10)

If we wish to get a minimum E, it is clear that we want $T = 0$ because this will zero out the $\| T \|^2$. Note that we could also get this result by computing $\partial E / \partial T$ and setting this partial derivative to zero. So, finally, we see that we want to find the rotation matrix R that will minimize

$$E = \frac{1}{2} \sum_{i=1}^{N} \| Rx^{(i)} - y^{(i)} \|^2.$$ (9.11)

Recall from the previous chapter that we want to find the matrix R that will minimize E, but it must be subject to the rotation constraint that $R^T R = I$. Lagrange multipliers can take care of this. But before we derive a Lagrangian, there is still another simplification that can be made. Because $R^T R = I$, we can write

$$
\begin{aligned}
E &= \frac{1}{2} \sum_{i=1}^{N} \| Rx^{(i)} - y^{(i)} \|^2 = \frac{1}{2} \sum_{i=1}^{N} (Rx^{(i)} - y^{(i)})^T (Rx^{(i)} - y^{(i)}) \\
&= \frac{1}{2} \sum_{i=1}^{N} (x^{(i)T} R^T - y^{(i)T})(Rx^{(i)} - y^{(i)}) \\
&= \frac{1}{2} \sum_{i=1}^{N} (x^{(i)T} R^T Rx^{(i)} - x^{(i)T} R^T y^{(i)} - y^{(i)T} Rx^{(i)} + y^{(i)T} y^{(i)}) \\
&= \frac{1}{2} \sum_{i=1}^{N} \left(\| x^{(i)} \|^2 - x^{(i)T} R^T y^{(i)} - y^{(i)T} Rx^{(i)} + \| y^{(i)} \|^2 \right) \\
&= \frac{1}{2} \sum_{i=1}^{N} \left(\| x^{(i)} \|^2 + \| y^{(i)} \|^2 \right) - \sum_{i=1}^{N} y^{(i)T} Rx^{(i)}.
\end{aligned}
$$ (9.12)

Notice that the first sum is independent of R. Consequently, we can minimize E by maximizing H where

$$H = \sum_{\gamma=1}^{N} y^{(\gamma)T} Rx^{(\gamma)}.$$ (9.13)

R is the 3×3 matrix:

$$R = \begin{bmatrix} r_1^1 & r_1^2 & r_1^3 \\ r_2^1 & r_2^2 & r_2^3 \\ r_3^1 & r_3^2 & r_3^3 \end{bmatrix} \tag{9.14}$$

subject to $R^T R = I$, a constraint that we will rewrite as

$$\sum_{\gamma=1}^{3} r_\gamma^\alpha r_\gamma^\beta = \delta_\alpha^\beta = \begin{cases} 0 & \text{if } \alpha \neq \beta \\ 1 & \text{if } \alpha = \beta. \end{cases} \tag{9.15}$$

Because $\alpha = 1, 2, 3$ and $\beta = 1, 2, 3$, there are nine of these constraints.

The Lagrangian will be $G = H - F$ where

$$H = \sum_{\gamma=1}^{N} y^{(\gamma)T} R x^{(\gamma)} \quad \text{and} \quad F = \frac{1}{2} \sum_{\alpha=1}^{3} \sum_{\beta=1}^{3} \lambda_\beta^\alpha \left[\left(\sum_{\gamma=1}^{3} r_\gamma^\alpha r_\gamma^\beta \right) - \delta_\alpha^\beta \right]. \tag{9.16}$$

Recall how multiple constraints are set up in a Lagrangian: Use a linear combination of all the constraints. The λ_β^α represent the nine Lagrange multipliers. We have chosen to index them with α and β. This notation is a convenience that will be useful later when representing the equations in matrix form. Note that the constraint does not change when we interchange α and β. This symmetry implies $\lambda_\beta^\alpha = \lambda_\alpha^\beta$.

How does H depend on the components of R? Noting that $R x^{(\gamma)}$ is just a 3-D vector:

$$R x^{(\gamma)} = \left[\sum_{\beta=1}^{3} r_1^\beta x_\beta^{(\gamma)} \quad \sum_{\beta=1}^{3} r_2^\beta x_\beta^{(\gamma)} \quad \sum_{\beta=1}^{3} r_3^\beta x_\beta^{(\gamma)} \right]^T \tag{9.17}$$

we can fully expand H to get

$$H = \sum_{\gamma=1}^{N} \sum_{\alpha=1}^{3} y_\alpha^{(\gamma)} \sum_{\beta=1}^{3} r_\alpha^\beta x_\beta^{(\gamma)}. \tag{9.18}$$

We will need to take the partial derivatives of G with respect to all nine components of the R matrix. First working with H:

$$\frac{\partial H}{\partial r_i^j} = \sum_{\gamma=1}^{N} \frac{\partial}{\partial r_i^j} \left(\sum_{\alpha=1}^{3} y_\alpha^{(\gamma)} \sum_{\beta=1}^{3} r_\alpha^\beta x_\beta^{(\gamma)} \right) = \sum_{\gamma=1}^{N} y_i^{(\gamma)} x_j^{(\gamma)}. \tag{9.19}$$

The simplification to the final sum follows from the fact that when taking derivatives, only those terms with $\beta = j$ and $\alpha = i$ will contribute.

To simplify our equations, we define a matrix C as follows:

$$\sum_{\gamma=1}^{N} y_i^{(\gamma)} x_j^{(\gamma)} = c_i^j \quad \Rightarrow \quad C = \sum_{\gamma=1}^{N} y^{(\gamma)} x^{(\gamma)\mathrm{T}}. \tag{9.20}$$

There will be nine of these c_i^j values, all derived from the input data.

The partial derivatives of F are a bit more complicated because the R matrix entries appear in a quadratic fashion: $r_\gamma^\alpha r_\gamma^\beta$. We have to apply the product rule when taking derivatives.

$$\frac{\partial F}{\partial r_i^j} = \frac{1}{2} \sum_{\alpha=1}^{3} \sum_{\beta=1}^{3} \lambda_\beta^\alpha \left[\left(\sum_{\gamma=1}^{3} \frac{\partial}{\partial r_i^j} r_\gamma^\alpha r_\gamma^\beta \right) - \delta_\alpha^\beta \right]$$

$$= \frac{1}{2} \sum_{\beta=1}^{3} \lambda_\beta^j r_i^\beta + \frac{1}{2} \sum_{\alpha=1}^{3} \lambda_j^\alpha r_i^\alpha = \sum_{\beta=1}^{3} r_i^\beta \lambda_\beta^j. \tag{9.21}$$

In the last line, the first sum comes from contributions made when $\alpha = j$ and $\gamma = i$. The second sum is from contributions made when $\beta = j$ and $\gamma = i$. The final simplification utilizes the symmetry: $\lambda_\beta^j = \lambda_j^\beta$.

Finally, because the Lagrangian is $G = H - F$:

$$\frac{\partial G}{\partial r_i^j} = 0 \quad \Rightarrow \quad \frac{\partial H}{\partial r_i^j} = \frac{\partial F}{\partial r_i^j} \quad \Rightarrow$$

$$\sum_{\gamma=1}^{N} y_i^{(\gamma)} x_j^{(\gamma)} = c_i^j = \sum_{\beta=1}^{3} r_i^\beta \lambda_\beta^j \quad \forall i, j. \tag{9.22}$$

We can consider these variables to be entries in arrays R, λ, and C defined as

$$R = \begin{bmatrix} r_1^1 & r_1^2 & r_1^3 \\ r_2^1 & r_2^2 & r_2^3 \\ r_3^1 & r_3^2 & r_3^3 \end{bmatrix} \quad \lambda = \begin{bmatrix} \lambda_1^1 & \lambda_1^2 & \lambda_1^3 \\ \lambda_2^1 & \lambda_2^2 & \lambda_2^3 \\ \lambda_3^1 & \lambda_3^2 & \lambda_3^3 \end{bmatrix} \quad C = \begin{bmatrix} c_1^1 & c_1^2 & c_1^3 \\ c_2^1 & c_2^2 & c_2^3 \\ c_3^1 & c_3^2 & c_3^3 \end{bmatrix}. \tag{9.23}$$

By adopting this notation, we can rewrite this last equation in a very succinct fashion as

$$C = R\lambda. \tag{9.24}$$

We know C. How do we solve for λ and then R? We have used the equation $R^T R = I$ to do various simplifications before we created the Lagrangian, but this constraint has not yet been used as a constraint for the Lagrangian analysis itself. So, note:

$$C^T C = \lambda^T R^T R \lambda = \lambda^T \lambda. \tag{9.25}$$

Because $C^T C$ is a square symmetric matrix, we can do an eigendecomposition:

$$\lambda^T \lambda = C^T C = V S^2 V^T. \tag{9.26}$$

If we can use this to find an appropriate λ, then we set $R = C\lambda^{-1}$ and we are done. What is meant by an "appropriate" λ? Selection of λ must be made with due attention to two issues that have not yet been addressed:

1. The rotation matrix must not introduce a reflection that changes chirality.

   • Angles and lengths are preserved, but this is not a guarantee that reflections are avoided.

2. Although we wanted to minimize E, there is nothing in the Lagrange strategy that guarantees this. The procedure could also lead to an R that maximizes E.

   • The Lagrange strategy only gets you critical rotations that produce extreme values of E.

To ensure a proper rotation, we insist that the determinant of the rotation matrix is +1, that is, $\det(R) = 1$. To be sure that E is minimized by our choice of R, we must look more deeply into the quantity

$$H = \sum_{\gamma=1}^{N} y^{(\gamma)T} R x^{(\gamma)} \tag{9.27}$$

to see how its value is determined by the choice of R. In particular, the construction of λ will involve the determination of signs of the square roots of the three entries on the diagonal matrix within VS^2V^T. A very elegant strategy for the computation of R starts with the singular value decomposition (SVD) of C. The theory for an SVD tells us that we can write

$$C = USV^T \quad \Rightarrow \quad C^TC = \lambda^T\lambda = VS^2V^T \tag{9.28}$$

where $S^2 = \text{diag}(s_1^2, s_2^2, s_3^2)$. It is easy to show that $\lambda = V\text{diag}(\sigma_1 s_1, \sigma_2 s_2, \sigma_3 s_3)V^T$

With $\sigma_i = \pm 1$, $i = 1,2,3$. Because $R\lambda = C$, we can write

$$R = C\lambda^{-1} = USV^TV\text{diag}\left(\sigma_1 s_1^{-1}, \sigma_2 s_2^{-1}, \sigma_3 s_3^{-1}\right)V^T$$
$$= U\text{diag}(\sigma_1, \sigma_2, \sigma_3)V^T. \tag{9.29}$$

In summary,

$$C = USV^T \quad \Rightarrow \quad R = U\text{diag}(\sigma_1, \sigma_2, \sigma_3)V^T. \tag{9.30}$$

Now we put this into our H quantity:

$$H = \sum_{\gamma=1}^{N} y^{(\gamma)T}Rx^{(\gamma)} = \sum_{\gamma=1}^{N} y^{(\gamma)T}U\text{diag}(\sigma_1, \sigma_2, \sigma_3)V^Tx^{(\gamma)}$$

$$= \sum_{\gamma=1}^{N}\sum_{k=1}^{3} \sigma_k y^{(\gamma)T}u^{(k)}v^{(k)T}x^{(\gamma)} = \sum_{\gamma=1}^{N}\sum_{k=1}^{3} \sigma_k(y^{(\gamma)} \bullet u^{(k)})(v^{(k)} \bullet x^{(\gamma)})$$

$$= \sum_{\gamma=1}^{N}\sum_{k=1}^{3} \sigma_k(u^{(k)} \bullet y^{(\gamma)})(x^{(\gamma)} \bullet v^{(k)}) = \sum_{\gamma=1}^{N}\sum_{k=1}^{3} \sigma_k u^{(k)T}y^{(\gamma)}x^{(\gamma)T}v^{(k)}$$

$$= \sum_{k=1}^{3} \sigma_k u^{(k)T}\left[\sum_{\gamma=1}^{N} y^{(\gamma)}x^{(\gamma)T}\right]v^{(k)} = \sum_{k=1}^{3} \sigma_k u^{(k)T}Cv^{(k)}. \tag{9.31}$$

The "big dots" indicate the dot product computations. Note that this derivation makes use of a useful formula:

$$U\text{diag}(\sigma_1, \sigma_2, \ldots, \sigma_l)V^T = \sum_{k=1}^{l} \sigma_k u^{(k)}v^{(k)T}. \tag{9.32}$$

The SVD of C tells us that $Cv^{(k)} = u^{(k)}s_k$, and so we get a very concise value for H:

$$H = \sum_{k=1}^{3} \sigma_k u^{(k)\mathrm{T}} Cv^{(k)} = \sum_{k=1}^{3} \sigma_k u^{(k)\mathrm{T}} u^{(k)} s_k = \sum_{k=1}^{3} \sigma_k s_k. \qquad (9.33)$$

Recall: E was minimized when H was maximized, so the best E occurs when $\sigma_i = +1$, $i = 1,2,3$. This gives us

$$R = U\mathrm{diag}(\sigma_1, \sigma_2, \sigma_3)V^{\mathrm{T}} = UV^{\mathrm{T}}. \qquad (9.34)$$

Now that we have the final elegantly simple result that $R = UV^{\mathrm{T}}$, are we finally done? Not quite yet. Remember that we stated that we must have $\det(R) = 1$. It is possible that the matrix C has a singular value decomposition that leads to $\det(UV^{\mathrm{T}}) = -1$. This is called an ***improper rotation***, and it introduces a reflection. We can still get a proper rotation by defining R as

$$R = U\mathrm{diag}(1,1,-1)V^{\mathrm{T}}. \qquad (9.35)$$

Why does this work? The determinant of R has changed sign because the determinant of the diagonal matrix has value -1. Now it is a proper rotation. However, the value of H is $s_1 + s_2 - s_3$, and so it is not as large as $s_1 + s_2 + s_3$. So, we have somewhat compromised E to get a proper rotation.

Note: To get the minimal E under these circumstances, we make sure that s_3 is the smallest of the three values. That is to say, we are assuming s_1, s_2, s_3 are in descending order.

Figure 9.1 shows the results of a structural alignment of 1MBN and 1JEB. Both sets of coordinates were given to the Chimera molecular modeling system, which then created a visualization showing the overlap of secondary structures in rounded ribbon form. Only one of the proteins provided the heme structure.

9.5.5 Summary

Steps for 3-D alignment of proteins P and Q:

1. Determine the subsequences of alpha carbons to be used in the 3-D alignment:

$$M(P) = (p^{(\alpha_1)}, p^{(\alpha_2)}, \ldots, p^{(\alpha_N)})$$
$$M(Q) = (q^{(\beta_1)}, q^{(\beta_2)}, \ldots, q^{(\beta_N)}), \qquad (9.36)$$

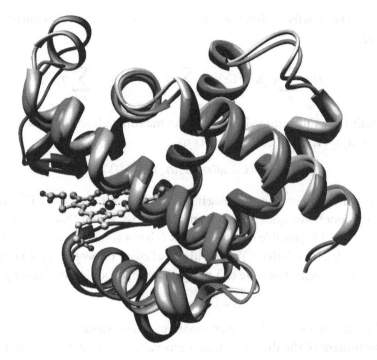

FIGURE 9.1 (See color insert following page 72). Structure alignment of 1MBN and 1JEB. (Courtesy of Shuo (Alex) Xiang, a graduate student at the University of Waterloo.)

2. Calculate centroids $p^{(c)}$ and $q^{(c)}$.

3. Shift the proteins so that centroids are at the origin. We are then working with $x^{(i)}$ and $y^{(i)}$ coordinate sets.

4. Calculate the C matrix and compute its SVD. This gives $C = USV^T$. If necessary, reorder the singular values so that $s_1 > s_2 > s_3$.

5. Compute the rotation matrix

$$R = UV^T. \tag{9.37}$$

6. Check to see if $\det(R) = 1$. If this determinant is negative, then we must redefine the rotation matrix to be

$$R = U\mathrm{diag}(1,1,-1)V^T. \tag{9.38}$$

7. Apply the rotation matrix to the $x^{(i)}$ coordinates.

9.5.6 Measuring Overlap

9.5.6.1 Calculation of the Root Mean Square Deviation (RMSD)

After the rotation matrix is applied to the $x^{(i)}$ coordinates, we can compute the squared distance between each $x^{(i)}$ point and its corresponding $y^{(i)}$ point:

$$d^{(i)2} = \|Rx^{(i)} - y^{(i)}\|^2 . \tag{9.39}$$

Having done this, we can compute the RMSD for that set of corresponding points:

$$\text{RMSD}(P,Q) = \sqrt{\frac{\sum_{i=1}^{N} d^{(i)2}}{N}} . \tag{9.40}$$

One should be careful about the interpretation of this calculation. When P and Q represent different conformations of the same protein and N is the number of atoms of this protein, then the RMSD represents the conventional measure of the similarity of P and Q as measured by degree of overlap. If the RMSD is zero (or very close to zero), we regard P and Q as being identical. If the RMSD is between 1 and 3 Å, then we consider P and Q to be very similar. If the RMSD is over 3 Å, then we typically consider P and Q to have similarity that is low or not present.

If P and Q have the same lengths but some differences in primary sequence, then we cannot compute $d^{(i)2}$ for atoms in corresponding but different residues. This is true because it is impossible to get a correspondence between the atoms in these residues. In this case, one may compute the RMSD with respect to the alpha carbons only. This is still reasonable in that we would have an evaluation of the similarity of the backbone conformations of P and Q.

If P and Q have different lengths, then the RMSD calculation can still be applied to the atoms in the equivalence set. In this case, one must bear in mind that it is **not** measuring overlap across the entire extents of P and Q.

9.5.6.2 RMSD Issues

In a simple calculation of the RMSD, all atoms are treated equally. However, this is not always reasonable because residues in the loops have more positional variation than residues in the hydrophobic core. This consideration

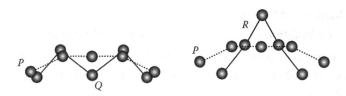

FIGURE 9.2   RMSD and structure alignment.

can apply to the computation of the translation and rotation matrices as well. The reader may work through Exercise 2 to derive the equations that would handle nonuniform weighting of the atoms in the equivalence set.

Another issue is that the significance of the RMSD may vary with protein length. For example, two lengthy proteins P and Q (say 500 residues in length) with a vague evolutionary relationship may produce an RMSD that is over 4 Å, whereas two shorter proteins (say 100 residues in length) may produce an RMSD that is less than 3 Å even though they have the same evolutionary distance between them.

Finally, the best structural alignment is not always achieved by the lowest RSMD. Consider Figure 9.2, which has been constructed to make a point. Suppose we have "2-D molecules" P, Q, and R, and we wish to use an RSMD calculation to determine which one of Q or R is most similar to P. In Figure 9.2, P is distinguished by having its bonds drawn with dotted lines. The overlap of P and Q may be almost exact because the corresponding atoms have a similar physical alignment. As illustrated in the figure, the overlap of P and R has a higher RMSD, but the overall shape of R is more similar to P. One might imagine the situation in 3-D with Q being a helix and R being a loop. Although this figure is somewhat contrived, it gives a rough demonstration that in some circumstances the lowest RSMD may not give the best alignment for the needs of the application. This is especially important for situations involving the structural comparison of distantly related proteins.

9.5.7 Dealing with Weaker Sequence Similarity

The previous topics covered the standard translate and rotate algorithm that can be used when there is a clear and extensive sequence alignment that defines the pairs of atoms that are to be placed into maximal overlap. The problem of similarity evaluation becomes much more of a challenge when a local sequence alignment becomes sketchy, for example, when the two proteins have a distant evolutionary relationship. As we know that structure is more conserved than sequence, it is reasonable to strive for algorithms that determine structural alignment with little or no help from a preliminary sequence alignment.

A local sequence alignment with a low score poses difficulties for the structural alignment. The strategy for overcoming these difficulties will depend on the reason for the low score:

1. ***Lengthy sequences of mismatches.*** In some cases, mismatches may reside in loop regions, whereas the hydrophobic core contains residues that show a higher percentage of matches in the sequence alignment. In such cases, it may be reasonable to simply remove the loop region from consideration and try to maximize the overlap of atoms in the hydrophobic core.

2. ***Presence of gaps in the alignment.*** Gaps in the sequence alignment pose a difficult problem because they indicate a break in continuity of the structural alignment. There are various strategies that can be employed:

   a. By breaking up a protein into fragments, we can try to derive separate structural alignments of fragments on either side of the gap.

   b. We can redefine the pairs of atoms that are to be put into maximal overlap.

The second approach is at the heart of a clever algorithm devised by Russell and Barton (see [RB92]). Their algorithm called STAMP (Structural Alignment of Multiple Proteins) uses a sequence alignment, but only as an initial step. It is one of several algorithms that establish a structural alignment with little or no reliance on a sequence alignment.

9.5.8 Strategies Based on a Distance Matrix

The following is a typical strategy that utilizes a distance matrix to derive a structural alignment:

1. Perform a local alignment of P and Q to get a set of atom pairs that will be used to define the overlap function E (as in Equation 9.4). Work with alpha carbons from a sequence of aligned positions with no gaps.

2. Derive the translation and rotation matrices that ensure maximum overlap for this set of atom pairs.

3. Construct a distance matrix D with axes corresponding to residue positions in each protein and cell $D(i, j)$ holding the Euclidean distance between alpha carbon i in protein P and alpha carbon j in protein Q.

4. Compute a similarity matrix for P and Q. Each entry of the similarity matrix is derived from $D(i, j)$ using some type of formula that produces large positive values of $V(i, j)$ when $D(i, j)$ is small. It produces small or zero values of $V(i, j)$ when $D(i, j)$ is large.

5. Use dynamic programming with a suitable recursion to compute a high score path through the matrix.

The optimal path defines a new alignment for P and Q. Continue with Steps 2, 3, 4, and 5, repeating until convergence is reached (when there is no change in the path computed in Step 5).

It is important to understand that three matrices are involved: the distance matrix D, the similarity matrix V, and the score matrix S that is computed during the execution of the dynamic program algorithm. Let us compare this situation with the Needleman–Wunsch sequence alignment algorithm covered in Chapter 6. There is no counterpart for the D matrix, so we consider V and S. Matrix V corresponds to the matrix that would be obtained by filling a matrix with $|P|$ rows and $|Q|$ columns so that *similarity* value $B(p_i, q_j)$ appears in row i and column j. The score matrix S corresponds to the NW matrix computed using the recursion Equation 6.23.

Several variations of this algorithm can be developed by using different approaches for the conversion of difference $D(i, j)$ into similarity $V(i, j)$. This must then be followed by a suitable dynamic programming algorithm to compute the high score path. For example, Barton and Sternberg [Bs88] use a simple calculation for $V(i, j)$. It is $D(i, j)$ subtracted from the maximum value in D. The dynamic program for Step 5 is essentially the same as Needleman–Wunsch. They report that no gap penalty was necessary. In the paper [Rb92], Russell and Barton use a much more complicated conversion from D to V, and the dynamic program is a modified Smith–Waterman.

There are other possibilities, for example, in the dynamic programming step, we might alter the recursion so that more weight is given to the part of an alignment that corresponds to helix or strand regions.

9.6 ALGORITHMS COMPARING RELATIONSHIPS WITHIN PROTEINS

Several algorithms have been developed to compare geometric relationships within the proteins being compared. The advantage of these *intramolecular* methods is that they avoid the need for an explicit superposition of the two structures. In all cases the algorithms essentially compare some local structures of a protein P with local structures of a protein Q. The notion of *local structure* varies from one algorithm to the next.

We will briefly describe some of the main ideas:

A. *Dali* (**Di**stance **ali**gnment). It does an optimal pairwise structural alignment of protein structures based on the similarity of local patterns extracted from distance maps (see [Hs93]).

B. SSAP (Secondary Structure Alignment Program). SSAP produces a structural alignment using double dynamic programming to generate an alignment of local "views" that are common to both proteins.

9.6.1 *Dali*

Before discussing the computations of the *Dali* algorithm, we should look at an example of a *distance map*. The map is a square matrix of cells that are indexed by the residues of the protein being studied. In simple versions of a distance map, the cell $D[i, j]$ at row i and column j is colored black if the distance between alpha carbon $[i]$ and alpha carbon $[j]$ is less than some particular threshold (say 10 Å), otherwise it is left as white. Note that cell $D[i, j]$ will have the same coloration as cell $D[j, i]$, and so the matrix is symmetric. A more informative map is shown in Figure 9.3.

The colorful triangular image above the main diagonal was obtained from the distance map generated by the server at http://i.moltalk.org. The protein is chain A of a Bovine Rabex-5 fragment cocrystallized with ubiquitin (PDB code: 2FIF). It was chosen because it is a small protein that has a good selection of secondary structural elements: helices and a beta sheet made up of both parallel and antiparallel strands. The distance map is a scatter plot of the distances between all the alpha carbon atoms in the backbone. The plot was done with a distance threshold of 12 Å. The bright yellow pixels of the scatter plot represent very close contacts (less than one half of the threshold distance), whereas the red pixels represent more distant contacts that are still below the threshold. The background

FIGURE 9.3   (See color insert following page 72) Distance map for 2FIF.

"tartan" appearance of the plot arises from the colors given to the secondary structure assignment for each residue. These pixels are colored using the STRIDE colors of yellow for strands, red for helices, and blue for turns (for information on STRIDE see [FA95]). Notice that these colors appear in the complete pixels that are immediately above the diagonal line. Below the diagonal, we have inserted the primary sequence of the protein (not done by the Moltalk server) along with the corresponding residue indices ranging from 1 to 73 inclusive. Below that there is the "strung-out" version of 2FIF with the Chimera visualization of the protein in the lower left corner. These secondary structures have also been given STRIDE colors. The secondary structure elements have been labeled as A, B, C, ..., H. Notice the contributions made by these secondary structures in the distance map. The areas labeled as AB, EF, and EH represent antiparallel beta strands. Their orientation in the distance map is, of course, very similar to the stem structures that we have seen in Chapter 5 on RNA secondary structure. The area labeled BC is caused by the B strand coming near the C helix. Note that this is also seen as an antiparallel interaction and if you follow the direction of increasing alpha carbon indices in the helix, you will see why this is so. There is a sudden backbone turn between helix C and helix D to the extent that D comes up against C thus producing the red area in the distance map labeled with CD. As can be seen in the structure image, strands A and H run parallel to one another, and this is seen in the map as area AH, which runs parallel to the main diagonal but some distance away from it. Finally, it should be noted that the helices C, D, and G produce yellow areas along the diagonal, a typical distance map signature for a helical structure where residues are interacting with other nearby neighboring residues.

The **Dali** algorithm goes through the following steps:

1. Distance maps (stored as matrices) are computed for both P and Q.

2. Extract a full set of overlapped hexapeptide submatrices from each matrix. Each submatrix is a square 6×6 array taken from the distance map. These are also called **contact patterns**.

3. Each of the $(|P| - 5)^2$ contact patterns obtained from the distance matrix for P is compared with the $(|Q| - 5)^2$ contact patterns obtained from the distance matrix for Q. Each contact pattern of P is paired with its most similar partner in Q. This produces a **pair list**.

4. The list is sorted with respect to strength of contact pattern pair similarity. Pairs with low similarity are eliminated.

5. Contact patterns from P are connected to form chains, and those from Q are also connected to form chains. Chain-forming connections are not made arbitrarily. The connections are generated, so that the two chains represent a more extended structural alignment of P and Q.

The last step of the algorithm is quite sophisticated, and all the details are beyond the scope of this text. More information can be found in [Hs93] and [Hs96].

We limit the discussion to an overview of the approach. Figure 9.4 shows two columns of 6×6 contact maps, one column for each of P and Q. Note that the contact maps may describe any type of local contact that we have seen in Figure 9.3: helices, neighboring beta strands, helix–helix contacts, helix–strand contacts, loop–helix contacts, and loop–strand contacts. This gives the algorithm a considerable amount of meaningful data, but it also complicates the detection of a structural alignment. Figure 9.4 has arrows to indicate the pairing of contact patterns considered to be similar.

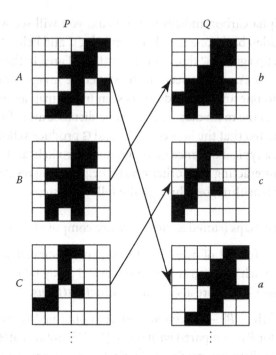

FIGURE 9.4 *Dali* contact patterns.

Figure 9.5 is a very simplified example of the type of processing involved in Step 5 of the algorithm. Within the entire distance map for *P*, adjacent contact patterns *A* and *C* form a short chain, and the adjacent contact patterns *a* and *c* within *Q* also form a chain. However, the concatenation of the patterns has not been done in an arbitrary manner. The similarity of *A* and *a* along with the similarity of *C* and *c* now give us the similarity of

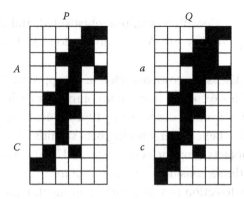

FIGURE 9.5 Combining *Dali* contact patterns.

chain AC and chain ac, this longer similarity representing a more extended structural alignment. This example has shown chain formation as the continuation of adjacent contact patterns, but in practice, chain formation may be realized through the continuation of overlapped contact patterns.

The complexity of the processing arises from the possibility of generating chains in many different ways, not all of them useful in the pursuit of a structural alignment. To increase the chances of meaningful chain formation, the algorithm favors the utilization of contact patterns that are ranked high in the sorting step. Nonetheless, the algorithm must search a very large space of possibilities, and this is facilitated by means of a Monte Carlo algorithm. The Monte Carlo optimizing strategy involves a type of random walk exploration of the search space containing all the chaining possibilities. Moves in this space are randomly chosen. A move corresponds to a change of chain formation and can be evaluated by means of a scoring function. The probability p of accepting a move is given by

$$p = \exp(\beta(S' - S)) \tag{9.41}$$

where S' is the new score and S is the old score. The variable β is a parameter that must be carefully chosen (see [MR53]). Moves with higher scores are always accepted. Moves with poor scores are sometimes accepted, and this helps the algorithm to get out of local minima that may trap the procedure resulting in less satisfactory structural alignments.

Another significant feature of **Dali** is the ability to detect the similarity of two hydrophobic cores even though the secondary structure components of a core have been "rewired." Figure 9.6 (adapted from Figure 2 of [Hs93]) illustrates this. **Dali** will analyze the extended contact pattern and recognize that an interchange of fragments c and b in Q will give a better alignment. After this is done, both sets of patterns are collapsed to a representation that reveals the final structural alignment of the three strands in P with the three strands in Q.

Dali is the structure comparison method used in the FSSP (fold classification based on structure–structure alignment of proteins) database.

9.6.2 SSAP

9.6.2.1 Motivation

If the proteins that are to be compared have primary sequences that are the same or very similar, then the basic superposition algorithm using a rotation matrix can be used for the structure comparison. If the structures

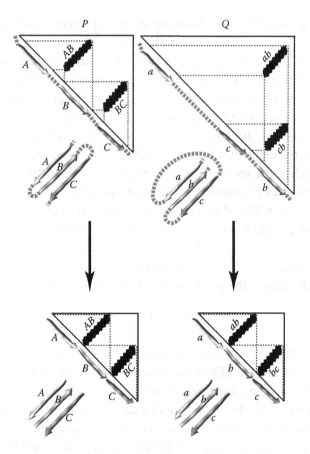

FIGURE 9.6 Reorganizing contact patterns.

are very similar, then the approach still works well, even when there is low sequence similarity. For example, members of the globin family may have low sequence similarity but can show a significant amount of structure alignment when the STAMP algorithm is applied. There may be some missing segments in the alignment, but there is still a large percentage of structural overlap as seen in Figure 9.1.

When the structural similarity of the proteins is less obvious, it becomes much more difficult to specify the equivalent residues in the comparison. For example, there may be two domains in P that are structurally similar to two domains in Q even though the sequence similarity is weak. If the physical separation of these two domains in P is quite different from the separation of the corresponding domains in Q, or if the domains have a very different spatial orientation, then the superposition strategy will not do well because there will be a poor overall fit between the topologically equivalent substructures.

9.6.2.2 Introduction to SSAP

We need a strategy that works with various *local structural* alignments just as the local sequence alignment algorithm depends on local sequence matching. An approach to the structural alignment problem, which cleverly handles insertions and deletions of residues, was developed by Taylor and Orengo [To89] in 1989. Their SSAP algorithm relies on the notions of *local views*. These views are used to create an overall structural alignment by means of *double* dynamic programming.

To motivate the idea of a view, consider Figure 9.7, which contains the backbone drawings for 1MBN and 1JEB. In each case the local view is relative to the proximal histidine just below the heme group. Each view is a set of vectors originating from the alpha carbon of the histidine and going to all other alpha carbon atoms of the protein.

In the figure, only 12 of these vectors have been drawn for the sake of clarity. A visual comparison of the arrows (vectors) in these two diagrams gives an immediate confirmation of the similarity of the local environments "seen" by these two atoms. This type of similar comparison could have been demonstrated for several other pairs of alpha carbons; the proximal histidine is chosen because this residue is highly conserved, and so it is likely that the surrounding atoms have a similar configuration. Table 9.1

TABLE 9.1   Alpha Carbon Atoms for Each View in Figure 9.7

	1JEB		1MBN	
	HIS	87	HIS	93
1	LYS	61	THR	67
2	VAL	62	VAL	68
3	ASN	97	TYR	103
4	VAL	96	LYS	102
5	PRO	95	ILE	101
6	LEU	83	LEU	89
7	LYS	82	PRO	88
8	TYR	89	THR	95
9	ILE	90	LYS	96
10	LEU	91	HIS	97
11	TYR	42	LYS	42
12	THR	39	THR	39

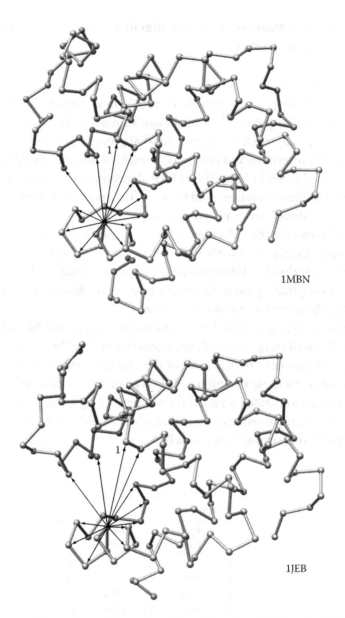

FIGURE 9.7 Views from the proximal histidine in 1MBN and 1JEB.

shows that this similarity in the local views is not due to percentage of residue identity. The table lists the residues in each view starting at the "one o'clock" position. Out of the 12 vector pairs, only three show an identical residue match. Such similarity in views does not persist across the entire protein. For example, substructures in the upper left corner differ in each protein, and so corresponding atoms of P and Q in the helix just below and to the right would have reasonably different views.

The idea of double dynamic programming (see [To89]) is to use a **high-level** dynamic programming algorithm to find a structural alignment that is comprised of the largest number of pairs of similar views. However, the evaluation of the similarity of two particular views is itself an optimization problem. This is solved by using a **low-level** dynamic programming strategy. The score matrix computed at this low level will be derived from a **view level matrix**. There will be many of these matrices. The single high-level matrix is called the **consensus** matrix.

9.6.2.3 Overview of SSAP

Let the sequence of coordinates for the alpha carbon atoms in proteins P and Q be represented by

$$\{p^{(i)}\}_{i=1}^{|P|} \qquad \{q^{(j)}\}_{j=1}^{|Q|}$$

respectively. The steps for the SSAP algorithm are as follows:

1. **Calculate the views.** Calculate a view for each alpha carbon atom of both P and Q. For a particular alpha carbon $p^{(i)}$, the view is a list of vectors. Each vector in the list goes from $p^{(i)}$ to another alpha carbon of the same protein. Formally, the view for $p^{(i)}$ is the set

$$\{p^{(i,r)}\}_{r=1}^{|P|}$$

where $p^{(i,r)}$ designates the vector with origin at $p^{(i)}$ and terminus at $p^{(r)}$. The same notation holds for the view corresponding to $q^{(j)}$:

$$\{q^{(j,s)}\}_{s=1}^{|Q|}.$$

2. **Build the consensus matrix.** For each combination of $p^{(i)}$ and $q^{(j)}$, $1 \leq i \leq |P|$ $1 \leq j \leq |Q|$, compare vector views using a dynamic

programming strategy that fills in a view level matrix with values that are based on the similarity of vectors. The dynamic program will compute an optimal path with a total path score defined as the sum of all view similarity evaluations encountered along the path minus any gap penalties. If the total path score is above a specified threshold, then the alignment scores on the path are added to accumulating similarity evaluations of the consensus matrix.

3. **Compute an optimal path in the consensus matrix.** Using dynamic programming, derive a set of equivalent residues by finding an optimal path in the consensus matrix.

These three steps have described an overview of the algorithm. There are many variations of these steps that address issues such as computational efficiency and the incorporation of structural description. We now review these steps, adding more details and relevant comments.

9.6.2.4 Calculating the Views

Recall that the view for an alpha carbon $p^{(i)}$ is the list of vectors $\{p^{(i,r)}\}_{r=1}^{|P|}$. To give an example of vectors in a view, we can note that the list of atoms in the second column of Table 9.1 includes 12 of the entries that would be found in the view of $p^{(87)}$.

We will eventually discuss how the comparison of two views will involve the comparison of two vectors. This may involve vector lengths, angles, etc. Variations of the algorithm may involve the inclusion of other measures associated with the atom at the terminus of a vector, for example, secondary structure assignment, solvent accessibility, conformational angles, hydrophobicity evaluations, and hydrogen bonding patterns. Some research studies have worked with views that are defined using the beta carbon atoms of a residue.

9.6.2.5 Building the Consensus Matrix

The procedure described in Step 2 has been simplified to state that all combinations of views for $p^{(i)}$ and $q^{(j)}$ with $1 \le i \le |P|$ $1 \le j \le |Q|$ are utilized to generate $|P||Q|$ pairs of views. As described in [TA99], this number can be greatly reduced by filtering the potential pairs. For example, we would not compare the views of $p^{(i)}$ and $q^{(j)}$ if $p^{(i)}$ is in a helix and $q^{(j)}$ is in a strand. However, this secondary structure test requires that we have a reliable assessment of the secondary structure status of a residue.

To avoid the discretization error that this implies, one may resort to a filtering strategy that compares the dihedral angles on either side of $p^{(i)}$ with the corresponding dihedral angles of $q^{(j)}$. Another filtering technique that may be used along with dihedral angle selection is residue exposure (also described in [TA99]). This strategy brings in the notion of residue burial: residues with a similar amount of solvent exposure are considered to be similar. Solvent exposure evaluation is discussed by Aszodi and Taylor in [AT94].

After filtering the entire set of $p^{(i)}$ and $q^{(j)}$ pairs, we will be dealing with considerably fewer than $|P||Q|$ pairs of views. Recall that for each of the surviving $p^{(i)}$ and $q(j)$ pairs, we construct a view level matrix $V^{(i,j)}$. An entry in cell (r, s) of this matrix will be denoted by $V_{r,s}^{(i,j)}$, and it will be a value that specifies the similarity of the vectors $p^{(i,r)}$ and $q^{(j,s)}$.

A straightforward approach is to compare the lengths of the two vectors $p^{(r)} - p^{(i)}$ and $q^{(s)} - q^{(j)}$. However, complications arise when the same positive length is calculated for two vectors going off in opposite directions. To avoid this, the difference vector $p^{(r)} - p^{(i)}$ is established in a frame of reference with the alpha carbon $p^{(i)}$ at the origin. There are several ways to build this reference frame. For example, the three atoms C_α, along with the C and N atoms bonded to it, define a plane. With the origin at C_α, as shown in Figure 9.8, we let the $C_\alpha - N$ bond define the direction of the x-axis. Then the y-axis is in the plane and perpendicular to the direction of the x-axis. Choose the positive direction of y to lie in the same direction as C. Finally, the z-axis is

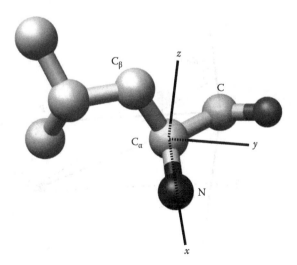

FIGURE 9.8   Local frame set up at the alpha carbon of leucine.

perpendicular to the plane and in the same direction as the C_β atom of the residue attached to C_α. We use the hydrogen atom that replaces C_β in the case of glycine. Each axis is represented by a normalized vector.

By calculating the inner product of $p^{(r)} - p^{(i)}$ with each these orthonormal vectors, we get the coordinates of $p^{(r)} - p^{(i)}$ in this local frame of reference. In the same fashion, a local frame of reference is constructed for $q^{(j)}$, and the coordinates of $q^{(s)} - q^{(j)}$ are calculated with respect to this frame of reference. We can now treat these newly computed coordinates as if they are relative to the **same** frame of reference. By doing this, we have essentially superimposed the $p^{(i)}$ alpha carbon and the $q^{(j)}$ alpha carbon although no rotation or translation operation was used. Suppose $p^{(r)} - p^{(i)}$ in this computed frame of reference is represented by the column vector

$$^P d^{(i,r)} = \left({}^P d_1^{(i,r)}, \quad {}^P d_2^{(i,r)}, \quad {}^P d_3^{(i,r)} \right)^{\mathrm{T}}. \tag{9.42}$$

A visualization of these vector components is shown in Figure 9.9. In a similar fashion, we can let $q^{(s)} - q^{(j)}$ in this frame of reference be represented by the column vector

$$^Q d^{(j,s)} = \left({}^Q d_1^{(j,s)}, \quad {}^Q d_2^{(j,s)}, \quad {}^Q d_3^{(j,s)} \right)^{\mathrm{T}}. \tag{9.43}$$

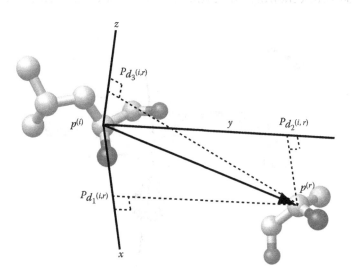

FIGURE 9.9 Coordinates with respect to local frame.

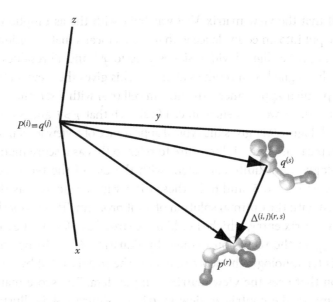

FIGURE 9.10   Difference of vectors in a view.

Then we define

$$\Delta^{(i,j)(r,s)} = {}^{P}d^{(i,r)} - {}^{Q}d^{(j,s)}$$

$$= \left({}^{P}d_1^{(i,r)} - {}^{Q}d_1^{(j,s)}, \quad {}^{P}d_2^{(i,r)} - {}^{Q}d_2^{(j,s)}, \quad {}^{P}d_3^{(i,r)} - {}^{Q}d_3^{(j,s)} \right)^{\mathrm{T}}. \quad (9.44)$$

The significance of $\Delta^{(i,j)(r,s)}$ is shown in Figure 9.10. Note that a small norm for $\Delta^{(i,j)(r,s)}$ would indicate that the view at $p^{(i)}$ is similar to the view at $q^{(j)}$ at least as far as the vectors $p^{(r)}$ and $q^{(s)}$ are concerned. Our goal is to establish more indications of vector similarity for several other values of r and s. The dynamic program at this view level will essentially extract an alignment of the primary sequences in such a way as to maximize the number of similar pairs of vectors $p^{(r)}$ and $q^{(s)}$.

In the original paper [To89], the authors decided to avoid the square roots involved in the calculation of a Euclidean distance in order to save computation time. Consequently, the norm squared is used in place of the vector length. They also needed to convert a measure of difference into a measure of similarity. This was done by using the following hyperbolic formula for the view matrix entries. The entry at cell (r, s) in view matrix $V^{(i,j)}$ is given by

$$V_{r,s}^{(i,j)} = \frac{a}{b + \| \Delta^{(i,j)(r,s)} \|^2}. \quad (9.45)$$

Experiments have determined that $a = 50$ and $b = 2$ give good results. Note that, with these parameter settings, $V_{i,j}^{(i,j)}$ is always $a/b = 25$.

Recall that the view matrix $V^{(i,j)}$ was built with the assumption that $p^{(i)}$ would be put into an equivalence with $q^{(j)}$. This means that all residues prior to $p^{(i)}$ can only be aligned with residues prior to $q^{(j)}$, and all residues after $p^{(i)}$ can only be aligned with residues after $q^{(j)}$. This gives the view matrix $V(i,j)$ a rather peculiar appearance. An entry in cell (r, s) with $r < i$ and $s > j$ will be undefined. As well, an entry in cell (r, s) such that $r > i$ and $s < j$ will be undefined. Figure 9.11 gives the typical appearance of a view level matrix.

It is typical in research literature to refer to this as a score matrix. This terminology is not quite consistent with the use of the term "score" as used in Chapters 4, 5, and 6. In these discussions, a score was the value associated with the optimal solution of a subproblem. This is not the same as a view matrix entry, which should be regarded as playing the same role as $B(p_i, q_j)$ in the recursion defined by Equation 6.23. To be consistent with this terminology, the score matrix is the matrix filled by applying a recursion that uses the view matrix as input data. The score matrix corresponding to the contrived view matrix of Figure 9.11 is illustrated in Figure 9.12.

The dynamic program to compute an optimal path in the view matrix is an adaptation of the dynamic program used for the global sequence alignment problem. We essentially start at location (i, j) of the score matrix $S^{(i,j)}$ and fill in the bottom right submatrix using the following recursion, which is valid for $r > i$ and $s > j$.

Base cases:

$$S_{i,s}^{(i,j)} = \frac{a}{b} - g(s - j)$$

$$S_{r,j}^{(i,j)} = \frac{a}{b} - g(r - i).$$

(9.46)

	W	A	T	E	R	M	A	N
N	1	1	2	2				
E	14	2	1	2				
E	1	4	12	1				
D					25			
L						3	1	2
E						8	2	1
M						11	1	3
A						1	11	3
N						2	3	14

FIGURE 9.11　A typical view matrix.

	W	A	T	E	R	M	A	N
N	39	31	25	19	13			
E	(43)	35	29	23	17			
E	25	(29)	(33)	26	21			
D	9	13	17	(21)	(25)	21	17	13
L					(21)	28	24	20
E					(17)	29	30	26
M					13	(28)	30	33
A					9	24	(39)	35
N					5	20	35	(53)

FIGURE 9.12   A score matrix for the view matrix of Figure 9.11.

Recurrence for cell (r, s):

$$S_{r,s}^{(i,j)} = \max \begin{cases} S_{r-1,s-1}^{(i,j)} + V_{r,s}^{(i,j)} \\ S_{r-1,s}^{(i,j)} - g \\ S_{r,s-1}^{(i,j)} - g. \end{cases} \qquad (9.47)$$

Then we start at location (i, j) of the score matrix $S^{(i,j)}$ and fill in the top left submatrix "going in the opposite direction" using the following recursion, which is valid for $r < i$ and $s < j$:

Base cases:

$$S_{i,s}^{(i,j)} = \frac{a}{b} - g(j - s)$$

$$\qquad (9.48)$$

$$S_{r,j}^{(i,j)} = \frac{a}{b} - g(i - r).$$

Recurrence for cell (r, s):

$$S_{r,s}^{(i,j)} = \max \begin{cases} S_{r+1,s+1}^{(i,j)} + V_{r,s}^{(i,j)} \\ S_{r+1,s}^{(i,j)} - g \\ S_{r,s+1}^{(i,j)} - g. \end{cases} \qquad (9.49)$$

After the matrix is filled, we locate the maximum element in the upper left submatrix and initiate a trace-back from this cell to cell (i, j). Similarly, starting at the maximum element is the lower right submatrix, we

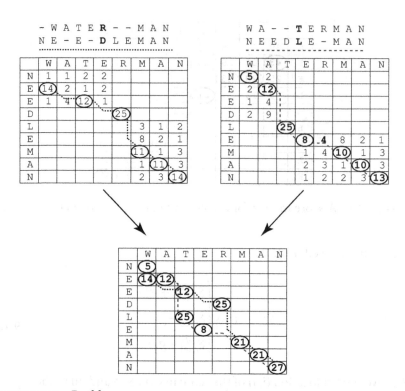

FIGURE 9.13 Building consensus.

initiate another trace-back that ends in cell (i, j). In Figure 9.12, each score in the trace-back path is set in bold font and enclosed by an ellipse. In the examples given, we are working with a gap penalty of $g = 4$.

The score for this path is the sum of the two maximum scores (96 in the score matrix of Figure 9.12). If this path score is above a preselected threshold, then the path elements $V_{r,s}^{(i,j)}$ that represent matches in the alignment are added to the corresponding elements of the consensus matrix. This is shown in Figure 9.13. The first matrix represents $V^{(4,5)}$, the view for $p^{(4)}$ and $q^{(5)}$. The alignment corresponding to the optimal path (marked by a black dotted line) is given just above the matrix. The second matrix represents $V^{(5,3)}$. Its optimal alignment is also shown above the matrix, and the path is designated with a gray dashed line. Of course, we will need many such additions to the consensus matrix before we are ready for the next step.

9.6.2.6 Compute the Optimal Path in the Consensus Matrix

After the consensus matrix has been constructed, an optimal path is derived using a conventional Smith–Waterman algorithm. The result of the SSAP algorithm is an alignment that gives an equivalence set for the various segments of *P* and *Q* that are presumed to have structural similarity.

Once again, it should be stressed that the description of SSAP that was just given is only one of several variations of the algorithm, all of which may employ the double dynamic programming strategy.

9.7 EXERCISES

1. To do this exercise you should write one or more programs that will compute the maximum superimposition of two given proteins. Show the capabilities of your software on the following two proteins:

   PDB code: 1E1G   Human prion protein

   PDB code: 1B10   Syrian hamster prion protein

   The following steps should be accomplished:

   a. Derive a sequence alignment for the proteins so that you get a list of amino acid pairs, each pair consisting of the amino acids that are to be "matched" in the superimposition step. The sequence alignment can be done with a server such as LALIGN.* Use your own judgment to determine whether you require a local or global alignment.

   b. Working with the coordinates of the matching alpha carbons in the backbone of the protein, compute the centroids of each protein. Be sure to use only those alpha carbons that are in the amino acids that are involved in the pairings extracted from the sequence alignment.

   c. Do the translate operation so that both proteins have their alpha carbon centroids at the origin.

   d. Use the algorithm described in this chapter to define a rotation matrix that will bring the alpha carbons of one protein into superimposition (as much as possible) with the corresponding alpha carbons of the other protein.

   e. Once you get the rotation matrix, apply it to all the atoms in the protein.

   f. Compute the RMSD of the superimposition.

---

* http://www.ch.embnet.org/software/LALIGN_form.html.

g. Provide a graphic that shows the extent to which the backbone atoms of the proteins have been overlapped. You could use an application such as MATLAB to simply draw a 3-D view of the chain of backbone bonds of both proteins. A more compelling picture would be to use Chimera to visualize both backbone ribbons in the same scene, each colored with a different color (for example: cornflower blue for 1E1G and magenta for 1B10.)

2. In Section 9.5.4, all the alpha carbons in the structural alignment were given equal weighting. In other words, we were required to find a rotation matrix R and a translation vector T such that the E score defined in Equation 9.4 was minimized. However, some researchers prefer to do a superimposition in which some atoms are given a higher weighting than others. For example, they might place more weight on residues in the hydrophobic core and less weight on residues in the loops. In this general type of structural superimposition, one is required to find a rotation matrix R and a translation vector T such that

$$E_w = \frac{1}{2} \sum_{i=1}^{N} w_i \, \|Rp^{(\alpha_i)} + T - q^{(\beta_i)}\|^2$$

is minimized where one is given a vector w with components w_i, $i = 1, 2, \ldots, N$ each a nonnegative real number between 0.0 and 1.0 inclusive. In this problem, you are required to derive the formulas that are used in the computation of R and T such that E_w is minimized. In particular, your derivation should go through the same steps as those done in Section 9.5.4 for the minimization of E, covering issues such as

- The computation of the centroid (be careful here!)

- The reformulation of E_w that leads to the appropriate definition of H_w

- The formulation of the Lagrangian and subsequent calculation of partial derivatives

- The equation that shows how R is calculated

- A summary that presents the computational steps needed to find R

3. Write a program that implements structural alignment based on a distance matrix. Use the following data sets:

   a. $P \leftarrow$ 1MBN, $Q \leftarrow$ 1JEB.

   This pair will be the easiest to structurally align. A translate and rotate should produce a structural overlap similar to that seen in Figure 9.1.

   b. $P \leftarrow$ 6PAX, $Q \leftarrow$ 1PDN.

   This pair presents more of a challenge and will require a more powerful strategy such as the STAMP algorithm.

   c. $P \leftarrow$ 4FXN, $Q \leftarrow$ 3CHY.

   d. $P \leftarrow$ 1MBC, $Q \leftarrow$ 1CPC (Chain A).

   The last two pairs have been used in various research studies (see, for example, [TA97]). The proteins in both pairs have similar folds for their hydrophobic cores. However, in each case the sequence similarity is rather low. A simple translate and rotate strategy will do very poorly for both of these pairs. Part of this exercise will be to see how well you can do with c and d using your algorithm. Both c and d will be back in the last exercise and at that time you will be applying double dynamic programming to do the comparison. For data set b, try to get two different structural alignments, one for each of the two different domains in the proteins. For data sets c and d, generate at least two different structural alignments with the objective of generating structural alignments that allow you to make reasonable statements about what is structurally conserved. For example, specify the beta strands and helices of protein P that are overlapped with corresponding beta strands and helices of protein Q for some structural alignment initiated by means of some particular local sequence alignment.

4. Section 9.6.2.5 gives a description of the computation of $\Delta^{(i,j)(r,s)}$. However, some of the details have been left out. Provide these details. In particular, how do you calculate the vectors representing the orthonormal axes at $p^{(i)}$ and $q^{(j)}$ when given the coordinates of these atoms and the other atoms that are bonded to them. Finally, having done this, write a program that can compute $\Delta^{(i,j)(r,s)}$ when given the needed input data.

5. Program the SSAP algorithm. Test the program using the data set described in Exercise 3. Explain any differences in what is observed when comparing the output of your double dynamic program with that of your algorithm from Exercise 3.

REFERENCES

[At94] A. Aszódi and W. R. Taylor. Secondary structure formation in model polypeptide chains. *Protein Engineering, 7* (1994), 633–644.

[Bs88] G. J. Barton and M. J. E. Sternberg. LOPAL and SCAMP: Techniques for the comparison and display of protein structures. *J. Molecular Graphics, 6* (1988), 190–196.

[Bo96] N. P. Brown, C. A. Orengo, and W. R. Taylor. A protein structure comparison methodology. *Computers and Chemistry, 20* (1996), 359–380.

[Ds04] A. V. Diemand and H. Scheib. iMolTalk: An interactive, Internet-based protein structure analysis server. *Nucleic Acids Res., 32* (2004), W512–516.

[Ej00] I. Eidhammer, I. Jonassen, and W. R. Taylor. Structure comparison and structure patterns. *Journal of Computational Biology, 7* (2000), 685–716.

[Fa95] D. Frishman and P. Argos. Knowledge-based protein secondary structure assignment. *Proteins, 23* (1995), 566–579.

[Hs93] L. Holm and C. Sander. Protein structure comparison by alignment of distance matrices. *J. Mol. Biol., 233* (1993), 123–138.

[Hs96] L. Holm and C. Sander. Mapping the protein universe. *Science, 273* (1996), 595–602.

[Ho97] G. 't. Hooft. *In Search of the Ultimate Building Blocks*, Cambridge University Press, Cambridge, U.K., 1997.

[Ka76] W. Kabsch. A solution for the best rotation to relate two sets of vectors. *Acta Crystallographica, A32* (1976), 922–923.

[Ka78] W. Kabsch. A discussion of the solution for the best rotation to relate two sets of vectors. *Acta Crystallographica, A34* (1978), 827–828.

[Mo07] S. Meier and S. Özbek. A biological cosmos of parallel universes: does protein structural plasticity facilitate evolution? *BioEssays, 29* (2007), 1095–1104.

[Mr53] N. Metropolis, A. W. Rosenbluth, M. N. Rosenbluth, A. H. Teller, and E. Teller. Equation of state calculations by fast computing machines. *J. Chem. Phys., 21* (1953), 1087–1092.

[Oj03] C. A. Orengo, D. T. Jones, and J. M. Thornton. *Bioinformatics: genes, proteins and computers,* BIOS Scientific Publishers, London, 2003.

[Rb92] R. B. Russell and G. J. Barton. Multiple protein sequence alignment from tertiary structure comparison: Assignment of global and residue confidence levels. *Proteins, 14* (1992), 309–323.

[SB06] M. SAGERMANN, W. A. BAASE, AND B. W. MATTHEWS. Sequential reorganization of β-sheet topology by insertion of a single strand. *Protein Science,* **15** (2006), 1085–1092.

[TA97] W. R. TAYLOR. Multiple sequence threading: An analysis of alignment quality and stability. *J. Mol. Biol.,* **269** (1997), 902–943.

[TA99] W. R. TAYLOR. Protein structure comparison using iterated double dynamic programming. *Protein Science,* **8** (1999), 654–665.

[TO89] W. R. TAYLOR AND C. A. ORENGO. Protein structure alignment. *J. Mol. Biol.,* **208** (1989), 1–22.

[VB96] I. R. VETTER, W. A. BAASE, D. W. HEINZ, J. P. XIONG, S. SNOW, AND B. W. MATTHEWS. Protein structural plasticity exemplified by insertion and deletion mutants in T4 lysozyme. *Protein Science,* **5** (1996), 2399–2415.

[YG05] Y. YE AND A. GODZIK. Multiple flexible structure alignment using partial order graphs. *Bioinformatics,* **21** (2005), 2362–2369.

[YB05] X. YUAN AND C. BYSTROFF. Non-sequential structure-based alignments reveal topology-independent core packing arrangements in proteins. *Bioinformatics,* **21** (2005), 1010–1019.

CHAPTER 10

Machine Learning

... the reality might be hard or even impossible to model (and understand) accurately, but there still can be ways to make the good decisions. One consequence of this idea is that, when one infers a model from data, one should judge the quality of the model not from the point of view of how it fits the "true model," but from the point of view of how good are the decisions one makes based on this model, and the goodness of a decision is not measured as how different it is from an hypothetical "ideal decision," but should be measured by how much return one gets, compared to the best possible return.

DR. OLIVIER BOUSQUET

10.1 MOTIVATION

A guiding principle of biology is that sequence defines structure, which in turn defines functionality. We start this chapter with the observation that these three topics—sequence, structure, and functionality—are progressively more difficult to model when computational problems are to be solved.

The *representation* of a biological sequence can be done with utmost precision. It is easy to construct data structures that represent the sequence of nucleotides in a DNA molecule. The same may be said for protein sequences. The mathematical modeling of structure is more of a challenge. We deal with coordinates of atoms that are derived from a procedure compromised by some error. For example, the positions of atoms in protein loops are often more difficult to assess than those that are in core helices. In addition to this, there is the issue of protein flexibility and the need to represent in some way the dynamic range of possible protein conformations. Consequently, our data structures become much more complicated, and simplifying assumptions are frequently made to facilitate an algorithmic solution even though these assumptions weaken the reality of the model. The assumption of molecular rigidity in the docking of a ligand with a protein is a good example of such a simplification. Because of our need to make "assumptions of convenience," we produce a mathematical model that is further divorced from the true behavior of the biological system under study.

The dependency of ligand docking on molecular structure and flexibility is only a simple example of a much larger challenge: How do we adequately model the functional behavior of molecules in a cell? The complexity of this problem ranges from the somewhat difficult issue of ligand docking to the much more complicated one of cellular modeling, which attempts to describe the simultaneous molecular interactions taking place in thousands of different biological pathways.

Let us consider the impact of this complexity on our problem-solving efforts. Recall from Chapter 1 that functional simplicity allows us to build a powerful methodology characterized by transparency and high accuracy. Using our planetary system as an example, we saw that gravitational laws could be used to formulate models with equations that described planetary motion with exquisite accuracy. The transparency attribute refers to the fact that we could look into the model and fully understand how each part of it contributes to the solution of some problem such as the prediction of planetary positions in the future. The success of space exploration vehicles going to distant planets is a convincing testament that such problems are indeed solvable. Unfortunately, biological systems have such an onerous complexity that we cannot build models based on relatively simple postulates and natural laws. For these systems, the "best" formulation of the Science block in Figure 1.1 (Chapter 1) poses a severe challenge. The luxury of high accuracy, fast computation, and model transparency are, to a significant extent, beyond our grasp.

Machine learning offers us a strategy to build models that facilitate prediction, even though such models are supported by little, if any, scientific justification in the form of postulates and laws. The content of the Science block consists of a *learning* model that can be trained. Training is done by a learning algorithm that takes the observed inputs and output of the natural process, performs calculations on this data, and then sets parameters within the model. If there is enough data and if the natural process is sufficiently "well behaved," the algorithm generates a model that has the power to *generalize*. A high level of generalization is achieved when, given a value that represents the

x input, the model can give a reasonable prediction of the output value for $f(x, x')$. Before getting into the details of this strategy, we review some of the issues that make modeling of biological systems complicated.

10.2 ISSUES OF COMPLEXITY

Difficulties in the mathematical modeling of a complex system can arise due to one or more of the following challenges:

1. Computational scalability

2. Intrinsic complexity

3. Unavailable knowledge

10.2.1 Computational Scalability

We might have a very accurate mathematical model that describes how a natural process works when it is a small system or some localized domain within a larger system. It might even be possible to have algorithms that can utilize this model and within a reasonable amount of time, predict the behavior of the system. We encounter scalability problems when this mathematical model fails to work for larger systems. This will happen if the required computational time rises so quickly relative to system size that it becomes prohibitive.

An example would be the assessment of molecules to determine whether they will properly bind to some active site in a protein. We could get a reasonable answer to this question by using an algorithm that attempts to dock the molecule in the binding site. Suppose we have a docking procedure that takes one second (a very optimistic estimate) to dock one molecule. This will be a good strategy for a few molecules. However, it is not reasonable when applied to a library of 30 million molecules because there are only 31,556,926 seconds in a year, and so this processing will take almost a full year. Therefore, a different strategy is needed.

10.2.2 Intrinsic Complexity

Biological systems typically have an overwhelming amount of detail with thousands of different interactions taking place simultaneously. Tracking the huge number of protein–protein and protein–ligand interactions taking place in a cell over its lifetime is a severe challenge. The characterization of these events and their appropriate organization into a meaningful context is an important but very difficult task.

Even if we narrow our attention to a very small system, such as the interaction between a protein-binding site and some ligand, we still get a considerable amount of complexity: avoidance of steric clashes, the analysis of charge distributions, hydrogen bond formation, solvation effects, entropic energy changes, etc. In an attempt to make a reasonable simplification, a docking algorithm may blithely disregard water molecules in a binding site when, in fact, there are examples in which the proper placement of a water molecule is a key step in a successful binding activity (see [Pc97]).

In some situations, it is very difficult to make a reasonable simplifying assumption. A process may depend on a large number of independent variables (often referred to as *features*), and in the interest of reducing the complexity of a mathematical model, it may be necessary to eliminate some of the variables that have little or no impact on the model. There are many ways to undertake feature selection, and it must be done very carefully. If the relationship between some observed output and the many features is a linear one, then there are techniques such as principal components analysis that can be applied. If the relationship is nonlinear, it is possible that some seemingly insignificant feature is actually the most critical in determining the output of a process and then its elimination would be detrimental.

10.2.3 Inadequate Knowledge

In some situations, we simply do not have enough detailed knowledge about a biological system to build a mathematical model that faithfully represents it. This may be due to the difficulty in obtaining the data. For example, getting data that describes the exact conformation of a membrane protein is very important for the pharmaceutical industry because at least half of the membrane proteins make worthwhile drug targets. However, protein solubilization prior to crystallization typically modifies the structural conformation and makes it different from the native conformation that is determined to some degree by the membrane in which it is embedded. While an estimated 30% of a typical cell's proteins are membrane proteins, they comprise less than 1 in 100 of the structures deposited in the Protein Data Bank (PDB; as of 2008).*

---

* For readers interested in membrane proteins, a good starting point is http://blanco.biomol. uci.edu/Membrane_Proteins_xtal.html.

10.3 PREDICTION VIA MACHINE LEARNING

Because the detailed modeling of complex systems is very difficult, we are forced to adopt another approach to the problem. We consider the natural process to be a "black box." By saying that we cannot see into the box, we are essentially admitting that we do not fully understand the exact behavior of the natural processes at work inside it. However, we do have the power to gather measurements or outputs from the box, and depending on the nature of the problem, we may be allowed to provide the box with some input and then observe the corresponding output.

For example, we might assume that the black box contains the natural processes that are responsible for a complex biological process such as protein folding. We would give the black box the primary sequence of a protein and it provides, as output, the coordinates of all atoms in the tertiary structure. In this case, the box is assumed to contain the natural processes that can fold a protein and then report the same results that we can get from the PDB. Similar to the PDB, it cannot tell us anything about the biophysics of the natural folding process that produced the tertiary structure.

In machine learning, our objective is to design and implement a learning algorithm that can observe the inputs and outputs of the black box and "learn" its behavior. The strategy is described in Figure 10.1, which has been adapted from Figure 1.1. (Chapter 1). We see that the main component of the SCIENCE box is an *artificial* model that contains the hypothesis function of Figures 1.1 and 1.3 in Chapter 1. The choice of such a model is usually stipulated by applying some machine-learning strategy. Several such methods are available: neural networks, decision trees, classifiers, support vector machines, expectation-maximization algorithms, and random forests, to name a few.

As seen in Figure 10.1, the learning algorithm may be designed to take advantage of some form of domain knowledge. This typically amounts to extra data or some set of calculations that specializes the algorithm for the learning task that is to take place. For example, if it is to predict the binding affinity of a ligand in a protein-binding site, the learning algorithm may work with various biophysical parameters that are incorporated into the calculations that formulate the prediction. To mention another example, the researcher may know that some natural processes involve a dependency that can be modeled as an exponential function. It may be possible to incorporate such a function as part of a regression strategy. Very often, such domain knowledge may be limited or very complicated, and so there is often no attempt to build it into the artificial model.

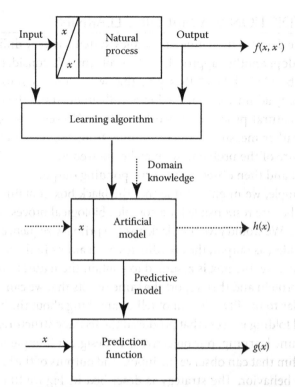

FIGURE 10.1 Machine learning.

The central idea is to have an artificial model that contains an infinite number of hypothesis functions. If the space is rich enough, it will contain one or more functions with a very small hypothesis error (see Figure 1.2, Chapter 1). In practice, we will never actually discover the best $h(x)$ function in this space. In Figure 10.1, we represent this fact by using dotted arrows for input and output of the artificial model. More precisely, the artificial model will be given a sample set of x vectors. However, as explained earlier, domain knowledge may be difficult to incorporate. Consequently, it may be impossible to find an exact representation of the optimal function $h(x)$. Our strategy is to use a learning algorithm that will calculate $g(x)$, an approximation to the optimal, but unknowable, $h(x)$. By calculating various parameters, the learning algorithm will specify some particular $g(x)$ that is a member in the space of all possible $h(x)$ functions. After the learning phase is completed, the prediction function $g(x)$ can be used on hitherto unseen values of x and, for each, it will generate a $g(x)$ value that is an approximation of the optimal $h(x)$ which in turn is an approximation of $f(x,x')$. As already described in Section 1.6, this approximation of an approximation involves two sources of error (see Figure 1.3).

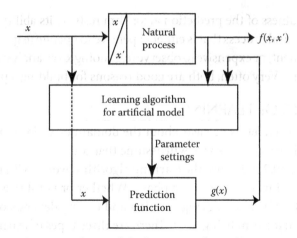

FIGURE 10.2   Training phase of the learning algorithm.

10.3.1   Training and Testing

Initially, the learning algorithm goes through a *training* phase. This is illustrated in Figure 10.2. In this figure, we assume that the learning algorithm is specialized for a particular artificial model. Its main objective is to set parameters that will specify $g(x)$ the best predictor for the optimal hypothesis function hx within the infinite hypothesis space that is proffered by the artificial model. During training, it has access to a training set comprised of a (hopefully) large selection of data samples of the form $\{(x^{(i)}, f^{(i)}) \mid i = 1, 2, ..., l\}$. During training, the algorithm has access to these data samples, and it will typically pass x values to the prediction function, get back $g(x)$ predictions, and make adjustments to the parameters after comparing the $g(x)$ values with the corresponding $f(x,x')$ outputs.

After the training phase is completed, the success of the algorithm can be evaluated in a *test* phase that compares predictions against the outputs of the black box. Testing is done by giving the learning algorithm inputs that are part of a dataset designated as a test suite. Data in the training suite should never reappear in the test suite. We are attempting to measure the learning success of the algorithm by measuring its ability to generalize, not its ability to recall previously seen examples.

If we are satisfied with the test results of the learning algorithm, it can be used as a facility that provides predictions corresponding to future inputs that have never been seen during training or testing phases. Therefore, we should consider the testing phase as providing a measure of confidence that indicates the likely success of the prediction function when it handles previously unseen x inputs.

The usefulness of the prediction algorithm rests on its ability to cheaply replace a natural process that is either expensive to operate (e.g., some scientific experiment) or expensive to observe (e.g., doing x-ray analysis of crystallized protein). Very often, both are good reasons for building a predictor.

10.4 TYPES OF LEARNING

From the outset, let us be clear about the domain of x. For most applications, x will be a vector, so we will assume that $x \in \mathbb{R}^n$.

As Figure 10.2 indicates, the learning algorithm works with the incoming x vectors during the training phase. Whether or not it sees the corresponding $f(x, x')$ values during training and testing depends on the type of learning algorithm being used. There are three types of learning strategies, described as follows:

1. Supervised learning: During the training phase, the supervised learning algorithm observes both the x vector and the corresponding $f(x, x')$ value as shown in Figure 10.2. During the testing phase, the generated predictions or $g(x)$ values are compared with the "true" $f(x, x')$ values as shown in Figure 10.3.

2. Semisupervised learning: During the training phase, the supervised learning algorithm observes both the x vectors and the corresponding $f(x, x')$ values as shown in Figure 10.2. In addition to this, it is given the x vectors that will be used during the testing phase. The algorithm is *not* given the $f(x, x')$ test results during its training phase. During the testing phase, the $g(x)$ predictions are compared with the true $f(x, x')$ values as shown in Figure 10.3. Although it is a fascinating topic, we will not cover semisupervised learning in this text.

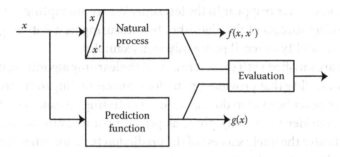

FIGURE 10.3   Testing the learning algorithm.

3. Unsupervised learning: In this type of learning, there is no training phase. The algorithm is given all the x vectors, which are now
considered as points in $\mathbb{R}^n$, along with a metric that can evaluate
distances between these points. It then undertakes a procedure that
uses the metric to partition these points into clusters, the success of
which can be evaluated by considering the average intercluster distances and comparing them with the average intracluster distances.
No $f(x,x')$ prediction is involved in clustering and so there is no need
to provide these data to the algorithm.

10.4.1 Types of Supervised Learning

Our immediate concern will be supervised learning because it most clearly
attempts to learn the behavior of the black box process with the objective
of generalization, that is, predicting outputs for previously unseen x values. Two scenarios fall into this type of learning:

Regression: In regression problems, the $f(x, x')$ outputs of the natural
process are considered to be real values and the learner is required to
provide a prediction $g(x)$ that matches them as closely as possible.

Classification: In classification problems, the $f(x, x')$ outputs are discrete values that are members of a finite set of labels. In binary classification, these labels are often represented as the values +1 and −1.

In a sense, both scenarios represent a type of pattern analysis problem. The
learning algorithm must find regularities or some relationship among the
components of the x vectors that will enable it to learn how it can compute
the best prediction. For most learning algorithms, this will amount to the
determination of a set of parameters that are used in some systematic way
to compute the prediction when given an x vector. In doing so, the algorithm is really generating a reasonable hypothesis function that attempts
to predict the output of the natural process.

10.4.2 Supervised Learning: Notation and Formal Definitions
Training

The input to our learning machine will be a finite set of samples known as
the *training set*. We will use a parenthesized superscript on the x variable
to provide an index for elements of the training set. There will be l vectors
in the set, so the training set is represented by the notation

$$\{x^{(i)}\} \quad i = 1, 2, \ldots, l.$$

When a sample vector $x^{(i)}$ is applied to the input of the physical process under study, it produces an output value y_i. For regression problems, y_i is a real value; for classification problems, y_i is a class label. The learning algorithm will accept the training set as input, and it will generate a function $g(x)$ that predicts the y values.

Testing

We then use another set called the *test set* along with a loss function $L(y, g(x))$ that measures the discrepancy of the predictor $g(x)$. The function L maps $Y \times Y$ to $\mathbb{R}_0^+$, the nonnegative numbers of the real line. A typical loss function is

$$L(y, g(x)) = [y - g(x)]^2. \qquad (10.1)$$

The objective of the training phase is the creation of a $g(x)$ function that will minimize L for the test data. More precisely, we typically want a $g(x)$ that minimizes the sum of the L evaluations across the entire test set (in this case, "minimizing sum of squares"). Of course, when the learning algorithm is processing data to generate $g(x)$, it will not have access to the test data, and so the training phase goal is the minimization of the sum of squares across all the training data. You can get a preview of this requirement by looking at Equation 10.4. Our prediction strategy relies on the reasonable hope that a successful predictor for the training data will also be a successful predictor for the test data if both test data and training data come from the same source. As noted earlier, this ability to work successfully with the test data after processing only the training data is called generalization.

10.5 OBJECTIVES OF THE LEARNING ALGORITHM

So far, we have discussed training, testing, and loss functions. We have not considered how much training data we will need and, most importantly, we have yet to discuss any of the algorithms that work in this setting. Before doing so, some goals should be kept in mind as we cover these topics. These are listed as follows:

Computational scalability: This was mentioned at the beginning of the chapter. Our algorithm must be capable of handling large data sets in a timely fashion. A more precise expression of this requirement is that the algorithm should be efficient, that is, it must have an execution time that is a polynomial function of the size of the training set.

Resistance to noisy data: As the training set usually arises from observations of a physical process, it is subject to measurement errors and incorrect values (outliers in the data set). The learning algorithm should be robust in the sense that it generates a reliable predictor even if the training set is somewhat compromised by noise

Stability: As discussed earlier, the learning algorithm must generate an accurate prediction for new data vectors that it has never seen during the training phase. Also, the most significant property of the learning algorithm will be its ability to "generalize, not just memorize." This ability is often referred to as *statistical stability*. Here is another way to describe it: Suppose our physical process can produce several training sets. Let us assume they have data that is independent and identically distributed (i.i.d.). Under these conditions, the learning algorithm should ideally produce the same predictor for each training set. We then say that the algorithm is *stable*. In other words, we want learning to be sensitive to the properties of the data source (the process under study) but not sensitive to a particular training data set.

The last goal is very important. Why is this? Good statistical stability should give a high level of success for our predictor. Assuming the test set has the same statistical properties as the training set, stability will help to ensure that low error on the training data carries over to low error on the test set. From a practical perspective, this means that we will formulate a predictor that will minimize the loss function (i.e., minimize error) on the training set. We then rely on statistical stability to ensure that the prediction error on the test set (plus any other future data inputs following the same i.i.d. properties as the training set) is also low.

10.6 LINEAR REGRESSION

We will start with linear regression as the first topic in supervised learning. Compared to classification, the mathematical underpinnings for linear regression are somewhat easier to explain, and so we will deal with classification later. Linear regression will also introduce kernel methods, which will be our main strategy for supervised learning when the predictor must be a nonlinear function. Finally, it has applicability to quantitative structure–activity relationship (QSAR) problems. For QSAR problems, we want to predict affinities of molecules in binding sites when given data that represent their molecular descriptors.

Without loss of generality, we will assume that the data have been centered at the origin. This can be done by shifting the data so that their centroid is at the origin. In other words, we have replaced each $x^{(i)}$ with $x^{(i)}$ minus the "average" $x^{(i)}$:

$$x^{(i)} \leftarrow x^{(i)} - \frac{1}{l}\sum_{j=1}^{l} x^{(j)}. \tag{10.2}$$

Because the data is centered, our predictor will be a homogenous real-valued function $g(x) = \langle x, w \rangle = x^T w$. We want the g(x) that gives the best interpolation of a given training set R = $\{(x^{(1)}, y_1), (x^{(2)}, y_2), \ldots, (x^{(l)}, y_l)\}$, where each column vector $x^{(i)} \in \mathbb{R}^n$ and each $y_i \in \mathbb{R}$. We construct the matrix X from the $x^{(i)}$ column vectors:

$$X = [x^{(1)} \quad x^{(2)} \quad \ldots \quad x^{(l)}]. \tag{10.3}$$

If $l = n$ and X is invertible, then we could simply find w by using the matrix, vector equation $X^T w = y$, and solving to get $w = (X^T)^{-1} y$. However, it is typical that $l \gg n$, and we need a "best fit" w to obtain a hyperplane that goes through the data. We solve this minimization problem with some calculus. We start with a loss function L:

$$
\begin{aligned}
L(w, D_R) &= \sum_{i=1}^{l} [y_i - g(x^{(i)})]^2 \\
&= \sum_{i=1}^{l} [y_i - x^{(i)T} w]^2 = \|y - X^T w\|^2 \\
&= y^T y - y^T X^T w - w^T Xy + w^T XX^T w.
\end{aligned}
\tag{10.4}
$$

Taking the derivative of L with respect to w (see Appendix D if you need help with this), we get

$$\frac{\partial L}{\partial w} = 0 \Rightarrow XX^T w = Xy. \tag{10.5}$$

This gives us the usual *normal equations for a least squares fit*. If XX^T is invertible, then $w = (XX^T)^{-1} Xy$. To interpolate (predict) a new value, we would simply use

$$g(x) = x^T w = x^T (XX^T)^{-1} Xy. \tag{10.6}$$

There is a computational issue with this strategy. Normal equations can be ill conditioned, their determinants being close to zero, and hence, the inverse of XX^T may be subjected to drastic changes when the training data is slightly changed. This leads to a lack of stability and poor generalization. The next algorithm attempts to alleviate this problem.

10.7 RIDGE REGRESSION

This is a similar approach, but we now try to minimize the w norm $\|w\|^2 = w^T w$ as well as the sum of the squared differences between the y_i values and $g(x)$ values. This is done by using a Lagrange multiplier approach. We proceed with the following minimization problem:

$$\min_w L_\lambda(w, D_R) = \min_w \left\{ \lambda w^T w + \sum_{i=1}^{l} (y_i - x^{(i)T} w)^2 \right\} \qquad (10.7)$$

where $\lambda > 0$ is a parameter that controls the amount of *regularization*. Higher values of λ imply that we are promoting the minimization of $\|w\|^2$. Taking derivatives, we get

$$\frac{\partial L}{\partial w} = 0 \Rightarrow (XX^T + \lambda I) w = Xy. \qquad (10.8)$$

The significant issue here is that $XX^T + \lambda I$ is always invertible if $\lambda > 0$. This is true because XX^T is symmetric, and so we can find an orthogonal matrix Q such that $XX^T = QDQ^T$ where D is a diagonal matrix with non-negative entries. Then we can write

$$XX^T + \lambda I = Q(D)Q^T + \lambda QQ^T = Q(D + \lambda I)Q^T. \qquad (10.9)$$

It is clear that $XX^T + \lambda I$ has the same eigenvectors as XX^T but with eigenvalues that are never zero. Because the determinant of a matrix is the product of its eigenvalues, we see that $XX^T + \lambda I$ will not be ill conditioned when $\lambda > 0$ is suitably chosen.

So, without any worries about the lack of an inverse, we can write

$$w = (XX^T + \lambda I)^{-1} Xy. \qquad (10.10)$$

This is called the *primal solution*. Interpolation is done as before: $g(x) = x^T w = x^T (XX^T + \lambda I)^{-1} Xy$. We now consider another approach to the problem. It will lead to an alternate solution called the *dual solution*, which has some very useful properties.

We start by regrouping the terms in the equation defining w to get

$$w = \lambda^{-1} X(y - X^T w). \qquad (10.11)$$

This seems to be a strange manipulation of the equation because we have not really solved for any particular variable. However, the form of the equation suggests that w can be written as a linear combination of column vectors in X. Consequently, we can write

$$w = X\alpha = \sum_{i=1}^{l} \alpha_i x^{(i)} \tag{10.12}$$

if $\alpha = \lambda^{-1}(y - X^T w)$. Replacing w in this last equation with $X\alpha$ and multiplying through by λ, we get $\lambda\alpha = y - X^T X\alpha$. Finally, solving for α we obtain

$$\alpha = (X^T X + \lambda I)^{-1} y = (G + \lambda I)^{-1} y \tag{10.13}$$

where

$$G = X^T X. \tag{10.14}$$

The matrix G is called the Gram matrix and has the typical entry

$$G_{ij} = \langle x^{(i)}, x^{(j)} \rangle. \tag{10.15}$$

We can now use our expression for α to define the $g(x)$ predictor as

$$g(x) = x^T w = \left\langle x, \sum_{i=1}^{l} \alpha_i x^{(i)} \right\rangle = \sum_{i=1}^{l} \alpha_i \langle x, x^{(i)} \rangle. \tag{10.16}$$

It is a common strategy in linear algebra to replace a sum by a calculation involving entire matrices or vectors. We can do this for the last sum if we consider $\langle x, x^{(i)} \rangle$ to be the ith component of a column vector k^x. The superscript x will emphasize the dependency of this vector on the test data vector x. Then with the understanding that

$$k_i^x = \langle x, x^{(i)} \rangle \tag{10.17}$$

we can rewrite the last sum as

$$\sum_{i=1}^{l} \alpha_i \langle x, x^{(i)} \rangle = \sum_{i=1}^{l} \alpha_i k_i^x = \alpha^T k^x \tag{10.18}$$

to finally obtain

$$g(x) = x^{\mathrm{T}}w = \alpha^{\mathrm{T}}k^x = y^{\mathrm{T}}(G+\lambda I)^{-1}k^x. \qquad (10.19)$$

This is known as the dual solution and the values of α_i $(i=1,2,\ldots,l)$ are referred to as the *dual variables*. There is a very important observation to be made at this time: *Whenever the dual solution calculation for g(x) uses the training set vector $x^{(i)}$, it appears as an argument within an inner product.*

10.7.1 Predictors and Data Recoding

Let us go back to our previous general discussion about prediction. As discussed earlier, learning algorithms typically extract a predictor from a parameterized family of functions often called the *hypothesis space*, usually designated as *H*. Here are some examples of a hypothesis space:

- Linear functions

- Polynomials of some preassigned degree

- Gaussian functions

- High-dimensional rectangles

We have just seen a linear function being used for regression problems. Later we will see linear functions acting as *separators* in the input space of training vectors. This topic will be covered when we consider binary classification.

How do polynomials fit into our previous discussion that describes the construction of $g(x)$? Consider the following extension of the previous analysis: We take a training set vector $x^{(i)} \in \mathbb{R}^n$ and double the number of its components by bringing in the squares of the components. So, $x^{(i)} = (x_1^i, \ x_2^i, \ \cdots, \ x_n^i)^{\mathrm{T}}$ becomes a vector in $\mathbb{R}^{2n}$:

$$\left(x_1^{(i)}, \ x_2^{(i)}, \ \cdots, \ x_n^{(i)}, \ \left(x_1^{(i)}\right)^2, \ \left(x_2^{(i)}\right)^2, \ \cdots, \ \left(x_n^{(i)}\right)^2\right)^{\mathrm{T}}.$$

This step essentially "recodes" the training data so that it includes nonlinear contributions in $g(x)$. Despite the fact that we have such nonlinearities, the derivation of the predictor $g(x)$ *remains the same* as that done in the previous section, only this time our *w* vector contains 2*n* entries instead of *n* entries. We now do the same calculations in $\mathbb{R}^{2n}$ that we did for $\mathbb{R}^n$. In fact, we can go directly to the dual solution and simply replace all the inner product calculations done in $\mathbb{R}^n$ with the analogous inner product calculations done in $\mathbb{R}^{2n}$.

Again, any particular prediction function is determined by setting parameters. In the regression application, we had to determine the particular w that was the minimizing value of a Lagrange problem. In general, for any supervised learning strategy, parameters are chosen so that the predictor function minimizes the sum of the loss function values when they are evaluated and accumulated across the entire training set.

10.7.2 Underfitting and Overfitting

As just described, the predictor is a parameterized function that is taken from some infinite family of functions. In most cases, it is important to make a wise choice in the selection of the function family that will be used. Roughly speaking, the "flexibility" made available in the family should match the needs of the problem. As an example, suppose we have a data set that is generated by a natural process and $f(x)$ is a quadratic function with some noise in the data. If our family is the set of linear functions, we will get underfitting, that is, the hypothesis is unable to account for data complexity and we get poor test results (see Figure 10.4a). If the family is the set of cubic polynomials (recoding the data so that it has cubic as well as quadratic terms and working in $\mathbb{R}^{3n}$), it is very likely that it overreacts to noise in the data. It becomes so specialized for training data that it gives poor predictions when given the test data. This is a case of overfitting (see Figure 10.4b). If the user has some domain knowledge, he or she may understand that the natural process is intrinsically quadratic, and so he or she would go with the quadratic family (see Figure 10.4c).

10.8 PREAMBLE FOR KERNEL METHODS

How do we select the family of functions that make up the hypothesis (H) space? As just noted, we could use domain knowledge if available. Some knowledge of the nature of the physical process may allow us to utilize some particular H space. In general, this may present difficulties, because expert knowledge may not yield a judicious H space choice as the physical process may be too complicated. Further, each domain would require a special set of algorithms.

The idea behind kernel methods is to have a systematic methodology for incorporating nonlinearities into the predictor function. The previous example showing how we could extend a training set vector by including

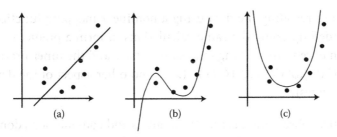

FIGURE 10.4 Fitting predictors. (a) Underfitting, (b) overfitting, and (c) best fit.

quadratic terms has two procedural techniques that are also seen when kernel methods are used:

1. The training data was a set of vectors in $\mathbb{R}^n$ but we ended up doing the analysis in a higher dimensional space $\mathbb{R}^{2n}$. We will refer to this new space as the *feature space,* and we will imagine that there is a mapping ϕ that maps a training vector in $\mathbb{R}^n$, now called the *input space,* to an image vector in this feature space.

2. The training data were used to do feature space computations but *only* with inner products.

Let us look at these two techniques more closely. The mapping function is formally defined as $\phi : R^n \rightarrow F$ where F represents the feature space containing the images

$$\phi(x^{(i)}) \quad i = 1, 2, \dots, l.$$

This means that we are essentially recoding the training set as

$$\left\{ \left(\phi(x^{(i)}), y_i \right) \right\}_{i=1}^{l}.$$

This is often described as *embedding* the training set in a feature space.

Recall the sequence of events in our previous discussion:

- We derived a regression analysis that worked with vectors in the input space.

- The final predictor only used the training data in two calculations: $\langle x^{(i)}, x^{(j)} \rangle$ and $\langle x, x^{(i)} \rangle$.

- After recoding the data using a nonlinear mapping function ϕ, we ended up doing the same calculations to form a predictor, but this time we were working in a feature space and the inner products are $\langle \phi(x^{(i)}), \phi(x^{(j)}) \rangle$ and $\langle \phi(x), \phi(x^{(i)}) \rangle$ (no other aspect of the derivation changes).

The critical observation is that the analysis and computations done in the feature space are linear, but the predictor defined in the lower dimensional input space is nonlinear.

To take further advantage of this technique, let us return to the final computations for ridge regression in the original $\mathbb{R}^n$ space now called the input space. We computed α using the equations $\alpha = (X^T X + \lambda I)^{-1}$ and $y = (G + \lambda I)^{-1} y$. After recoding the $x^{(i)}$ vectors, we get

$$\alpha = (G^\phi + \lambda I)^{-1} y \tag{10.20}$$

where

$$G_{ij}^\phi = \langle \phi(x^{(i)}), \phi(x^{(j)}) \rangle \tag{10.21}$$

and

$$g(x) = \langle w, \phi(x) \rangle = y^T (G^\phi + \lambda I)^{-1} k^\phi \tag{10.22}$$

with

$$k_i^\phi = \langle \phi(x^{(i)}), \phi(x) \rangle. \tag{10.23}$$

Note the use of the superscript ϕ to denote calculations involving vector images in the feature space. If the reader is skeptical about these final lines, he or she should go over the derivation of the dual solution to verify that the data recoding effort (replacing $x^{(i)}$ with $\phi(x^{(i)})$) does not substantially change our linear analysis and we do get these final equations.

Why is all of this so significant? Because even though we are doing linear operations in the feature space, the nonlinear ϕ mapping makes these linear computations appear as nonlinear computations in the input space. We are still in the convenient realm of linear algebra when working in the feature space, but we are actually doing nonlinear manipulations in the input space. An example in the next section will illustrate this.

A reasonable question to ask at this point is, how do I get the right ϕ mapping? This question cannot be adequately answered right now. We first need to explore the wonderful world of kernel functions.

10.9 KERNEL FUNCTIONS

We start with the formal definition of a kernel function.

> *Definition:* A kernel is a function k such that for all x and z in the input space $\mathcal{X}$, we have
>
> $$k(x,z) = \langle \phi(x), \phi(z) \rangle \tag{10.24}$$

where ϕ is a mapping from $\mathcal{X}$ to a (inner product) feature space F.

Almost any tutorial or text introducing kernels will use the following example to illustrate a simple kernel, and we will do likewise. Consider an explicit ϕ function:

$$\phi : x = (x_1, x_2)^T \in \mathbb{R}^2 \rightarrow \phi(x) = \left(x_1^2, x_2^2, \sqrt{2} x_1 x_2 \right)^T \in F = \mathbb{R}^3. \tag{10.25}$$

As we did earlier, let $g(x) = \langle w, \phi(x) \rangle$ so

$$g(x) = w_1 x_1^2 + w_2 x_2^2 + w_3 \sqrt{2} x_1 x_2. \tag{10.26}$$

The computation for $g(x)$ is done in the feature space as a linear calculation. This is true because the coordinate axes are labeled as $x_1^2, x_2^2, \sqrt{2} x_1 x_2$, and so the calculation of $g(x)$ is a simple inner product as indicated. To illustrate this behavior, suppose $w = (1, 3, 2)^T$ and consider the equation $g(x) = 8$. In the feature space, our inner product computation for $g(x)$ gives $x_1^2 + 3x_2^2 + 2\sqrt{2} x_1 x_2 = 8$, which is the equation of a plane as the axes are $x_1^2, x_2^2,$ and $\sqrt{2} x_1 x_2$. However, in the input space with axes x_1 and x_2, $x_1^2 + 3x_2^2 + 2\sqrt{2} x_1 x_2 = 8$ represents the quadratic (i.e., nonlinear) equation of an ellipse.

Now consider

$$\langle \phi(x), \phi(z) \rangle = \left\langle \left(x_1^2, x_2^2, \sqrt{2} x_1 x_2 \right), \left(z_1^2, z_2^2, \sqrt{2} z_1 z_2 \right) \right\rangle$$

$$= x_1^2 z_1^2 + x_2^2 z_2^2 + 2 x_1 x_2 z_1 z_2 \tag{10.27}$$

$$= (x_1 z_1 + x_2 z_2)^2$$

$$= \langle x, z \rangle^2.$$

So, if we define a function k using the equation

$$k(x,z) = \langle x,z \rangle^2 \tag{10.28}$$

then the preceding derivation tells us that it is indeed a kernel because there is a function ϕ such that $k(x,z) = \langle \phi(x), \phi(z) \rangle$ as required by our definition.

Note the two ways to calculate the kernel function:

1. Knowing ϕ explicitly we can calculate $\langle \phi(x), \phi(z) \rangle$. However, this may be computationally impractical if F is a very high-dimensional space.

2. The calculation of $\langle x,z \rangle^2$ can be done in the input space, and it is an easier computation (especially when compared to an explicit inner product computation in a high-dimensional feature space).

Considering these two approaches, we will want to go with the second choice. This strategy is described in the next section.

10.9.1 The "Kernel Trick"

Let us return to the nonlinear regression analysis discussed earlier. We needed to do two significant calculations:

$$G_{ij}^{\phi} = \langle \phi(x^{(i)}), \phi(x^{(j)}) \rangle \tag{10.29}$$

and

$$k_i^{\phi} = \langle \phi(x^{(i)}), \phi(x) \rangle. \tag{10.30}$$

The previous section tells us that these two dot product calculations will be possible (even for an infinite dimensional feature space) if we make them *kernel calculations*. That is to say, we will compute k^{ϕ} and G^{ϕ} using

$$G_{ij}^{\phi} = \langle \phi(x^{(i)}), \phi(x^{(j)}) \rangle = k(x^{(i)}, x^{(j)}) \tag{10.31}$$

and

$$k_i^{\phi} = \langle \phi(x^{(i)}), \phi(x) \rangle = k(x^{(i)}, x). \,^* \tag{10.32}$$

---

* Note the slight overloading of notation here. We are using the variable name k for both the name of the kernel function and the name of the vector needed in the regression analysis. However, the lack of superscripts and subscripts on the variable k will make it clear that we are dealing with the kernel function k.

The benefits of the last two equations are as follows:

- The kernel calculation is done using data in the input space, giving us an efficient computation.

- Nevertheless, we are calculating a dot product in the feature space and we do not need to compute ϕ explicitly! If we have the kernel function $k(.,.)$, then we can "skip" the computation of the feature space inner product that uses an explicit ϕ.

The clever avoidance of work mentioned in the second point is the essence of the "kernel trick."

10.9.2 Design Issues

Recall the question that was posed earlier: How do I get the right ϕ mapping? By using kernels, we can avoid explicit computations with data images $\phi(x)$ and $\phi(x^{(i)})$. When describing the theory behind an algorithm, the ϕ mappings appear as arguments of feature space inner products but in practise the kernel function $k(x,z)$ is an alternate way to calculate a feature space inner product that does not explicitly rely on ϕ. Therefore, we do not have to answer the question. Instead, we have to pose the question: "How do I get the right kernel function?" In effect, this is almost the same question because the kernel will implicitly define the ϕ mapping. Currently, researchers working with kernel methods cannot answer this question in a definitive manner. Selection of a kernel is often a difficult decision although, at times, there are some aspects of the prediction problem that provide guidance. For example, if the data is taken from a source characterized by a Gaussian distribution, then it would be advisable to use the Gaussian kernel (defined in the second line of Equations 10.33).

People who write machine-learning applications that require a kernel usually select one of the following:

$$k(x,z) = (<x,z>+c)^d \qquad c>0, d>0$$

$$k(x,z) = e^{(-\|x-z\|^2/\sigma^2)}$$

$$k(x,z) = \tanh(c<x,z>+d) \qquad c>0, d>0 \qquad (10.33)$$

$$k(x,z) = \frac{1}{\sqrt{\|x-z\|^2+c^2}}.$$

All these functions are valid kernels according to the kernel definition stated earlier. Although outside the scope of this text, it should be noted that there are many other kernel functions, some dealing with input data that involves data structures other than $\mathbb{R}^n$ vectors such as sequences, graphs, text, and images.

Therefore, although the choice of the most appropriate kernel is not always clear, changing application code so that it uses a different kernel is a very simple procedure. This is a very compelling reason to use kernels; the application software will have a type of design modularity that is very useful across a wide variety of applications. Once you decide on the algorithm to be implemented (e.g., some form of regression or classification), it can be programmed and this program will not have to be modified later to suit some future input space. Nonetheless, it is possible to adapt the algorithm to a specific task by changing the kernel, and in doing so, utilize a different implicit ϕ mapping.

We finish this section with some important observations. As we do *not* use an explicit mapping into the feature space, we do not "see" the actual images of points there. Instead, algorithms dealing with the feature space use inner products of these images. At first, this might seem to be a very severe constraint. Surely, such a limitation will hamper feature space algorithms in general, despite the regression algorithm in which it was sufficient to use inner products when working with the training data. However, it turns out that this restriction is not that onerous, and researchers have indeed taken many algorithms and "kernelized" them to produce very useful tools, all of which observe the "inner products only" rule. Lastly, as noted earlier, the feature space inner products are typically computed in an efficient manner (even for feature spaces of infinite dimensions) by using a kernel function that works with the input data.

10.9.3 Validation Data Sets

Most kernels have parameters such as c, d, and σ, as seen in Equations 10.33. The usual strategy for the calculation of such parameters is to run a number of small experiments that will evaluate a sequence of possible values for each parameter to determine the parameter setting that provides the best prediction. This is done by extracting a *validation set* from the training set, and using this validation set to evaluate parameter settings. Caution: This extraction should *never* be done using data in the test set. This would bias the parameters to do very well with the test set and you would get an overoptimistic evaluation of the predictor.* There are various strategies for working with the validation set, as given in the following sections.

10.9.3.1 Holdout Validation

A reasonable selection of data for the validation set would be a random extraction of, say, 20% of the training set. The remaining 80% is used to train the predictor with a particular parameter setting for c, d, or σ. Then the predictor is evaluated using the validation set. This procedure is repeated for several other settings of the parameter, and the best value of the parameter is taken to be the one that gives the smallest prediction error on the validation set. Typically, the prediction error falls to some minimum and then rises again as the parameter value is increased. For kernels that have two parameters, the situation is more complicated and time consuming because the number of possibilities is greatly increased. The validation evaluations are run over several different pairs of parameter settings.

10.9.3.2 N-Fold Cross Validation

We can utilize the training data more effectively by doing N-fold cross validation. The given training set is randomly partitioned into N disjoint subsets of equal size (or as close to equal as possible) and then one of these sets is chosen to be the validation set, which is used in the same manner as the validation set in hold-out validation. This procedure is repeated N times, each of the N subsets taking a turn as the validation set. For any setting of a parameter, this procedure gives N different evaluations of the prediction error and the average of these errors becomes the final error assessment of the predictor for that parameter setting. The entire procedure is repeated for different parameter settings. It is important that learning always starts from scratch for each validation set. In other words, it is not allowed to have any parameter setting information carry over from one validation subset experiment to another. As expected, we choose the parameter setting that has the lowest average error across all N validation subsets. The value of N really depends on how much computation time you can afford to expend. More accurate results can be achieved for larger values of N.

---

* Even though this is proper procedure, you may find the occasional research paper in which it is clear (or perhaps not so clear) that the author has simply run the algorithm repeatedly on the test data with different model parameters and then reported the best results, considering this a worthy assessment of how well the algorithm performs.

With sufficient computation time, one can chose N to be the number of samples in the training set. In other words, the size of the validation subset is 1. This extreme variant of N-fold cross validation is called *leave-one-out* cross validation. An advantage of this is that training is done each time with the largest possible training set while still doing validation. Consequently, you get the most reliable and most stable training but, of course, the computation expense greatly increases.

After cross validation has been completed and the parameter values defined, one last training session can be done using all the given training data. Only then are we allowed to work with the test data in order to do a final evaluation of the prediction capabilities of the learning algorithm.

10.10 CLASSIFICATION

Classification is a core problem in artificial intelligence with many applications in structural bioinformatics and biology in general. In a typical classification exercise, we are given a set of entities, each described by a set of attributes, and associated with each entity is a particular category or *class* that is predefined by the needs of the application. For example, we might be given a set of drug candidates with data that serves to describe each candidate. The data describing a drug candidate might be various physicochemical properties such as its molecular weight, number of rotatable bonds, number of hydrogen bond donors, number of hydrogen bond acceptors, polar surface area, etc. The classification might be the ability of a drug candidate to gain access to synapses in the brain. In this scenario, our classification software would process the list of physicochemical properties of a drug candidate and generate a prediction such as "after ingestion, more than 50% of this drug will cross the blood–brain barrier". This type of prediction is phrased as a **predicate**. It is either true or false, and so we are working with a **binary** classifier. In some applications, a classifier may deal with k classes where $k > 2$ but, as we will see, this requires some extra care and more computational expense.

In general, an entity is specified by a set of **attribute** values that are stored in a vector with n components. Some texts will refer to these attributes as **features**, and the various classes are specified by means of **labels**. So, the classifier can be seen as a function $y = f(x)$ that maps features stored in a vector x to a value y considered to be a member of some set of labels. Later, we will see that it is very convenient to assume that $y \in \{-1, +1\}$ when we are dealing with binary classification.

In some situations, the classifier may be some simple function that operates on the feature vector and does an easy binary classification specified by a predicate such as "the molecule has a molecular weight greater than 500." In this chapter, we will ignore such trivial classifiers and instead concentrate on classification problems for which there is no known simple a priori calculation. Usually, we are assuming that there is some biological process involved with the generation of the set of entities. Further, this process is typically so complicated that it is not possible to formulate a simple explicit function that does the classification by processing the features using, for example, some clever analysis based on biophysical theory.

10.10.1 Classification as Machine Learning

Because the classifier cannot be formulated from first principles, we adopt a learning strategy. The description of this strategy will be similar to that presented in Section 10.3, but in classification we predict a label instead of some numerical value.

The classification program will be given several examples of correctly classified feature vectors, and it will be required that the program *learns* the function that will do the class prediction. We will accept the possibility that learning will not be perfect, and so predictions will be made with a certain amount of unavoidable error, and it will be necessary to design a learning algorithm that attempts to minimize this learning error. In the simplest situation, the classifier learns its classification function by first processing properly labeled data that comprises a training set. To determine how well our learning algorithm is doing, we can apply the classifier to more labeled vector data designated as a test set. By comparing the classifier predictions with the labels in the test set, we can evaluate the percentage of prediction errors, and if it is low, it will give us some confidence in the abilities of the classifier when it is given unlabeled data in the future. As noted earlier we may need a ***validation set***, but for now the two sets (training and testing) will suffice.

A mathematical formalization of the classification problem is as follows: We are given a training set R of vectors with labels $\{x^{(i)}, y_i\}_{i=1}^N$. Each label is designated y_i with $y_i \in C = \{c_1, c_2, \ldots, c_k\}$ and a particular c_i represents one of the k class labels. To test the learning algorithm, we utilize a test set $\{x^{(i)}, y_i\}_{i=1}^M$.

In the remainder of this chapter, we will look at various simple classification strategies of increasing sophistication, eventually leading to support vector machines.

10.10.2 Ad Hoc Classification

In some cases, the features of a training vector may be used to formulate a simple classification predictor based on empirical observation with perhaps little or no physical theory. In 1997, Christopher Lipinski et al. gathered data on 2245 compounds. The objective was to identify testable parameters that were likely related to absorption or permeability (see [LL97]). After an inspection of the data, they found that although there were some exceptions, poor absorption or permeation was more likely when

- The molecular weight was over 500
- The CLogP value was over 5
- There were more than 5 hydrogen bond donors
- The number of hydrogen-bond acceptors was more than 10

The CLogP value measures lipophilicity and involves the ratio of octanol solubility to aqueous solubility. Hydrogen bond donors were counted by evaluating the number of OH and NH groups, whereas hydrogen bond acceptors were expressed as the count of N and O atoms.

Because all thresholds were either related to five or a multiple of five, the points in the list became "The Rule of 5" and later on "Lipinski's Rule of 5". Even though possibly more relevant physicochemical properties could have been used to predict absorption permeability, the authors deliberately chose to use rules that were simple and motivated by chemical structure, as they put a high value on visual communication of the observed results. Because drug design research papers abound with references to The Rule of 5, it would seem that the authors were correct in their anticipation of the popularity of such a predictor. In fact, many people regard it as a quick and useful rule of thumb predictor that helps drug developers avoid costly experimentation with candidates that are likely to fail.

We consider these rules as providing an ad hoc classification because they do not attempt to provide a classifier that is optimal in the sense of minimizing errors. Nonetheless, this classifier does provide us with a starting point for discussion because it represents an elementary classification based on a vector of features: molecular weight (MWT), CLogP, H-bond donor count, and H-bond acceptor count.

10.11 HEURISTICS FOR CLASSIFICATION

In the previous section, it was demonstrated that classification could be done by checking a vector of physicochemical features to verify that each was within a particular range of values. The Rule of 5 essentially defines a hyperrectangle in a 4-D space, and if a point (in this case, corresponding to a chemical compound) is within the rectangle, then the compound is considered to have favorable attributes. Intuitively, we can understand that these tests are in some sense "overly rectilinear." Natural processes are more plastic and more organic, and we would expect decision surfaces to be more curvilinear, for example, a surface defined by some threshold imposed on a Gaussian distribution.

In drug research, experiments might be done to find an "ideal drug" for some protein-binding site. The optimal drug might correspond to a single point in some high-dimensional space but other nearby points could also represent drugs that are valid candidates, perhaps exhibiting other qualities of choice such as having a safer pathway for elimination by the renal system. The keyword in the last sentence is "nearby." To have the notion of a nearby drug, we must have a measure of distance in this high-dimensional space. A mathematical modeling of this requirement puts us in the realm of metric spaces (see the review in Appendix A). If we maintain the notion of an entity description by means of a feature vector, then we could measure distance using a Euclidean metric such as

$$d(x,y) = \sqrt{\sum_{i=1}^{n} (x_i - y_i)^2}. \qquad (10.34)$$

Later in this chapter, we will be using distance metrics in our classification algorithms.

10.11.1 Feature Weighting

At this time, it should be noted that a naïve application of the last distance metric might be ill advised. It may be necessary to use a more general distance measure such as

$$d(x,y) = \sqrt{\sum_{i=1}^{n} \alpha_i (x_i - y_i)^2} \qquad (10.35)$$

in which the positive α_i values provide appropriate weighting of the various features.

If we wanted to reformulate the Rule of 5 classification by devising a metric that defines a distance between molecule A and molecule B, we could start with a formula such as

$$d(A,B) = \sqrt{\alpha_m(m_A - m_B)^2 + \alpha_c(c_A - c_B)^2 + \alpha_d(d_A - d_B)^2 + \alpha_a(a_A - a_B)^2}$$

(10.36)

where the variables m, c, d, and a designate values for the Rule of 5 features (MWT, CLogP, H-bond donor count, and H-bond acceptor count, respectively) and, as expected, the subscripts A and B designate the molecule associated with the feature. With this as a starting point, we would then strive to specify appropriate values for the weights $\alpha_m, \alpha_c, \alpha_d$, and α_a so that the distance function $d(A,B)$ gives us the appropriate separation during classification. Strategies to determine the required weights are referred to as feature weighting. For now, let us assume that these weights have been defined, and so we have a metric that can assess the distance between two points in our "space of molecules." The availability of a metric allows us to approach classification with extra powers of discrimination, as we will see in the strategies that follow.

10.12 NEAREST NEIGHBOR CLASSIFICATION

The idea underlying nearest neighbor classification is really quite straightforward. Given a new point of unknown classification (let us call it the *query* point), we compute its distance from each point in the training set and find point x^* in the set in that is closest to the query point. The predicted classification for the query point is simply that of x^*. The situation can be visualized quite nicely if we consider an input space with two dimensions. For a start, we consider only two points in the training set, so $R = \{x^{(1)}, x^{(2)}\}$, and we will assume that each has a different class assignment, say $y_1 = +1$ and $y_2 = -1$. The perpendicular bisector of the dashed line segment that connects $x^{(1)}$ and $x^{(2)}$ (see Figure 10.5) acts as a discrimination line separating all points in the plane into two sets: P_+ with points that would be closer to $x^{(1)}$ than to $x^{(2)}$, and P_- with points that would be closer to $x^{(2)}$ than to $x^{(1)}$. When the number of training points is increased, we can still use the perpendicular bisector strategy to draw discrimination lines among the points belonging to different classes, although it is a more complicated separator as it is only piecewise linear.

For purposes of explanation, the piecewise linear discrimination function is shown in isolation but it is really a subset of a much larger set of line segments shown in Figure 10.7.

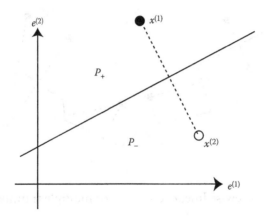

FIGURE 10.5   Separating line for two training points.

10.12.1 Delaunay and Voronoi

Figure 10.7 illustrates two networks of lines. The dashed lines (both thick and thin) represent a Delaunay triangulation. This is a triangulation formed by putting in edges to join neighboring points in R. We put in as many edges as possible with the constraint that no given point of R resides within any triangle formed by any other point in R. The set of solid lines (both thick and thin) represent a Voronoi tessellation, which can be derived by constructing perpendicular bisectors for each edge in the Delaunay triangulation. Each bisector separates two points in the plane, and all such bisectors form a set of convex polygons each one containing a single point from R. As a rule of construction, any bisector is trimmed so that its remaining points are closer to a point in R than any other point on any other bisector. At times, this trimming is done in such a way as to remove a portion of the bisector that actually crosses the Delaunay line segment defining that bisector. On carefully comparing Figures 10.6 and 10.7, you can see that Figure 10.6 can be derived from Figure 10.7 simply by removing any Delaunay edges that join any R points with the *same* classification, removing as well any bisectors of such edges.

Finally, it should be appreciated that the separation lines do not necessarily form a single jagged line. Consider Figure 10.7 again. If you were to change the classification of the point that is closest to the bottom of the figure, then the separation lines would include the two Voronoi lines that separate it from the two points immediately above it.

Although illustrated here in a 2-D space, it should be clear that the notion of a Voronoi tessellation can be generalized to an *n*-dimensional

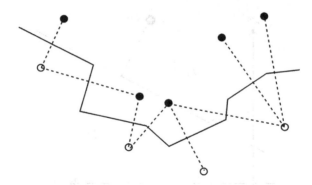

FIGURE 10.6 Piecewise linear separator for multiple training points.

space. Once you have an inner product, you can define a distance function and it is possible to deal with bisectors and notions of perpendicularity. For applications with $n = 3$, many research studies have dealt with Voronoi tessellations in physical space, for example, putting Voronoi convex polyhedrons around the atoms of a protein to study various types of packing in the conformation. For higher-dimensional extensions, the boundaries defining Voronoi neighborhoods are hyperplanes.

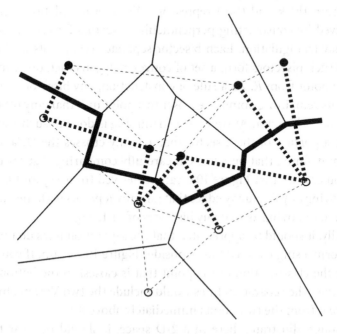

FIGURE 10.7 Delaunay and Voronoi lines.

10.12.2 Nearest Neighbor Time and Space Issues

The nearest neighbor classification algorithm described earlier has time and space characteristics as follows:

- Each query point requires $O(n)$ operations to accomplish a class prediction, where n is the number of existing classified points in the training set.

- The storage requirements are also $O(n)$ (necessary to store all the data in the training set).

For nearest neighbor algorithms, the labeled data is a training set in a rather loose sense of this terminology because no processing of the training data is done prior to the presentation of query points. The reader should understand that the preceding Voronoi and Delaunay discussion presents an analysis of how the space is carved up by the separation planes. The actual classification algorithm is quite simple and does not require the construction of the Voronoi tessellation. In the next section, we consider an improvement of nearest neighbor classification that will improve these time and space requirements to some extent.

10.13 SUPPORT VECTOR MACHINES

The topic of support vector machines (SVMs) has become very popular in bioinformatics over the last few years. Several papers have been written that use SVMs for prediction of protein related functionality. See for example: protein-protein interactions [Bw05]; identifying DNA binding proteins [KG07]; subcellular localization of proteins [CJ04], [DH06], [GL06], [Hs01], and [Sc05]; prediction of G-protein coupled receptor subfamilies [GL05]; and prediction of the functional class of lipid-binding proteins [LH06]. The SVM is another classification algorithm, and it can be implemented using a kernel methods approach. We start with the basic ideas behind linear discrimination.

10.13.1 Linear Discrimination

The easiest classification scenario is when the labeled data can be separated using a simple linear discriminator. Reviewing our learning environment, we are given a training set, $D_R = \{(x^{(i)}, y_i)\}_{i=1}^{l}$, in which each vector $x^{(i)} \in \mathbb{R}^n$ and each y_i is a class label. We assume $y_i \in \{+1, -1\}$. We are required to learn a linear function $g(x)$ that predicts the class label when

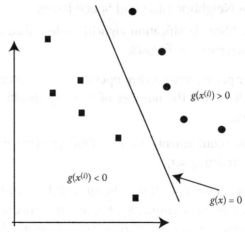

FIGURE 10.8 Linear separator.

given a new x value. On the training set or test set, perfect predictions would give

$$g(x^{(i)}) > 0 \quad \text{if} \quad y_i = +1,$$
$$g(x^{(i)}) < 0 \quad \text{if} \quad y_i = -1. \tag{10.37}$$

Linear separability implies that we can implement the prediction function as

$$g(x) = w^T x + w_0 \tag{10.38}$$

where w is a weight vector and w_0 is the *bias* or *threshold*. As implied by the term "separability," all points with the same class label are on the same side of the line, whereas points with different labels are on opposite sides of the separator as shown in Figure 10.8.

If we consider x^α and x^β to be two points on the line $g(x) = w^T x + w_0$, then

$$\left(w^T x^\alpha - w_0\right) - \left(w^T x^\beta - w_0\right) = g(x^\alpha) - g(x^\beta) = 0 \Rightarrow w^T (x^\alpha - x^\beta) = 0. \tag{10.39}$$

So, w is normal to the vector $x^\alpha - x^\beta$, which is the line determined by x^α and x^β. This means that w determines the orientation of the separator line (see Figure 10.9).

We now calculate the distance d from the origin to the separator along the normal vector w. As in Figure 10.10, we let an extension of the w vector meet the separator at x^γ. Then, x^γ is the unit vector along w scaled

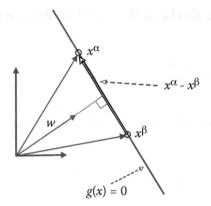

FIGURE 10.9   Significance of *w*.

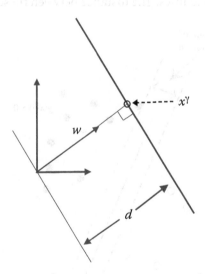

FIGURE 10.10   Distance from origin to separator along *w*.

by the value of d. We also know that $g(x^\gamma)=0$ because x^γ is on the line. Consequently,

$$x^\gamma = \frac{w}{\|w\|} d$$

$$g(x^\gamma)=0 \quad \Rightarrow w^T \left(\frac{w}{\|w\|} d \right) + w_0 = 0 \quad \Rightarrow \qquad (10.40)$$

$$d = \frac{-w_0}{\|w\|}.$$

10.13.2 Margin of Separation

Support vector machines rely on the notion of *maximum margin* for a linear separator.

If we consider a set of linear separators with the same orientation (they are all parallel to one another), we can evaluate the distance between the two separators that are furthest apart. In Figure 10.11, these two separators are drawn as dashed lines. The distance between these two lines is the

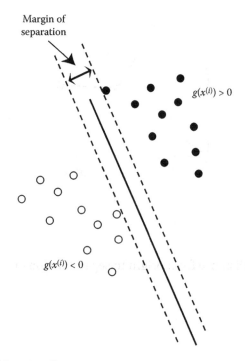

FIGURE 10.11   Margin of separation.

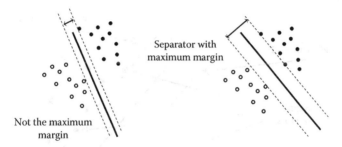

FIGURE 10.12   Maximum margin of separation.

margin of separation for the set. The central line in the set (the solid line in Figure 10.11) is assigned this margin of separation.

Not all margins are created equal. Among the infinite number of linear separators, typically one has a maximum margin of separation (see Figure 10.12).

10.13.3 Support Vectors

The main idea behind the SVM algorithm is to find the $g(x) = 0$ linear separator that has the maximum margin of separation. Therefore, we need w_0 and the w orientating normal vector that does this. The data points that define the boundary lines with maximum separation are called the *support vectors* (see Figure 10.13).

It can be shown that a classifier with a margin that is smaller than the maximum margin runs a higher expected risk of misclassification for future data. We will not include a formal proof for this, but the intuitive notion is that the maximal margin will generate a linear separator that makes decisions on the training points in such a way as to guarantee that they are maximally robust to noise or "feature jitter." Another very important aspect of this linear separator is that it is defined only by the support vectors that are usually much fewer in number than the size of the entire training data set. The other members of the training set do not participate in the equation that defines the separator. This means that we have a *sparse solution*, and so the predictor can be calculated in a very rapid fashion.

Notice that the equations for lines going through the support vectors (dashed lines in Figure 10.13) are given as $w^T x + w_0 = +1$ and $w^T x + w_0 = -1$. Why not $w^T x + w_0 = +c_1$ and $w^T x + w_0 = -c_2$? If $w^T x + w_0 = 0$ is the centerline, then it is easy to see that $c_1 = c_2 = c$. However, if we use $w^T x + w_0 = +c$ and $w^T x + w_0 = -c$, then without loss of generality we can assume that w is

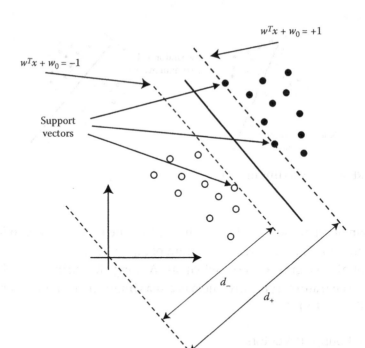

FIGURE 10.13 Support vectors.

scaled so that $c = +1$. This value of c is very useful when we formulate the constraints for the problem.

So, what is the margin of separation? Looking at the figure, we see that we need to calculate margin $r = d_+ - d_-$. Recall that for $w^T x + w_0 = 0$, the distance to the origin is $d = -w_0/\|w\|$. So for $w^T x + w_0 = -1$, we get $d_- = -(w_0 + 1)/\|w\|$ (just by replacing w_0 with $w_0 + 1$), and similarly for $w^T x + w_0 = +1$, we get $d_+ = -(w_0 - 1)/\|w\|$. Consequently, the margin is

$$r = d_+ - d_- = \frac{2}{\|w\|}. \tag{10.41}$$

10.13.4 The SVM as an Optimization Problem

This calculation tells us that we can maximize the margin by minimizing $\|w\|$. In summary, this is a minimization problem with constraints, stated formally as

$$\text{minimize } \|w\|$$

subject to the constraints:

$$y_k\left(w^T x^{(k)} + w_0\right) \geq 1 \quad \text{for all} \quad k = 1, 2, \ldots, l. \tag{10.42}$$

Notice how the y_k labels have been utilized to combine the new constraints: $w^T x + w_0 \leq -1$ when $y_k = -1$, and $w^T x + w_0 \geq +1$ when $y_k = +1$ into a single inequality. We are now reaping the benefit of having labels that are ± 1 while also having $c = +1$ in the earlier analysis.

This problem can be solved by reformulating it as a Lagrange multiplier problem (see Equation 10.43). We wish to minimize $L_P(w, w_0, \alpha)$ with respect to w and w_0, while simultaneously maximizing $L_P(w, w_0, \alpha)$ with respect to the Lagrange multipliers α_k where $k = 1, 2, \dots, l$. In other words, the stationary points that we want to find are actually *saddle points* of the hyperdimensional surface.

$$L_P(w, w_0, \alpha) = \frac{1}{2} w^T w - \sum_{k=1}^{l} \alpha_k \left[y_k \left(w^T x^{(k)} + w_0 \right) - 1 \right]$$

$$\text{(10.43)}$$

$$\alpha_k \geq 0 \quad k = 1, 2, \dots, l.$$

This expression for $L_P(w, w_0, \alpha)$ is known as the *primal form* of the Lagrangian. Taking derivatives to find the stationary points gives the following equations:

$$\frac{\partial L_P}{\partial w_0} = 0 \quad \Rightarrow \quad \sum_{k=1}^{l} \alpha_k y_k = 0$$

$$\text{(10.44)}$$

$$\frac{\partial L_P}{\partial w} = 0 \quad \Rightarrow \quad w = \sum_{k=1}^{l} \alpha_k y_k x^{(k)}.$$

The derivative with respect to w_0 produces a single equation, whereas the derivative with respect to w produces a vector equation that you can view as n equations (because w has n components). Note that the derivative of $L_P(w, w_0, \alpha)$ with respect to α simply produces the constraints. Before moving on, you should observe that the equation for w is expressing the requirement that w must be some linear combination of the training data.

Replacing w in $L_P(w, w_0, \alpha)$ with the sum just derived and doing some necessary simplification, we get

$$L_D(\alpha) = \sum_{k=1}^{l} \alpha_k - \frac{1}{2} \sum_{i=1}^{l} \sum_{j=1}^{l} \alpha_i \alpha_j y_i y_j x^{(i)T} x^{(j)}. \quad \text{(10.45)}$$

This is the *dual form* of the Lagrangian. We have not really solved the primal problem at this point, but we have put our optimization problem into a form that is much more "agreeable." We want to maximize $L_D(\alpha)$ with respect to the α_k subject to $\alpha_k \geq 0$ $k = 1, 2, \ldots, l$ and $\alpha_1 y_1 + \alpha_2 y_2 + \cdots + \alpha_l y_l = 0$.

The dual form is important because

- It is a conceptually straightforward maximization problem with respect to α (there are no saddle points).

- Even more significantly, $L_D(\alpha)$ is expressed in terms of inner products of the training data. Therefore, we can use kernel methods.

Unfortunately, the dual form is complicated by the fact that it is quadratic in α.

10.13.5 The Karush–Kuhn–Tucker Condition

Recall our two constraints for the SVM: $y_k(w^T x^{(k)} + w_0) - 1 \geq 0, \alpha_k \geq 0$ where $k = 1, 2, \ldots, l$. The KKT "complementarity condition"* can be stated as follows:

$$\alpha_k \left[y_k \left(w^T x^{(k)} + w_0 \right) - 1 \right] = 0 \quad k = 1, 2, \ldots, l. \tag{10.46}$$

Notice that it is essentially the product of the constraints but with the left-hand side set exactly to zero instead of having an inequality. The complementarity condition demands that for any $k = 1, 2, \ldots, l$, either α_k or $y_k(w^T x^{(k)} + w_0) - 1$ must be zero (but not both). We do not go into the proof of this result, but it is extremely important when discussing the sparseness of the SVM predictor (see [KT51]).

10.13.6 Evaluation of w_0

After the dual form of the optimization problem is solved, you will have values for all the α_k where $k = 1, 2, \ldots, l$. The KKT condition stipulates that most of these α_k variables will be zero, except for the few that are associated with support vectors. Once we have the α_k values, we can use

$$w = \sum_{k=1}^{l} \alpha_k y_k x^{(k)} \tag{10.47}$$

---

* The KKT complementarity condition was discovered in 1939 by W. Karush, a Master's student, who did not fully appreciate the significance of this condition (see [KA39]). Consequently, it was essentially ignored and was later rediscovered in 1951 by Kuhn and Tucker [KT51]].

to derive the value of w. Then a new vector z will be classified according to the sign of $\langle w, z \rangle + w_0$. However, to do this, we need the value of w_0 that can be evaluated as follows: Working from the KKT condition, we let SV be the set of indices for which $\alpha_k > 0$. Then $y_k(w^T x^{(k)} + w_0) - 1 = 0, \quad k \in SV$. Multiply through by y_k and solve for w_0 (noting that $y_k^2 = 1$) to get $w_0 = y_k - w^T x^{(k)}, \quad k \in SV$.

Note that we could do this for any $k \in SV$ and get a value for w_0. However, using a single support vector may introduce errors due to issues such as noise. Therefore, it is advisable to average over all support vectors as follows:

$$w_0 = \frac{1}{n_{SV}} \left[\sum_{k \in SV} y_k - w^T \sum_{k \in SV} x^{(k)} \right] \tag{10.48}$$

and because

$$w = \sum_{j \in SV} \alpha_j y_j x^{(j)} \tag{10.49}$$

we finally get

$$w_0 = \frac{1}{n_{SV}} \left[\sum_{k \in SV} y_k - \sum_{k \in SV} \sum_{j \in SV} \alpha_j y_j x^{(j)T} x^{(k)} \right] \tag{10.50}$$

where n_{SV} is the number of support vectors.

10.14 LINEARLY NONSEPARABLE DATA

The previous section assumes that the data is linearly separable. In this case, the SVM that is produced is said to have a *hard margin*. If there were no strategy to get around the linearly separable requirement, we would hear very little about SVMs. This is because most applications do not provide linearly separable data. Due to noise, outliers, or the intrinsic properties of the data, there may be some mixing of two data classes with a single class appearing on both sides of any linear separator. The problem then becomes one of choosing the straight line that will minimize the amount of "off-side" behavior.

The strategy is to utilize some *slack* or "forgiveness" that *allows misclassification* but only with a penalty. The forgiveness is accomplished by using *slack variables* $\xi_k \geq 0$, one for each training vector $x^{(k)}$.

The problem formulation is

$$w^T x^{(k)} + w_0 \geq 1 - \xi_k \qquad \text{for all } k \text{ such that } y_k = +1 \qquad (10.51)$$

$$w^T x^{(k)} + w_0 \leq -1 + \xi_k \qquad \text{for all } k \text{ such that } y_k = -1 \qquad (10.52)$$

$$\xi_k \geq 0 \qquad \text{for all } k = 1, 2, \ldots, l. \qquad (10.53)$$

To see how the slack variable works, consider the first inequality. In this case, the label is positive, and if the point $x^{(k)}$ did not require some slack, then we would simply have $w^T x^{(k)} + w_0 \geq 1$ as in the hard margin case with $\xi_k = 0$. For cases when $x^{(k)}$ is actually on the other side of $w^T x^{(k)} + w_0 = 1$, we still get the inequality satisfied by changing it so that the right-hand side is decreased by an amount $\xi_k \geq 0$. The value of ξ_k is chosen to be as small as possible, just enough to make sure that $w^T x^{(k)} + w_0 \geq 1 - \xi_k$. A similar argument can show how a slack variable handles an offside for the other boundary. To be sure that ξ_k is chosen to be a small positive quantity, we incorporate a penalty into the Lagrange formulation. Instead of minimizing $(w^T w)/2$ as in the hard margin case, we want to minimize

$$\frac{1}{2} w^T w + C \sum_{k=1}^{l} \xi_k \qquad (10.54)$$

where $C > 0$ is a parameter that controls the trade-off between margin width and misclassification penalty. As C increases, the algorithm will be less tolerant of misclassification. That is to say, we get closer to the generation of a hard margin solution. To achieve this, the solution may need to settle for a narrower margin. Typically, C is calculated by means of a validation data set. With this understanding of the slack variables, we can write the primal Lagrangian for the soft margin as

$$L_p(w, w_0, \alpha, \xi, \beta) = \frac{1}{2} w^T w + C \sum_{k=1}^{l} \xi_k$$

$$- \sum_{k=1}^{l} \alpha_k \left[y_k (w^T x^{(k)} + w_0) - 1 + \xi_k \right] - \sum_{k=1}^{l} \beta_k \xi_k. \qquad (10.55)$$

We now have two sets of Lagrange multipliers: $\alpha_k \geq 0$ and $\beta_k \geq 0$. The α_k variables serve the same purpose as for the hard margin case, whereas the β_k values are there to ensure that the constraints $\xi_k \geq 0$ where $k = 1, 2, \ldots, l$ are satisfied.

We wish to minimize $L_p(w, w_0, \alpha, \xi, \beta)$ with respect to w, w_0, ξ_k while maximizing it with respect to α_k, β_k. As before, we compute the stationary points by using partial differentiation:

$$\frac{\partial L_p}{\partial w_0} = 0 \implies \sum_{k=1}^{l} \alpha_k y_k = 0$$

$$\frac{\partial L_p}{\partial w} = 0 \implies w = \sum_{k=1}^{l} \alpha_k y_k x^{(k)} \tag{10.56}$$

$$\frac{\partial L_p}{\partial \xi} = 0 \implies C = \alpha_k + \beta_k.$$

Observe that the first two lines are the same as those produced in the hard margin case. The last equation $C = \alpha_k + \beta_k$ is new. When this value for C is enforced in the primal form of the Lagrangian, an interesting event takes place. All the coefficients of each ξ_k where $k = 1, 2, \ldots, l$ sum to zero! It is as if the ξ_k values are not in the primal form when we compute the dual form by replacing w with the sum defined by the second line in Equations 10.56. The result is that we get the *same* dual form that we derived for the hard margin case. Another observation is that $\alpha_k \geq 0$, $\beta_k \geq 0$, and $C = \alpha_k + \beta_k$ imply that

$$0 \leq \alpha_k \leq C. \tag{10.57}$$

So, even though the dual form of the Lagrangian is the same as in the hard margin case, the algorithm that does the quadratic optimization for the dual form must derive α_k values that obey this inequality, and so the α_k will be different *if* there are training vectors that require slack values greater than zero.

For this soft margin case, the KKT complementarity condition involves two equations:

$$\alpha_k \left[y_k \left(w^T x^{(k)} + w_0 \right) - 1 + \xi_k \right] = 0 \tag{10.58}$$

$$\beta_k \xi_k = 0.$$

Note that the last equation implies

$$(C - \alpha_k)\xi_k = 0. \tag{10.59}$$

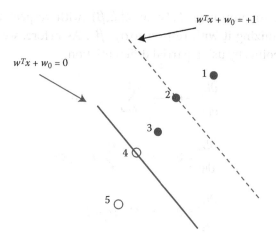

FIGURE 10.14 Different possibilities for a training point.

10.14.1 Parameter Values

It is instructive to look at the values of α_t, β_t, ξ_t associated with a training point $x^{(t)}$ that is in various positions with respect to the lines of separation. Without loss of generality, we will assume that $y_t = +1$. A similar analysis could be carried out for a training point with a negative label. As you read these, remember the KKT conditions $(C - \alpha_t)\xi_t = 0$ and $\alpha_t + \beta_t = C$. Figure 10.14 shows the five possible scenarios for $x^{(t)}$, with positions numbered to correspond to entries in the following list:

1. $x^{(t)}$ properly classified (no slack):

$$1 < w^T x^{(t)} + w_0 \quad \alpha_t = 0 \quad \beta_t = C \quad \xi_t = 0. \tag{10.60}$$

2. $x^{(t)}$ a support vector (no slack):

$$w^T x^{(t)} + w_0 = 1 \quad 0 < \alpha_t < C \quad 0 < \beta_t < C \quad \xi_t = 0. \tag{10.61}$$

3. $x^{(t)}$ properly classified but in the margin (slack > 0):

$$0 < 1 - \xi_t = w^T x^{(t)} + w_0 < 1 \quad \alpha_t = C \quad \beta_t = 0 \quad 0 < \xi_t < 1. \tag{10.62}$$

4. $x^{(t)}$ classification indeterminate (slack > 0):

$$w^T x^{(t)} + w_0 = 0 \quad \alpha_t = C \quad \beta_t = 0 \quad \xi_t = 1. \tag{10.63}$$

5. $x^{(t)}$ misclassified (slack > 0):

$$1-\xi_t = w^T x^{(t)} + w_0 < 0 \quad \alpha_t = C \quad \beta_t = 0 \quad 1 < \xi_t. \tag{10.64}$$

Note that $x^{(t)}$ is a support vector for all cases except the first one.

10.14.2 Evaluation of w_0 (Soft Margin Case)

Starting with the KKT conditions,

$$\alpha_k \left[y_k \left(w^T x^{(k)} + w_0 \right) - 1 + \xi_k \right] = 0 \quad k = 1, 2, \ldots, l \tag{10.65}$$

let SV be the set of indices such that $0 < \alpha_k < C$, $\xi_k = 0$.

Then $y_k(w^T x^{(k)} + w_0) - 1 = 0$ where $k \in SV$. Multiply through by y_k and solve for w_0 (noting that $y_k^2 = 1$) to get $w_0 = y_k - w^T x^{(k)}$ where $k \in SV$. As we did for the hard margin case, we average over all support vectors with $\xi_k = 0$ to get

$$w_0 = \frac{1}{n_{SV}} \left[\sum_{k \in SV} y_k - \sum_{k \in SV} w^T x^{(k)} \right]. \tag{10.66}$$

Let ASV be the set of indices for all support vectors (those for which $0 < \alpha_k \leq C$, $0 \leq \xi_k$), then

$$w = \sum_{j \in ASV} \alpha_j y_j x^{(j)}. \tag{10.67}$$

So, we finally get

$$w_0 = \frac{1}{n_{SV}} \left[\sum_{k \in SV} y_k - \sum_{k \in SV} \sum_{j \in ASV} \alpha_j y_j x^{(j)T} x^{(k)} \right] \tag{10.68}$$

where n_{SV} is the number of support vectors. Again, note that the training set in this computation only appears within inner products.

10.14.3 Classification with Soft Margin

A new vector z is classified according to the sign of $\langle w, z \rangle + w_0$. Considering the computations in the previous section, we would assign z to the +1 class if

$$\sum_{j \in ASV} \alpha_j y_j x^{(j)T} z + \frac{1}{n_{SV}} \left[\sum_{k \in SV} y_k - \sum_{k \in SV} \sum_{j \in ASV} \alpha_j y_j x^{(j)T} x^{(k)} \right] > 0 \tag{10.69}$$

otherwise it is assigned a class of –1. To conserve computations, we should observe that the sums within the square brackets and various products in the first sum could all be computed immediately after training is completed. Consequently, all future predictions involve evaluation of a first sum plus the addition of a precomputed value.

10.15 SUPPORT VECTOR MACHINES AND KERNELS

The kernel version of the SVM can be generated in a very straightforward fashion. We employ the same strategy as that used to derive the kernel version of regression. Because a training vector always appears within inner products, we can recode the data using the implicit mapping $x^{(i)} \rightarrow \phi(x^{(i)})$ by replacing all such inner products with kernel calculations. Thus, the discriminator that we derived in the previous section becomes

$$h(z) = \sum_{j \in ASV} \alpha_j y_j k(x^{(j)}, z) + \frac{1}{n_{SV}} \left[\sum_{k \in SV} y_k - \sum_{k \in SV} \sum_{j \in ASV} \alpha_j y_j k(x^{(j)}, x^{(k)}) \right] > 0.$$

(10.70)

If this inequality holds, that is, if $h(z) > 0$, then z is classified as having a +1 label, otherwise it is put into the –1 class.

It should be stressed that a nonlinear kernel generates a decision surface in the input space that is a nonlinear curve even though there is a linear separation being done in the feature space. Figure 10.15 illustrates this. It shows how the $h(z)$ curve winds through the data separating empty circles (negative labels) from filled circles (positive labels). The squares represent vectors that have been selected as support vectors. All vectors here are situated in a 2-D space, so it is easy to see the separation. In most applications, the vector data resides in a Euclidean space of high dimension, and this type of simple diagram is not possible to visualize.

10.16 EXPECTED TEST ERROR

This chapter mainly concentrates on the presentation of the SVM as a tool for prediction (both regression and classification). It is natural to ask: "How well does the SVM perform?" It would require at least one more chapter of rather strenuous mathematics to frame the question and to provide an adequate answer. Consequently, we will limit the depth of this discussion to the note provided by Alpaydin in his treatment of SVMs

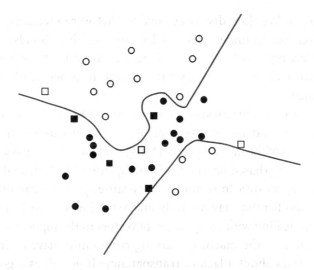

FIGURE 10.15 Nonlinear separation in the input space.

(page 225 of [Al04]): In 1995, Vapnik (see [Va95]) demonstrated that the *expected test error rate* is given by

$$E_l[P(\text{error})] \leq \frac{E_l[\text{number of support vectors}]}{l-1} \qquad (10.71)$$

where $E_l[.]$ denotes expectation over training sets of size l. The term $P(\text{error})$ represents the probability of committing a classification error on a test example. This inequality indicates that an upper bound on the error rate is inversely proportional to the number of samples in the training set. An even more significant observation is that the error rate depends on the number of support vectors and not on the dimension of the input space. Consequently, if the optimal hyperplane of separation can be constructed using a small number of support vectors (relative to l), then the generalization ability will be high *even though the input space may be of infinite dimension.*

10.17 TRANSPARENCY

Looking back over the various scientific models that we have encountered in this text, one can observe a general trend. As the complexity of the problems became more daunting, the realism of the model became less and less evident. Our best description of nature, quantum mechanics, was abandoned from the very start because the equations are intractable for large systems. We then discussed models that were mechanical in their formulation (for example, Hooke's law for modeling bonds). Although these models represent approximations, we can at least see how various terms contribute to final energy calculations. In other words, they have transparency.

In the case of machine learning, we may have a highly successful predictive model, but the details of its computations may give us no clue about the inner workings of the natural process being studied. After using a nonlinear implicit mapping (based on a kernel from Equation 10.33) that takes vectors in the input space over to vectors in the feature space, we can often derive a reliable classifier but may not fully understand why it works in terms of calculations dealing with components of vectors in the input space.

Researchers in the machine-learning community have a tendency to dismiss worries about a lack of transparency: If $g(x)$ does a good job in terms of prediction, why worry about how faithful it is to some true function $f(x, x')$, especially if the form of this function is so complex that it cannot be derived by conventional analysis? The quote at the start of this chapter typifies this pragmatic approach.

My view on the issue is very much similar to this but with some provisos: First, one should be aware that the artificial model used by machine learning is *artificial*, and so it makes no attempt to reveal inner function even though it may be a successful predictor. Second, we should not give up the challenge of developing more transparent models even though, initially, they may be less capable.

In the practice of machine learning, we are devising an algorithm that will take a training set and process it to derive a generalization that applies to a broader set of data instances. This methodology is basic to inductive reasoning, but does it really give us "truth." This has become a central debating issue in the more philosophical reaches of artificial intelligence (see [Gi96], for example). In a more applied setting, researchers working in the area have developed quite sophisticated statistical learning theory bounds to provide indications of successful learning. However, in other research papers that use these learning algorithms to handle some practical application, learning bounds are hardly ever mentioned! To underscore this, we quote Dr. Bousquet in closing this chapter:

> So, I think that the theoretical analysis is mainly a way to popularize an algorithm and to raise its visibility. The effect is then that more people try it out, and streamline it. So in the end, the

algorithm may be adopted, but a theoretical analysis rarely justifies the algorithm and never provides guarantees. Hence theory is a way to attract attention to an algorithm.

DR. OLIVIER BOUSQUET

10.18 EXERCISES

1. Recall Equation 10.6, we want a homogenous real-valued function $g(x) = x^T w$ that gives the best prediction for a given training set

$$D_R = \{(x^{(1)}, y_1), (x^{(2)}, y_2), \ldots, (x^{(l)}, y_l)\}$$

where each column vector $x^{(i)} \in \mathbb{R}^n$ and each $y_i \in \mathbb{R}$. We then derived a solution for w that was given as $w = (XX^T)^{-1} Xy$, where $y = [y_1, y_2, \ldots, y_l]^T$ and X is an $n \times l$ matrix made up from the $x^{(i)}$ column vectors. That is,

$$X = [x^{(1)} \quad x^{(2)} \quad \ldots \quad x^{(l)}].$$

In a practical application of these equations, we would get poor results seen as a large value of the loss function

$$L(w, D_R) = \sum_{i=1}^{l} [y_i - x^{(i)T} w]^2$$

because there is an optimistic assumption that $g(x) = x^T w$ is a reasonable predictor for linear regression. In actual use, $g(x) = x^T w$ is a valid form of the prediction only if the data undergoes some preprocessing. Without the preprocessing, our linear regression analysis would have to assume that g was given by $g(x) = x^T w + b$, where *both* the vector w and scalar b must be assigned optimal values if we are to achieve the minimization of L.

a. In this exercise, you are to derive the preprocessing calculations on the input data that would allow us to use $g(x) = x^T w$ instead of $g(x) = x^T w + b$. In other words, what must we do to the data if the linear regression analysis is to come up with the result that $b = 0$? Your answer should include the equations that describe the preprocessing along with a justification as to why they allow us to assume that $b = 0$.

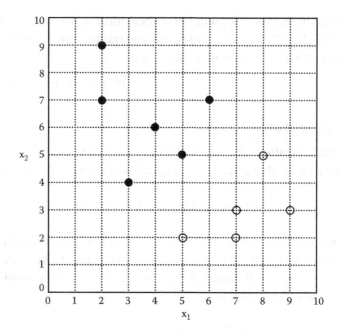

FIGURE 10.16   Training data for a linear SVM.

b. Suppose that you have derived $g(x)$ by means of a linear regression that works with the preprocessing that you have described in part (a). Given a test value x, describe how would you compute the corresponding y for x. You will need to state the equation that does this.

2. Consider the following labeled training data, also illustrated in Figure 10.16:

$((2\ 9), +1), ((2, 7), +1), ((3, 4), +1), ((4, 6), +1), ((5, 5), +1), ((6, 7), +1),$

$((5, 2), -1), ((7, 3), -1), ((7, 2), -1), ((8, 5), -1), ((9, 3), -1).$

On a copy of Figure 10.16, draw the two parallel margin lines that define the maximum margin and give their equations. What is the equation for the decision boundary? Specify the coordinates of the support vectors and mark their location on the graph.

3. Recall that a kernel is a function k such that for all x and z in the input space χ we have $k(x,z) = \langle\phi(x),\phi(z)\rangle$, where ϕ is a mapping from χ to an inner product feature space F. Show that $k(x,z) = (\langle x,z\rangle + 1)^2$ is a valid kernel.

4. Read the paper by Kumar, Gromiha, and Raghava (see reference [KG07]). This work is an excellent starting point for a project that

applies SVMs to a problem that is easily understood, namely: Given the primary sequence of a protein, does it bind to DNA? The goal of the project is to build an SVM that generates a prediction (binding, yes or no) when it is given the primary sequence of the protein.

The Web site for their DNAbinder server is at http://www.imtech.res.in/raghava/dnabinder/index.html.

The authors of the paper have kindly provided a download page, so data sets are available: http://www.imtech.res.in/raghava/dnabinder/download.html.

After you redo these experiments, try to improve the results by incorporating a different approach to the processing of the given primary sequences.

5. Essay question: The philosopher Karl Popper marks out *falsifiability* as a very important component of scientific methodology. For example, Albert Einstein proposed experiments that would falsify his general theory of relativity if the predictions made by the theory were not observed in the experiments. The observation that star light is bent by the gravitational field of the sun did not prove the general theory, but if this bending was *not* observed, then he would have to abandon his theory. Another famous example: When someone asked J. B. S. Haldane for an observation that would refute the theory of evolution, his terse answer was "the discovery of fossil rabbits in the Precambrian." The core benefit of such experiments and observations is that they lend credibility to the scientific theory. If the theory is packaged with an experiment that can prove it to be incorrect, then we can place more trust in the reality of the theory: It is not simply some dogmatic ad hoc statement that demands belief because it is outside the arena of criticism.

Write an essay that investigates the notion of falsifiability with respect to machine learning. Questions that you might consider

- Is the concept of falsifiability appropriate for a methodology that only strives to produce accurate predictions?

- Read Section 3.2 of Gillies book (see [G196]). He states:

   Now in all the machine-learning programs we considered, falsification played a crucial role. Typically the program generated a hypothesis from some data and the background knowledge. This hypothesis was then tested out against data and, if refuted by the data, was replaced by a new or modified hypothesis. Such repeated sequences of conjectures and refutations played a crucial role in all the machine-learning programs, and, in this respect, they agreed with Popper's methodology, except of course, that the conjectures were generated mechanically by the program rather than intuitively by humans.

Do you think that this is a reasonable statement for a machine-learning program that is working within an infinite-dimensional hypothesis space? Do you think that the falsification as described by Gillies really adds to the credibility of such a predictive methodology?

REFERENCES

[AL04]   E. ALPAYDIN. *Introduction to Machine Learning*. MIT Press, Cambridge, MA, 2004.

[Bw05]   J. R. BRADFORD AND D. R. WESTHEAD. Improved prediction of protein-protein binding sites using a support vector machines approach. *Bioinformatics*, **21**, (2005), 1487–1494.

[CJ04]   Q. CUI, T. JIANG, B. LIU, AND S. MA. Esub8: A novel tool to predict protein subcellular localizations in eukaryotic organisms. *BMC Bioinformatics*, **5**, (2004), http://www.biomedcentral.com/1471-2105/5/66.

[DH06]   L. DAVIS, J. HAWKINS, S. MAETSCHKE, AND M. BODÉN. Comparing SVM sequence kernels: A protein subcellular localization theme. *The 2006 Workshop on Intelligent Systems for Bioinformatics (WISB2006 Hobart, Australia)*, Conferences in Research and Practice in Information Technology (CRPIT), **73**, (2006) 39–47.

[GI96]   D. GILLIES. *Artificial Intelligence and Scientific Method*. Oxford University Press, London, 1996.

[GL05]   Y.-Z. GUO, M.-L. LI, K.-L. WANG, Z.-N. WEN, M.-C. LU, L.-X. LIU, AND L. JIANG. Fast Fourier transform-based support vector machine for prediction of G-protein coupled receptor subfamilies. *Acta Biochemica et Biophysica Sinica*, **37**, (2005), 759–766.

[GL06]   J. GUO AND Y. LIN. TSSub: eukaryotic protein subcellular localization by extracting features from profiles. *Bioinformatics*, **22**, (2006), 1784–1785.

[Hs01]   S. HUA AND Z. SUN. Support vector machine approach for protein subcellular localization prediction. *Bioinformatics*, **17**, (2001), 721–728.

[KA39]   W. KARUSH. *"Minima of Functions of Several Variables with Inequalities as Side Constraints."* M.Sc. Dissertation. Dept. of Mathematics, Univ. of Chicago, Chicago, 1939.

[KG07]   M. KUMAR, M. M. GROMIHA, AND G. P. S. RAGHAVA. Identification of DNA-binding proteins using support vector machines and evolutionary profiles. *BMC Bioinformatics*, **8**, (2007), 463.

[Kt51] H. W. KUHN AND A. W. TUCKER. *Nonlinear programming. Proceedings of 2nd Berkeley Symposium,* Berkeley: University of California Press (1951), 481–492.

[Lh06] H. H. LIN, L. Y. HAN, H. L. ZHANG, C. J. ZHENG, B. XIE, AND Y. Z. CHEN. Prediction of the functional class of lipid binding proteins from sequence-derived properties irrespective of sequence similarity. *Journal of Lipid Research,* **47,** (2006), 824–831.

[Ll97] C. A. LIPINSKI, F. LOMBARDO, B. W. DOMINY, AND P. J. FEENEY, Experimental and computational approaches to estimate solubility and permeability in drug discovery and development settings. *Advanced Drug Delivery Reviews,* **23** (1997), 3.

[Pc97] M. PASTOR, G. CRUCIANI, AND K. WATSON. A strategy for the incorporation of water molecules present in a ligand binding site into a three-dimensional quantitative structure-activity relationship analysis. *Journal Medicinal Chemistry,* **40** (1997), 4089–4102.

[Sc05] D. SARDA, G. H. CHUA, K.-B. LI, AND A. KRISHNAN. pSLIP: SVM based protein subcellular localization prediction using multiple physiochemical properties. *BMC Bioinformatics,* **6,** (2005), http://www.biomedcentral.com/1471-2105/6/152.

[Va95] V. VAPNIK. *The Nature of Statistical Learning Theory.* Springer, New York, 1995.

[Wc04] P. WALIAN, T. A. CROSS, AND B. K. JAP. Structural genomics of membrane proteins. *Genome Biology,* **5** (2004), Article 215.

Appendices

OVERVIEW

There are four appendices:

- Appendix A provides some basic definitions related to vector spaces. Not all of this material is needed in the book. Extra definitions have been included to increase general knowledge and sophistication in the area. This is important if you wish to understand some of the more abstract terminology contained in various research papers that have a mathematical setting.

- Appendix B is a basic review of linear algebra. The presentation style is somewhat informal, and although there are several theorems, none of them have proofs.

- Appendix C contains only the theorems mentioned in Appendix B, but with proofs.

- Appendix D defines the notion of taking a derivative with respect to a vector. This is useful in those parts of the text that discuss Lagrange multipliers.

As you go through Appendix B, try to appreciate the main pathway through the material. In particular, notice that more and more constraints are placed on the structure of the matrices being discussed. Singular value decomposition (SVD) is applied to general $m \times n$ matrices, but you have to learn all the preceding material before understanding why SVD works.

We assume that you are familiar with introductory topics related to real $m \times n$ matrices: matrix algebra, elementary matrices, vector spaces and

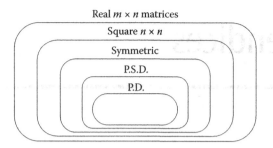

FIGURE A.1   Placing constraints on matrix structure.

subspaces, basis and dimension, rank of a matrix, linear transformations, matrix representations, inner products, and angles and orthogonality.

In these notes, we are mainly concerned with square matrices. Given a square matrix A, we can evaluate the determinant of A and, in particular, we can evaluate $\det(A - \mu I) = 0$; this allows us to calculate the eigenvalues and associated eigenvectors of the eigenvector equation $Av = \mu v$. The availability of the eigenvectors allows us to represent A in the form $A = P^{-1}DP$ as long as the eigenvectors form a complete set. This means that A is similar to a diagonal matrix (we say that A is diagonalizable). There are various applications—for example, computation of the inertial axis requires us to solve the eigenvector equation.

Adding more structure to a square matrix, we can suppose that A is symmetric: $A = A^T$. The main benefit is that A is not just diagonalizable; it is now orthogonally diagonalizable (Theorem 9). This means that there exists Q such that $A = QDQ^T$ with $Q^TQ = I$. Because the inner product between u and v is given as $u^T v$, we can see that the transpose has a special significance, and many proofs will combine the orthogonality of Q with the calculation of some inner product. For example, if A is symmetric, then the "generalized inner product" $u^T A v$ can be seen as $u^T A v = u^T QDQ^T v = (Q^T u)D(Q^T v)$. In other words, the inclusion of A in the inner product is equivalent to taking u and v, premultiplying both by Q^T (a rotation), and then forming an inner product that is similar to the usual inner product but with the terms in the sum weighted according to the entries on the main diagonal of D.

Finally, the insistence that A is a positive semidefinite (P.S.D) or positive definite (P.D.) matrix will give us more capabilities for A. For example, A positive definite implies that $x^T A x$ is always positive, and in fact it is easily seen that $x^T A x$ can act as a type of norm. We can also define the "square root" of a positive definite matrix.

APPENDIX A: A REVIEW OF VECTOR SPACES

A.1 Definition of a Vector Space

The familiar field of real numbers will be denoted by $\mathbb{R}$. We refer to the elements of $\mathbb{R}$ as *scalars*. By a vector space V we mean a nonempty set V with two mapping operations:

$(x, y) \mapsto x + y$ *addition* mapping $V \times V$ into V and

$(\lambda, x) \mapsto \lambda x$ *multiplication by a scalar* mapping $\mathbb{R} \times V$ into V such that all the following rules are satisfied:

1. Commutativity $\quad x + y = y + x$

2. Associativity $\quad (x + y) + z = x + (y + z)$

3. Additive identity $\quad$ There exists a vector $0 \in V$ such that $x + 0 = x$

4. Additive inverse

   For every $x \in V$ there exists $-x \in V$ such that $x + (-x) = 0$

5. Scalar associativity $\quad \alpha(\beta x) = (\alpha\beta)x$

6. Distributivity across scalar sums $\quad (\alpha + \beta)x = \alpha x + \beta x$

7. Distributivity across vector sums $\quad \alpha(x + y) = \alpha x + \alpha y$

8. Scalar identity $\quad 1x = x$

9. Scalar zero $\quad 0x = 0$

The first four rules make V an *additive group* with respect to the vectors that it contains.

A.2 Linear Combinations

Linear combinations are defined to be sums such as

$$\sum_{i=1}^{n} \alpha_i v^{(i)} \tag{A.1}$$

where $\alpha_i \in \mathbb{R}$, and $v^{(i)} \in V$ for $i = 1, 2, \ldots, n$.

A.3 Convex Combinations

A special linear combination is the *convex* combination

$$\sum_{i=1}^{n} \alpha_i v^{(i)} \quad \text{with} \quad v^{(i)} \in V, \text{all} \quad \alpha_i \geq 0 \quad \text{and} \quad \sum_{i=1}^{n} \alpha_i = 1. \tag{A.2}$$

A.4 Span of Vectors

Given a set of vectors $W = \{v^{(i)}\}_{i=1}^{m}$, the set of all possible linear combinations

$$\sum_{i=1}^{m} \alpha_i v^{(i)} \quad \text{with} \quad \alpha_i \in \mathbb{R} \quad \text{for all } i = 1, 2, \ldots, m. \tag{A.3}$$

is called the *span* of the set W.

A.5 Linear Independence

Given nonzero vectors $W = \{v^{(i)}\}_{i=1}^{m}$, consider the equation $\sum_{i=1}^{m} \alpha_i v^{(i)} = 0$. If this equation is true *only* when all $\alpha_i = 0$, then the set of vectors in W forms a *linearly independent* set. If this sum is zero with some $\alpha_j \neq 0$, then we can write $v^{(j)}$ as a linear combination of the other vectors, and we say that the set is linearly dependent.

A.6 Basis

Suppose we are given a set of vectors $B = \{v^{(i)}\}_{i=1}^{n}$ with the property that *any* vector in V can be written as a linear combination of vectors in B. In other words, the span of B is V. Such a set B is called a basis for the vector space V if it is also linearly independent. It can easily be shown that this linear combination is unique (there is only one way to define the set of α_j coefficients). Also, the number of vectors in a basis is always the same and is called the dimension of V.

A.7 Column Vectors

A common example of an n-dimensional vector space is $\mathbb{R}^n$, the infinite set of column vectors each having n real-valued components. We can write the typical vector in this space as $v = (v_1, \ldots, v_n)^{\mathrm{T}}$ where T denotes the transpose operation. All the conditions for a vector space are easily demonstrated if we assume that addition and scalar multiplication are defined in an elementwise fashion.

A.8 Standard Basis

The so-called *standard basis* for $\mathbb{R}^n$ is the set of n component vectors $\{e^{(i)}\}_{i=1}^{n}$ where $e^{(i)}$ is a column vector having all entries equal to 0 except for the ith entry, which is equal to 1. Using a notational convenience, we can write the jth entry of $e^{(i)}$ denoted by $e_j^{(i)}$ as $e_j^{(i)} = \delta_{ij}$ where δ_{ij} is the *Kronecker delta*, a symbol defined as

$$\delta_{ij} = \begin{cases} 1 & \text{if } i = j \\ 0 & \text{if } i \neq j. \end{cases} \quad\quad (A.4)$$

The reader must keep in mind that $\mathbb{R}^n$ is only one example of a vector space. The abstract definition given at the start of this section includes many other significant mathematical objects. For example, the set of all real-valued functions defined on a domain χ and denoted by $\mathbb{R}^\chi$ also constitutes a vector space with addition and scalar multiplication defined as

$$(f + g)(x) := f(x) + g(x) \quad \text{and} \quad (\alpha f)(x) := \alpha f(x) \quad \text{for all} \quad x \in \chi. \text{ (A.5)}$$

The other vector space properties are easily demonstrated.

A.9 Norms

A norm is a function $\|.\| : V \to \mathbb{R}_0^+$. That is, it maps a vector $v \in V$ to a nonnegative real number. For any $u, v \in V$, and $\alpha \in \mathbb{R}$, a norm must have the following properties:

$$\|v\| > 0 \quad \text{if } v \neq 0$$
$$\|\alpha v\| = |\alpha| \|v\| \quad \text{and}$$
$$\|u + v\| \leq \|u\| + \|v\|.$$

In general, once we have a norm, we can define a metric as

$$d(x, y) = \|x - y\|. \quad\quad (A.6)$$

A.10 Bilinear Forms

A bilinear form defined on a vector space V is a function F such that $F : V \times V \to \mathbb{R}$, with the following properties:

$$F(\alpha u + \beta v, w) = \alpha F(u, w) + \beta F(v, w) \quad\quad (A.7)$$

$$F(u, \alpha v + \beta w) = \alpha F(u, v) + \beta F(u, w) \quad\quad (A.8)$$

where u, v, and w are any vectors in V, and α and β are any numbers in $\mathbb{R}$.

Be sure to realize that the bilinear property extends to longer sums as in

$$F\left(\sum_{i=1}^{m}\alpha_i u^{(i)}, v\right) = \sum_{i=1}^{m}\alpha_i F(u^{(i)}, v). \tag{A.9}$$

A.10.1 Symmetric Bilinear Forms

If $F(u,v) = F(v,u)$ for all $u,v \in V$, then we say that F is a symmetric bilinear form.

A.11 Inner Product

For a vector space V, an inner product is a symmetric bilinear form that is strictly positive definite. Using the notation $\langle u,v \rangle$ to designate the inner product of u and v, we can restate these inner product properties as follows:

$$\langle u,v \rangle = \langle v,u \rangle \tag{A.10}$$

$$\langle u+v,w \rangle = \langle u,w \rangle + \langle v,w \rangle \tag{A.11}$$

$$\langle \alpha u,v \rangle = \alpha \langle u,v \rangle \tag{A.12}$$

$$\langle v,v \rangle \geq 0 \quad \text{and} \quad \langle v,v \rangle = 0 \Leftrightarrow v = 0. \tag{A.13}$$

Note that the first two equations grant us $\langle w, u+v \rangle = \langle w,u \rangle + \langle w,v \rangle$. It is also easily demonstrated that $\langle 0,v \rangle = \langle v,0 \rangle = 0$, and $\langle u,\alpha v \rangle = \alpha \langle u,v \rangle$.

A.12 Cauchy–Schwarz Inequality

For all $u,v \in V$,

$$\langle u,v \rangle^2 \leq \langle u,u \rangle \langle v,v \rangle. \tag{A.14}$$

The proof of this is simple but not necessarily immediately obvious. Here is the usual argument:

$$\langle \alpha u - v, \alpha u - v \rangle \geq 0 \tag{A.15}$$

and the bilinear properties give us

$$\langle \alpha u - v, \alpha u - v \rangle = \alpha^2 \langle u,u \rangle - 2\alpha \langle u,v \rangle + \langle v,v \rangle. \tag{A.16}$$

Setting $A = \langle u,u \rangle, B = \langle u,v \rangle$, and $C = \langle v,v \rangle$, we have $A\alpha^2 - 2B\alpha + C \geq 0$, and as $A \geq 0$, we can also write $A(A\alpha^2 - 2B\alpha + C) \geq 0$. However,

$$A(A\alpha^2 - 2B\alpha + C) = (A\alpha - B)^2 + AC - B^2. \qquad (A.17)$$

If this quantity is to be greater than or equal to zero for any value of α, for example, when $\alpha = B/A$, then we must have $AC - B^2 \geq 0$ and the inequality is proven.

A.13 Minkowski's Inequality

For all $u,v \in V$,

$$\sqrt{\langle u+v,u+v \rangle} \leq \sqrt{\langle u,u \rangle} + \sqrt{\langle v,v \rangle}. \qquad (A.18)$$

The proof is easily derived as an extension of the previous proof of the Cauchy–Schwarz inequality. We use the same definitions of A, B, and C. Then bilinear properties give

$$\langle u+v,u+v \rangle = A + 2B + C$$

$$\leq A + 2\sqrt{AC} + C \qquad (A.19)$$

$$= \left(\sqrt{A} + \sqrt{C} \right)^2$$

the last inequality due to Cauchy and Schwarz. Taking square roots of both sides gives us the needed result.

A.14 Induced Norms

If we have an inner product defined for a vector space, then we can define a norm for the space by using

$$\|v\| := \sqrt{\langle v,v \rangle}. \qquad (A.20)$$

This is usually described by saying that the inner product *induces* the norm. Using $\|.\|$ notation for this induced norm, we can rewrite the Cauchy–Schwarz and Minkowski inequalities as

$$\langle u,v \rangle \leq \|u\|\|v\| \qquad (A.21)$$

and

$$\|u+v\| \leq \|u\| + \|v\| \qquad (A.22)$$

respectively. The Minkowski inequality can now be seen as the familiar *triangle inequality*.

A.15 Metric Spaces

A metric space M is a set of points with a function d that maps any pair of points in M to a nonnegative real value that can be regarded as the distance between the two points. More formally, the metric space is a pair (M, d), where $d: M \times M \to [0, \infty)$, and the points in M can be almost any mathematical object as long as the mapping satisfies the metric properties listed as follows:

1. $0 \le d(x,y) < \infty$ for all pairs $x, y \in M$,

2. $d(x,y) = 0$ if and only if $x = y$,

3. $d(x,y) = d(y,x)$ for all pairs $x, y \in M$,

4. $d(x,y) \le d(x,z) + d(z,y)$ for all $x, y, z \in M$.

The last property is referred to as the *triangle inequality*.

As seen earlier, if we have a norm, we can use it to induce a metric on the normed vector space:

$$d(u,v) := \|u - v\|. \tag{A.23}$$

So, note the progression: With an inner product we can always induce a norm, and with a norm we can always induce a metric.

A.16 Inner Product, Norm and Metric for Column Vectors

The preceding discussion deals with the abstract definitions of inner product, norm, and metric. Let us now focus on our earlier example of an n-dimensional Euclidean space $\mathbb{R}^n$. Recall that a typical vector is represented as the column vector $v = [v_1, v_2, \ldots, v_n]^\mathrm{T}$. We can then write the inner product as the symmetric bilinear form

$$\langle u,v \rangle = \sum_{i=1}^{n} u_i v_i = u^\mathrm{T} v \tag{A.24}$$

and from this we can induce a norm as

$$\|v\| := \sqrt{\langle v,v \rangle} = \sqrt{\sum_{i=1}^{n} v_i^2} \tag{A.25}$$

and finally, the metric for $\mathbb{R}^n$ is

$$d(u,v) := \|u-v\| = \sqrt{\sum_{i=1}^{n}(u_i - v_i)^2}. \qquad \text{(A.26)}$$

The Cauchy–Schwarz and Minkowski inequalities become

$$\left(\sum_{i=1}^{n} u_i v_i\right)^2 \leq \left(\sum_{i=1}^{n} u_i^2\right)\left(\sum_{i=1}^{n} v_i^2\right) \text{ and } \sqrt{\sum_{i=1}^{n}(u_i + v_i)^2} \leq \sqrt{\sum_{i=1}^{n} u_i^2} + \sqrt{\sum_{i=1}^{n} v_i^2}$$

$$\text{(A.27)}$$

respectively.

APPENDIX B: REVIEW OF LINEAR ALGEBRA

B.1 The Gram–Schmidt Orthogonalization Procedure

Every inner product space has an orthonormal basis. Let $x^{(1)}, x^{(2)}, \ldots, x^{(n)}$ be a basis for an n-dimensional inner product space V. Define

$$u^{(1)} = \frac{x^{(1)}}{\|x^{(1)}\|}$$

$$\text{(B.1)}$$

$$u^{(2)} = \frac{g^{(2)}}{\|g^{(2)}\|}, \quad \text{where } g^{(2)} = x^{(2)} - \langle u^{(1)}, x^{(2)}\rangle u^{(1)}.$$

As $\{x^{(1)}, x^{(2)}\}$ is a linearly independent set, $u^{(2)} \neq 0$. It is also true that $\langle u^{(1)}, u^{(2)}\rangle = 0$ with $\|u^{(2)}\| = 1$. We can continue for higher values of the index on u by using

$$u^{(k)} = \frac{g^{(k)}}{\|g^{(k)}\|}, \quad \text{where } g^{(k)} = x^{(k)} - \sum_{i=1}^{k-1}\langle u^{(i)}, x^{(k)}\rangle u^{(i)}. \qquad \text{(B.2)}$$

It can be demonstrated that the set of vectors $\{u^{(i)}\}_{i=1}^{n}$ is orthonormal in the n-dimensional vector space V. Because an orthonormal set is linearly independent, it provides an orthonormal basis for V.

B.2 Matrix Column and Row Spaces

Given an $m \times n$ matrix A, with columns written as $\{c^{(i)}\}_{i=1}^{n} \subset \mathbb{R}^m$, the subspace $L(c^{(1)}, c^{(2)}, \ldots, c^{(n)})$ spanned by all the columns of A is called the column space of A and is written as $COLSPACE(A)$. Similarly, we can work with the rows of A: $\{r^{(i)}\}_{i=1}^{m} \subset \mathbb{R}^n$ and consider $L(r^{(1)}, r^{(2)}, \ldots, r^{(m)})$ to be the $ROWSPACE(A)$.

It can be shown that the dimension of $COLSPACE(A)$ equals the dimension $ROWSPACE(A)$, and we denote this as the rank of A. In short,

$$rank(A) = \dim COLSPACE(A) = \dim ROWSPACE(A). \quad \text{(B.3)}$$

If A is an $m \times n$ matrix, then all the solutions of $Ax = 0$ form a linear space called the null space of A, written as $NULLSPACE(A)$, with the dimension called *nullity*(A). Introductory linear algebra informs us that

$$rank(A) + nullity(A) = n. \quad \text{(B.4)}$$

If we are given two matrices A and B for which the product AB can be defined (say A is $m \times n$ and B is $n \times p$), then

$$COLSPACE(AB) = \{ABx \mid x \in \mathbb{R}^p\}$$
$$\subseteq \{Ay \mid y \in \mathbb{R}^n\} = COLSPACE(A) \quad \text{(B.5)}$$

because $Bx \in \mathbb{R}^n$ for any $x \in \mathbb{R}^p$. So,

$$\dim COLSPACE(AB) \leq \dim COLSPACE(A) \quad \text{(B.6)}$$

and hence

$$rank(AB) \leq rank(A). \quad \text{(B.7)}$$

Because

$$ROWSPACE(AB) = COLSPACE((AB)^{\mathrm{T}})$$
$$= COLSPACE(B^{\mathrm{T}} A^{\mathrm{T}})$$
$$\subseteq COLSPACE(B^{\mathrm{T}}) \quad \text{(B.8)}$$
$$= ROWSPACE(B)$$

we also have

$$rank(AB) \leq rank(B) \quad \text{(B.9)}$$

and so

$$rank(AB) \leq \min\{rank(A), rank(B)\}. \qquad \text{(B.10)}$$

Note that if A is square (say $n \times n$) and invertible, then we can make the more definite statement

$$rank(AB) = rank(B). \qquad \text{(B.11)}$$

This is true because if A is invertible, it has "full rank," and so $rank(A) = n$. Thus,

$$rank(AB) \leq \min\{n, rank(B)\} = rank(B). \qquad \text{(B.12)}$$

The last equality is true because $rank(B) \leq \min\{n, p\}$. However,

$$A^{-1}(AB) = B \implies$$

$$rank(B) = rank(A^{-1}(AB)) \qquad \text{(B.13)}$$

$$\leq rank(AB).$$

So, we are left with $rank(AB) = rank(B)$ if A is square and invertible.

B.3 Eigenvectors and Eigenvalues

Suppose A is a square $n \times n$ matrix. The eigenvector equation is

$$Av = \mu v \qquad \text{(B.14)}$$

where vector v is called the eigenvector, and the constant μ is called the eigenvalue. We want to obtain all the combinations of μ and v for which Equation B.14 is true. Note that if we have a vector v for which Equation B.14 is true, then any multiple of that vector will also satisfy the equation. So, it is best to view Equation B.14 as essentially specifying a direction instead of some particular vector. More precisely, the eigenvectors of A define those directions in which the effect of A on v is simply to change the length of v while leaving its direction unchanged: Multiplying an eigenvector v of A by A does not rotate v. Later, when we need particular eigenvectors, we will ensure that each has a unit length, that is, $\|v\| = 1$.

Equation B.1.4 can be rewritten as

$$(A - \mu I)v = 0 \qquad \text{(B.15)}$$

and matrix theory tells us that this equation has nonzero solution v if and only if $A - \mu I$ is a singular matrix, that is, when $\det(A - \mu I) = 0$. For an $n \times n$ matrix A, the left side of the last equation represents a polynomial of degree n in μ. The roots of this polynomial are the eigenvalues of (B.14).

It is possible that a particular root appears multiple times, and this represents an extra complication in an eigenvector analysis that we will discuss later. However, in most practical applications, the polynomial will have n different roots corresponding to n different eigenvectors.

B.3.1 The Eigenvector Matrix

If we consider all n eigenvector equations,

$$Av^{(i)} = v^{(i)}\mu_i \qquad i = 1, 2, \ldots, n \tag{B.16}$$

we can combine all of them by using the single matrix equation

$$AV = V \, diag[\mu_1, \mu_2, \ldots, \mu_n]. \tag{B.17}$$

where the diagonal matrix

$$diag[\mu_1, \mu_2, \ldots, \mu_n] = \begin{pmatrix} \mu_1 & & 0 \\ & \ddots & \\ 0 & & \mu_n \end{pmatrix} \tag{B.18}$$

and

$$V = [v^{(1)} \quad v^{(2)} \quad \cdots \quad v^{(n)}] = \begin{pmatrix} v_1^{(1)} & \cdots & v_1^{(n)} \\ \vdots & \ddots & \vdots \\ v_n^{(1)} & \cdots & v_n^{(n)} \end{pmatrix} \tag{B.19}$$

This is easily verified by considering the corresponding columns in Equation (B.17).

B.3.2 Motivation for Eigenvectors

When students are first introduced to the eigenvector Equation (B.14), they might wonder why this particular equation is so very important. What is so special about eigenvectors? We will show that eigenvectors tell us a lot about the behavior of a matrix A when it is considered to be a matrix operating on an arbitrary vector, say x, to produce $y = Ax$. Another topic of concern is the ability to write (B.17) as $A = V \, diag[\mu_1, \mu_2, \ldots, \mu_n] V^{-1}$. The invertibility of V will depend on certain properties of A. However, before considering this, we need more matrix theory, and this will be covered in the following theorems.

B.4 Definition: Similar Matrices

Suppose A and B are both $n \times n$ matrices; then we say that B is similar to A if there exists an invertible $n \times n$ matrix P such that

$$B = P^{-1}AP. \tag{B.20}$$

Theorem 1: Similar Matrices Have the Same Eigenvalue Spectrum

Let

$$B = P^{-1}AP \tag{B.21}$$

then $\{\lambda, x\}$ is an eigenpair of B, then $\{\lambda, Px\}$ is an eigenpair of A.

Theorem 2: Schur Decomposition

Let A be an $n \times n$ matrix with real eigenvalues. Then there exists an $n \times n$ orthogonal matrix X such that

$$X^T A X = \Delta \tag{B.22}$$

where Δ is an upper triangular matrix having the eigenvalues of A as its diagonal entries.

Definition: Diagonalizable

A matrix A is diagonalizable if it is similar to a diagonal matrix. That is, A is diagonalizable if there exists invertible P such that

$$D = P^{-1}AP \tag{B.23}$$

where D is a diagonal matrix.

Theorem 3: Diagonalizability

An $n \times n$ matrix A is diagonalizable if and only if it has n linearly independent eigenvectors.

Theorem 4: Distinct Eigenvalues and Linear Independence

Suppose A has k eigenvalues $\lambda_1, \lambda_2, \ldots, \lambda_k$ that are all distinct. Then the corresponding eigenvectors $v^{(1)}, v^{(2)}, \ldots, v^{(k)}$ are linearly independent.

Theorem 5: Distinct Eigenvectors $\Rightarrow$ Diagonalizability

If an $n \times n$ matrix A has n distinct eigenvalues, then it is diagonalizable.

Definition: Complete Set of Eigenvectors

A complete set of eigenvectors for an $n \times n$ matrix A is a set of n linearly independent eigenvectors of A.

Definition: Deficient Matrix

A matrix that does not have a complete set of eigenvectors is said to be deficient.

B.5 Diagonalizability and Distinctiveness of Eigenvalues

The preceding discussion says

n distinct eigenvalues $\Rightarrow$ complete set $\Rightarrow$ diagonalizability.

However, what if eigenvalues are not distinct? If there are repeated eigenvalues, then there is still the possibility that a complete set of eigenvectors can exist. We elaborate the possibilities as follows:

1. Distinct eigenvalues

2. A repeated eigenvalue and a complete set of eigenvectors

3. A repeated eigenvalue and a deficient set of eigenvectors

To distinguish the last two cases, we need more definitions and theorems.

Definition: Eigenspace

Suppose the $n \times n$ matrix A has an eigenvalue λ. Then the subspace containing the vector 0 and all the eigenvectors belonging to λ are called the eigenspace associated with λ.

Note that if u and v are both eigenvectors of A with the same eigenvalue, that is, $Au = \lambda u$ and $Av = \lambda v$, then

$$A(av + bu) = aAv + bAu = \lambda(av + bu) \tag{B.24}$$

and so we see that any linear combination of eigenvectors belonging to λ is also an eigenvector belonging to λ.

Theorem 6: Linear Independence of the Union of Bases

Suppose an $n \times n$ matrix A has k eigenvalues $\lambda_1, \lambda_2, \ldots, \lambda_k$ that are all distinct. Let B_i be a basis for the eigenspace associated with λ_i for $i = 1, 2, \ldots,$ k. Then $\cup_{i=1}^{k} B_i$ is a linearly independent set of eigenvectors of A.

Note that this theorem does not promise that the linear independent set of eigenvectors is complete because there is no guarantee that $\cup_{i=1}^{k} B_i$ contains n linearly independent eigenvectors.

B.6 Algebraic and Geometric Multiplicity

Definition: Algebraic and Geometric Multiplicity

For an $n \times n$ matrix A let

$$\det(A - \tau I) = (\lambda_1 - \tau)^{\alpha_1} (\lambda_2 - \tau)^{\alpha_2} \ldots (\lambda_m - \tau)^{\alpha_m} \tag{B.25}$$

where the λ_i are distinct. We call α_i the algebraic multiplicity of λ_i. Note that

$$\sum_{k=1}^{m} \alpha_k = n. \tag{B.26}$$

Let $\lambda_1, \lambda_2, \ldots, \lambda_m$ be the distinct eigenvalues of the $n \times n$ matrix A, and let γ_i be the dimension of the eigenspace associated with λ_i $i = 1, 2, \ldots, m$. Then γ_i is referred to as the geometric multiplicity of λ_i.

Theorem 7: Geometric Multiplicity Is Less Than or Equal to Algebraic Multiplicity

For any eigenvalue λ_i of A, we have $\gamma_i \leq \alpha_i$.

Theorem 8: Matrix A Diagonalizable if Algebraic Multiplicity Equals Geometric Multiplicity

An $n \times n$ matrix A with distinct eigenvalues $\lambda_1, \lambda_2, \ldots, \lambda_m$ $m \leq n$ is diagonalizable if and only if for all $i = 1, 2, \ldots, m$ $\gamma_i = \alpha_i$.

B.7 Orthogonal Diagonalization

Definition: Orthogonal Diagonalization

Consider an $n \times n$ matrix A such that there exists an orthogonal nonsingular matrix Q and

$$Q^{\mathrm{T}} A Q = D \tag{B.27}$$

for some diagonal matrix D. Then we say that A is orthogonally diagonalizable.

Definition: Symmetric Matrices

An $n \times n$ matrix A is symmetric if it has the property $A = A^T$.

Note: This additional property of A allows us to say more about eigenvalues, eigenvectors, and diagonalizability. There are two significant benefits that are gained when $A = A^T$. First, A is always diagonalizable. In fact, we will eventually see that the earlier issue of algebraic multiplicity possibly being not equal to geometric multiplicity goes away. Symmetry guarantees that $\gamma_i = \alpha_i$ for all $i = 1, 2, \ldots m$. Second, A is not just diagonalizable; it is orthogonally diagonalizable. This means that the eigenvectors are mutually orthogonal, and hence provide us with a very useful orthonormal basis.

Definition: Orthogonal Matrix

A square matrix Q is said to be an orthogonal matrix if

$$Q^T Q = QQ^T = I. \tag{B.28}$$

Note 1: If Q is nonsingular, then we have a wonderful simplicity for the inverse of the matrix Q; in fact, $Q^T Q = I$ implies $Q^{-1} = Q^T$.

Note 2: $Q^T Q = I$ if and only if the columns of Q are such that

$$q^{(i)T} q^{(j)} = \delta_{ij} := \begin{cases} 1 & i = j \\ 0 & i \neq j. \end{cases} \tag{B.29}$$

Note 3: $Q^T Q = I$, and Q nonsingular implies that Q^T is a left inverse of Q. As Q is square, the left inverse is also a right inverse of Q, that is, $QQ^T = I$. So the rows of Q are also orthonormal.

Use of Orthogonal Basis

If we use the orthogonal basis vectors to form a change-of-basis matrix, then we have the following benefit: Orthogonal operations on a vector (multiplying a vector by an orthogonal matrix) do not change the length of a vector.

$$\| x \|^2 = x^T x$$

$$= x^T Q^T Q x$$

$$= (Qx)^T Q x$$

$$= \| Qx \|^2 . \tag{B.30}$$

Also, we see that the angle between two vectors given by

$$\cos\theta = \langle u,v\rangle/(\|u\|\|v\|) \tag{B.31}$$

remains invariant after an orthogonal operation on both vectors because the vector lengths in the denominator are unchanged, and

$$\langle u,v\rangle = u^{\mathrm{T}}v = u^{\mathrm{T}}Q^{\mathrm{T}}Qv = (Qu)^{\mathrm{T}}Qv = \langle Qu,Qv\rangle. \tag{B.32}$$

Theorem 9: Symmetry iff Orthogonal Diagonalizability

Matrix A is symmetric if and only if A is orthogonally diagonalizable.

Corollary

Suppose $n \times n$ symmetric matrix A has k distinct eigenvalues $\lambda_1, \lambda_2, \ldots, \lambda_k$ with algebraic multiplicities $\alpha_1, \alpha_2, \ldots, \alpha_k$, respectively, and geometric multiplicities $\gamma_1, \gamma_2, \ldots, \gamma_k$, respectively. Then,

$$\sum_{i=1}^{k} \alpha_i = \sum_{i=1}^{k} \gamma_i = n \tag{B.33}$$

and

$$\gamma_i = \alpha_i \quad \text{for } i = 1, 2, \ldots, k. \tag{B.34}$$

Corollary

Suppose $n \times n$ symmetric matrix A has n (not necessarily distinct) eigenvalues written as $d_1, d_2, \ldots, d_n$. Then there exists $n \times n$ orthogonal Q such that

$$Q^{\mathrm{T}}AQ = diag(d_1, d_2, \ldots, d_n). \tag{B.35}$$

B.8 Spectral Decomposition

Definition: Spectral Decomposition of a Symmetric Matrix

Consider an $n \times n$ symmetric matrix A. The foregoing discussion asserts the existence of Q and D such that

$$A = QDQ^{\mathrm{T}}. \tag{B.36}$$

We can rewrite this last equation as

$$A = \sum_{i=1}^{n} d_i q^{(i)} q^{(i)\mathrm{T}} \tag{B.37}$$

where $q^{(i)}$ is the i-th column of Q.

This decomposition is unique (aside from the order of the d_i entries) if all multiplicities are equal to 1. However, if there is an eigenvalue of multiplicity greater than one, then the associated eigenspace admits eigenvectors that are linear combinations of two or more eigenvectors, and so the uniqueness property is lost. However, we can derive another decomposition with a summation over k entries (k = number of distinct eigenvalues), and it is unique.

Theorem 10: Spectral Decomposition

Consider $\lambda_1, \lambda_2, \ldots, \lambda_k$ to be the k distinct eigenvalues of $n \times n$ symmetric matrix A, with $v_1, v_2, \ldots, v_k$ representing the respective multiplicities of these eigenvalues. Assume the existence of Q and D such that

$$A = QDQ^{\mathrm{T}} \tag{B.38}$$

and so

$$A = \sum_{i=1}^{n} d_i q^{(i)} q^{(i)\mathrm{T}}. \tag{B.39}$$

Define

$$S_j = \{i \mid d_i = \lambda_j\} \tag{B.40}$$

as the set of indices i for which

$$d_i = \lambda_j. \tag{B.41}$$

Then A has the unique spectral decomposition given by

$$A = \sum_{j=1}^{k} \lambda_j E_j \tag{B.42}$$

where

$$E_j = \sum_{i \in S_j} q^{(i)} q^{(i)\mathrm{T}} \quad \text{for} \quad j = 1, 2, \ldots, k. \tag{B.43}$$

Note: We will see that this is a very useful representation of a symmetric matrix. It might seem that such a representation is of limited value because, after all, we need A to be symmetric. In practice this requirement is fulfilled for several situations. For example, if the n^2 entries of A are derived from all possible distances between n points in some metric space, $(a_{ij} = dist(x^{(i)}, x^{(j)}))$, then the symmetry of a distance calculation $dist(x^{(i)}, x^{(j)}) = dist(x^{(j)}, x^{(i)})$ leads to symmetry of A, that is, $a_{ij} = a_{ji}$. Also, for any arbitrary $m \times n$ matrix M, note that the $m \times m$ matrix MM^T and the $n \times n$ matrix $M^T M$ are both symmetric. This is easily demonstrated by doing a substitution into the identities: $(A^T)^T = A$ and $(AB)^T = B^T A^T$, to get

$$(MM^T)^T = (M^T)^T M^T = MM^T \quad \text{and} \quad (M^T M)^T = M^T (M^T)^T = M^T M.$$

$$(B.44)$$

The following theorem is useful in that it provides a condition for the invertibility of $M^T M$.

Theorem 11: $rank(M^T M) = rank(M)$

For any $m \times n$ matrix M, the symmetric matrix $M^T M$ has

$$rank(M^T M) = rank(M). \tag{B.45}$$

If we substitute C^T for M and note that

$$rank(M^T) = rank(M) \tag{B.46}$$

we can derive the next corollary.

Corollary

For any $m \times n$ matrix C, the symmetric matrix CC^T has

$$rank(CC^T) = rank(C). \tag{B.47}$$

B.9 Positive Definiteness

Definition: Positive Definiteness

The matrix product $M^T M$ has an interesting property. If we multiply on the left by x^T and multiply on the right by x, we get

$$x^T M^T M x = (Mx)^T Mx$$
$$= \| Mx \|^2 \geq 0.$$

$$(B.48)$$

If M has nullity > 0, then there are x vectors such that

$$Mx = 0. \qquad (B.49)$$

Otherwise,

$$\| Mx \|^2 > 0 \quad \text{for all } x. \qquad (B.50)$$

This property of the matrix $M^T M$ is given a name. We say that $M^T M$ is positive definite because

$$x^T M^T M x > 0 \quad \text{for all } x \neq 0. \qquad (B.51)$$

In a more general context we have the following definitions:

Definition: Definiteness

For $n \times n$ symmetric matrix A:

A is positive semidefinite if and only if $x^T A x \geq 0$ for all $x \in \mathbb{R}^n$ with $x \neq 0$.

A is positive definite if and only if $x^T A x > 0$ for all $x \in \mathbb{R}^n$ with $x \neq 0$.

A is negative semidefinite if and only if $x^T A x \leq 0$ for all $x \in \mathbb{R}^n$ with $x \neq 0$.

A is negative definite if and only if $x^T A x < 0$ for all $x \in \mathbb{R}^n$ with $x \neq 0$.

A is indefinite if $x^T A x$ can have both positive and negative values for all $x \in \mathbb{R}^n$ with $x \neq 0$.

Theorem 12: Positive Definiteness Is Invariant Under a Similarity Transformation

Suppose A is positive definite. Let

$$B = P^T A P \qquad (B.52)$$

with P invertible. Then B is also positive definite.

Theorem 13: Positive Definite if All Eigenvalues Positive

For symmetric matrix A, the following statements are equivalent:

(a) A is positive definite.

(b) All eigenvalues of A are positive.

We can derive, for positive definite matrix A, a matrix B that is very much like the square root of A.

Theorem 14: Possession of "Square Root" if Positive Definite

An $n \times n$ symmetric matrix A is positive definite if and only if there exists an invertible $n \times n$ matrix B such that

$$A = B^{\mathrm{T}} B. \tag{B.53}$$

Theorem 15: $A^{\mathrm{T}}A$ and AA^{T} Share Positive Eigenvalues

Suppose A is an $m \times n$ matrix with $rank(A) = r$. Then the positive eigenvalues of $A^{\mathrm{T}}A$ are equal to the positive eigenvalues of AA^{T}.

Note that this theorem only guarantees equivalence of the sets of eigenvalues and not eigenvectors (as made clear in the proof of the theorem).

B.10   Singular Value Decomposition

We now explore a very useful factorization that can be applied to *any* real matrix. Suppose we assume (without proof) that an $m \times n$ real matrix A can be written as

$$A = USV^{\mathrm{T}} \tag{B.54}$$

where both U and V are orthogonal matrices.

We get some interesting observations by computing the following:

$$AA^{T} = US^{2}U^{\mathrm{T}} \tag{B.55}$$

and

$$A^{T}A = VS^{2}V^{\mathrm{T}}. \tag{B.56}$$

From our previous discussion we know that both $A^{\mathrm{T}}A$ and AA^{T} are symmetric positive definite matrices and have the same eigenvalues. Because both $A^{\mathrm{T}}A$ and AA^{T} are symmetric matrices, we also know that they are diagonalizable and can be written as portrayed in the last two equations. With these observations, we can use the equations to compute U, V, and S.

Can we say anything about the eigenvectors of AA^{T} and $A^{\mathrm{T}}A$? There is an elegant relationship among them. If $v^{(i)}$ is a column vector of V, then it is an eigenvector. In fact,

$$A^{\mathrm{T}}Av^{(i)} = s_{i}^{2}v^{(i)} \tag{B.57}$$

where s_i^2 is the ith entry of the diagonal matrix S^2. Premultiplying this last equation by A, we get

$$AA^TAv^{(i)} = s_i^2 Av^{(i)} \tag{B.58}$$

and with some parentheses put in for clarity, we see that

$$(AA^T)(Av^{(i)}) = s_i^2(Av^{(i)}) \tag{B.59}$$

which says that we can get an eigenvector of AA^T by taking an eigenvector of A^TA and premultiplying it by A.

We now consider the $A = USV^T$ decomposition in a more formal fashion.

Theorem 16: The Singular Value Decomposition

If A is an $m \times n$ matrix with

$$rank(A) = r, \tag{B.60}$$

then there exists orthogonal $m \times m$ matrix U and orthogonal $n \times n$ matrix V such that

$$A = USV^T. \tag{B.61}$$

where S is a diagonal matrix with nonnegative entries on its diagonal, and

(i) S is a full rank $r \times r$ diagonal matrix if $r = m = n$.

(ii) S is $r \times n$ and has extra zero-filled columns if $r = m < n$.

(iii) S is $m \times r$ and has extra zero-filled rows if $r = n < m$.

(iv) S is $m \times n$ and has extra zero-filled rows and columns if $r < m$ and $r < n$.

APPENDIX C : LINEAR ALGEBRA THEOREM PROOFS

Theorem 1: Similar Matrices Have the Same Eigenvalue Spectrum

Let

$$B = P^{-1}AP. \tag{C.1}$$

If $\{\lambda, x\}$ is an eigenpair of B, then $\{\lambda, Px\}$ is an eigenpair of A.

Proof:

Note that similar matrices have the same characteristic polynomial. Recall that the eigenvalues of a matrix A are the roots of the characteristic polynomial

$$\det(\lambda I - A). \tag{C.2}$$

If $B = P^{-1}AP$, then

$$\det(\lambda I - B) = \det(\lambda I - P^{-1}AP)$$

$$= \det(\lambda P^{-1}IP - P^{-1}AP)$$

$$= \det(P^{-1}(\lambda I - A)P)$$

$$= \det(P^{-1})\det(\lambda I - A)\det(P) \tag{C.3}$$

$$= \frac{1}{\det(P)}\det(\lambda I - A)\det(P)$$

$$= \det(\lambda I - A).$$

Consequently, B and A have the same eigenvalues.

To show that $\{\lambda, Px\}$ is an eigenpair of A when $\{\lambda, x\}$ is an eigenpair of B, we note that

$$Bx = \lambda x \;\;\Rightarrow\;\; P^{-1}APx = \lambda x \;\;\Rightarrow\;\; A(Px) = \lambda(Px). \tag{C.4}$$

Theorem 2: Schur Decomposition

Let A be an $n \times n$ matrix with real eigenvalues. Then there exists an $n \times n$ orthogonal matrix X such that

$$X^T A X = \Delta \tag{C.5}$$

where Δ is an upper triangular matrix having the eigenvalues of A as its diagonal entries.

Proof:

Let $\lambda_1, \lambda_2, \ldots, \lambda_n$ be the (not necessarily distinct) eigenvalues of A, and let $v^{(1)}$ be an eigenvector of A associated with λ_1. Assume $v^{(1)}$ is normalized so that

$$v^{(1)T}v^{(1)} = 1 \tag{C.6}$$

Let V be any $n \times n$ orthogonal matrix having $v^{(1)}$ as its first column (this can be derived using the Gram–Schmidt orthogonalization procedure). Writing V in partitioned form with $v^{(1)}$ as the first column, we get

$$V = [v^{(1)} U] \tag{C.7}$$

with U an $n \times n - 1$ matrix containing all the other orthonormal columns. Note that

$$Av^{(1)} = \lambda_1 v^{(1)} \tag{C.8}$$

and

$$U^{\mathrm{T}} v^{(1)} = 0 \tag{C.9}$$

so

$$
\begin{aligned}
V^{\mathrm{T}} A V &= \begin{bmatrix} v^{(1)\mathrm{T}} A v^{(1)} & v^{(1)\mathrm{T}} A U \\ U^{\mathrm{T}} A v^{(1)} & U^{\mathrm{T}} A U \end{bmatrix} \\[2mm]
&= \begin{bmatrix} \lambda_1 v^{(1)\mathrm{T}} v^{(1)} & v^{(1)\mathrm{T}} A U \\ \lambda_1 U^{\mathrm{T}} v^{(1)} & U^{\mathrm{T}} A U \end{bmatrix} \\[2mm]
&= \begin{bmatrix} \lambda_1 & v^{(1)\mathrm{T}} A U \\ 0 & B \end{bmatrix}
\end{aligned} \tag{C.10}
$$

where the $(n-1) \times (n-1)$ matrix

$$B = U^{\mathrm{T}} A U. \tag{C.11}$$

Using Theorem 1 and the cofactor expansion formula for determinants, we get

$$
\begin{aligned}
\det(A - \lambda I) &= \det(V^{\mathrm{T}} A V - \lambda I) \\
&= (\lambda_1 - \lambda) \det(B - \lambda I)
\end{aligned} \tag{C.12}
$$

where the identity matrix in $B - \lambda I$ is $(n-1) \times (n-1)$.

As the eigenvalues of $V^{\mathrm{T}} A V$ are the same as those of A, the eigenvalues of B must be $\lambda_1, \lambda_2, \ldots, \lambda_n$.

The foregoing observations provide us with insights to generate an inductive proof. If $n = 2$, then B is a scalar, and must be λ_2. In this case, $V^T A V$ would be upper triangular, and we are done. For $n > 2$, we can show that the result holds for $n \times n$ matrices if it holds for $(n-1) \times (n-1)$ matrices. Because B is $(n-1) \times (n-1)$, we can assume the existence of Y such that

$$Y^T B Y = R \tag{C.13}$$

where R is an upper triangular matrix with $\lambda_1, \lambda_2, \ldots, \lambda_n$ as its diagonal entries. Now we define $n \times n$ matrix W as

$$W = \begin{bmatrix} 1 & 0^T \\ 0 & Y \end{bmatrix}. \tag{C.14}$$

We see that W is orthogonal because Y is orthogonal. Finally, we let $X = VW$. Note that

$$(VW)^T VW = W^T V^T VW = I \tag{C.15}$$

and so X is also orthogonal. Furthermore,

$$
\begin{aligned}
X^T A X &= W^T V^T A V W \\
&= \begin{bmatrix} 1 & 0^T \\ 0 & Y^T \end{bmatrix} \begin{bmatrix} \lambda_1 & v^{(1)T} A U \\ 0 & B \end{bmatrix} \begin{bmatrix} 1 & 0^T \\ 0 & Y \end{bmatrix} \\
&= \begin{bmatrix} \lambda_1 & v^{(1)T} A U Y \\ 0 & Y^T B Y \end{bmatrix} \\
&= \begin{bmatrix} \lambda_1 & v^{(1)T} A U Y \\ 0 & R \end{bmatrix}
\end{aligned}
\tag{C.16}
$$

where this matrix is upper triangular with diagonal elements $\lambda_1, \lambda_2, \ldots, \lambda_n$.

Theorem 3: Diagonalizability

An $n \times n$ matrix A is diagonalizable if and only if it has n linearly independent eigenvectors.

Proof:

($\Rightarrow$):

If A is diagonalizable, then there exists invertible P such that

$$P^{-1}AP = D \tag{C.17}$$

and if we write matrix P as a sequence of column vectors, that is,

$$P = [v^{(1)} \quad v^{(2)} \quad \ldots \quad v^{(n)}] \tag{C.18}$$

then

$$
\begin{aligned}
AP &= A[v^{(1)} \quad v^{(2)} \quad \ldots \quad v^{(n)}] \\
&= [Av^{(1)} \quad Av^{(2)} \quad \ldots \quad Av^{(n)}]
\end{aligned} \tag{C.19}
$$

and

$$PD = [d_1 v^{(1)} \quad d_2 v^{(2)} \quad \ldots \quad d_n v^{(n)}]. \tag{C.20}$$

Because $AP = PD$, we can equate corresponding columns to get

$$Av^{(k)} = d_k v^{(k)} \quad k = 1, 2, \ldots, n. \tag{C.21}$$

This means that the $v^{(k)}$ are eigenvectors of A with associated eigenvalues $d_k \quad k = 1, 2, \ldots, n$. Moreover, these eigenvectors must be linearly independent because, if they were not, then the matrix P would have $\det(P) = 0$ and P would not be invertible.

($\Leftarrow$):

If we have n linearly independent eigenvectors, we can construct the matrix system $AP = PD$. Then, because P is invertible (the n column vectors being linearly independent), we can write

$$P^{-1}AP = D \tag{C.22}$$

and necessity is proven.

Theorem 4: Distinct Eigenvalues and Linear Independence

Suppose A has k eigenvalues $\lambda_1, \lambda_2, \ldots, \lambda_k$ that are all distinct. Then the corresponding eigenvectors $v^{(1)}, v^{(2)}, \ldots, v^{(k)}$ are linearly independent.

Proof:

The proof is by induction on k. When $k = 1$, we have a single eigenvector $v^{(1)}$ which gives us a trivial linearly independent set. Assume now that we have $k-1$ linearly independent eigenvectors associated with $k-1$

distinct eigenvalues. If we consider k eigenvectors corresponding to k distinct eigenvalues, are they linearly independent? The answer is "yes" if any linear combination of these k eigenvectors set to zero implies that each cofficient in the linear combination must be zero. Consequently, we start with the assumption that

$$\sum_{i=1}^{k} c_i v^{(i)} = 0. \tag{C.23}$$

Multiply this equation by the matrix $(A - \lambda_1 I)$ to get

$$\sum_{i=2}^{k} c_i \left(A v^{(i)} - \lambda_1 v^{(i)} \right) = 0. \tag{C.24}$$

Terms corresponding to $i = 1$ do not appear in (C.24) because $v^{(1)}$ is an eigenvector and so

$$A v^{(1)} - \lambda_1 v^{(1)} = 0. \tag{C.25}$$

Note that (C.24) can be rewritten as:

$$\sum_{i=2}^{k} c_i (\lambda_i - \lambda_1) v^{(i)} = 0. \tag{C.26}$$

Since we have $k-1$ eigenvectors, in (C.26), we can use the induction hypothesis to claim that they are linearly independent. Hence, this linear combination must have all coefficieents equal to zero, that is

$$c_i (\lambda_i - \lambda_1) = 0 \quad \text{for all} \quad i = 2, \ldots, k. \tag{C.27}$$

Because the eigenvalues are distinct, $\lambda_i - \lambda_1 \neq 0$ for each $i = 2, \ldots, k$ and this implies that

$$c_i = 0 \quad \text{for all} \quad i = 2, \ldots, k. \tag{C.28}$$

Going back to the first sum, in (C.23) we get $c_1 v^{(1)} = 0$ since all the other terms are zero. Hence, our initial assumption (C.23) leads to the observation that

$$c_i = 0 \quad i = 1, 2, \ldots, k \tag{C.29}$$

and so the k eigenvectors are linearly independent.

Theorem 5: Distinct Eigenvectors ⇒ Diagonalizability

If an $n \times n$ matrix A has n distinct eigenvalues, then it is diagonalizable.

Proof:
The n distinct eigenvalues of A have, by Theorem 4, n linearly independent eigenvectors, and so, by Theorem 3, A is diagonalizable.

Theorem 6: Linear Independence of the Union of Bases

Suppose an $n \times n$ matrix A has k eigenvalues $\lambda_1, \lambda_2, \ldots, \lambda_k$ that are all distinct. Let B_i be a basis for the eigenspace associated with λ_i for $i = 1, 2, \ldots, k$. Then $\cup_{i=1}^{k} B_i$ is a linearly independent set of eigenvectors of A.

Proof:
For $i = 1, 2, \ldots, k$, let the basis vectors in B_i be indexed using number pairs (i, j) and listed as

$$B_i = \{v^{(i,j)}\}_{j=1}^{|B_i|} \tag{C.30}$$

where $|B_i|$ is the number of basis vectors for the eigenspace associated with λ_i. Assume that a linear combination of all these vectors is set to zero:

$$\sum_{i=1}^{k} \sum_{j=1}^{|B_i|} \alpha_{ij} v^{(i,j)} = 0. \tag{C.31}$$

Note that for each $i = 1, 2, \ldots, k$ the sum

$$\sum_{j=1}^{|B_i|} \alpha_{ij} v^{(i,j)} \tag{C.32}$$

is a linear combination of vectors in the eigenspace associated with λ_i, and so is itself an eigenvector of λ_i. Let

$$\sum_{j=1}^{|B_i|} \alpha_{ij} v^{(i,j)} = u^{(i)} \tag{C.33}$$

so

$$\sum_{i=1}^{k} u^{(i)} = 0. \tag{C.34}$$

The vectors $\{u^{(i)}\}_{i=1}^k$ are either 0 or are eigenvectors associated with distinct eigenvalues $\{\lambda_i\}_{i=1}^k$, in which case they are linearly independent, and hence the sum

$$\sum_{i=1}^k u^{(i)} = 0 \tag{C.35}$$

would require that

$$u^{(i)} = 0 \quad i = 1, 2, \ldots, k. \tag{C.36}$$

This means that

$$\sum_{j=1}^{|B_i|} \alpha_{ij} v^{(i,j)} = 0 \quad \text{for } i = 1, 2, \ldots, k. \tag{C.37}$$

Because these vectors are the basis B_i, they are linearly independent, and so $\alpha_{ij} = 0$ for $i = 1, 2, \ldots, k$ and $j = 1, 2, \ldots, |B_i|$. Consequently, all the α_{ij} have been shown to be 0, and so $\cup_{i=1}^k B_i$ is a linear independent set of eigenvectors.

Theorem 7: Geometric Multiplicity Is Less Than or Equal to Algebraic Multiplicity

For any eigenvalue λ_i of A, we have

$$\gamma_i \leq \alpha_i. \tag{C.38}$$

Proof:
Suppose the eigenspace for eigenvalue λ has dimension γ. This means that there is a linearly independent set

$$V = (v^{(1)}, v^{(2)}, \ldots, v^{(\gamma)}) \tag{C.39}$$

where each $v^{(i)}$ $i = 1, 2, \ldots, \gamma$ is an eigenvector of A belonging to λ. From linear algebra we know that the basis V can be extended to provide a basis B for $\mathbb{R}^n$. We collect these vectors to make a matrix B as follows:

$$B = [v^{(1)} \quad \cdots \quad v^{(\gamma)} \quad v^{(\gamma+1)} \quad \cdots \quad v^{(n)}]. \tag{C.40}$$

We can assume that these column vectors are orthogonal. If they are not, then the Gram–Schmidt orthogonalization procedure can be used to replace them with column vectors that are orthonormal. Furthermore, note that if Gram–Schmidt orthogonalization is used, we do not destroy the property that the first γ vectors are eigenvectors associated with λ because new orthonormal vectors are linear combinations of the given $v^{(i)}$ $i=1,2,\ldots,\gamma$ (see the definition of the eigenspace associated with an eigenvalue). Consequently, we can write B as

$$B=[V \quad W] \tag{C.41}$$

with $AV=\lambda V$ and $W^{\mathrm{T}}V=0$. Now consider

$$
\begin{aligned}
B^{\mathrm{T}}AB &= \begin{bmatrix} V^{\mathrm{T}} \\ W^{\mathrm{T}} \end{bmatrix} [AV \quad AW] \\[2mm]
&= \begin{bmatrix} V^{\mathrm{T}}AV & V^{\mathrm{T}}AW \\ W^{\mathrm{T}}AV & W^{\mathrm{T}}AW \end{bmatrix} \\[2mm]
&= \begin{bmatrix} \lambda I & V^{\mathrm{T}}AW \\ 0 & W^{\mathrm{T}}AW \end{bmatrix}.
\end{aligned}
\tag{C.42}
$$

As similar matrices have the same eigenvalue spectrum, we know that the eigenvalues of A are the roots of the following polynomial in r:

$$
\det\left(\begin{bmatrix} \lambda I & V^{\mathrm{T}}AW \\ 0 & W^{\mathrm{T}}AW \end{bmatrix} -rI\right)=\det\left(\begin{bmatrix} (\lambda-r)I & V^{\mathrm{T}}AW \\ 0 & W^{\mathrm{T}}AW-rI \end{bmatrix}\right) \tag{C.43}
$$

$$=(\lambda-r)^{\gamma}\det(W^{\mathrm{T}}AW-rI)$$

the last equality due to the cofactor expansion of determinants. It is possible that the geometric multiplicity of λ —call it α —might be larger than γ, for example, if the polynomial equation

$$\det(W^{\mathrm{T}}AW-rI)=0 \tag{C.44}$$

also has $r=\lambda$ as a root. However, it is clear from C.43 that $\gamma\leq\alpha$.

Theorem 8: Matrix A Diagonalizable if Algebraic Multiplicity Equals Geometric Multiplicity

An $n \times n$ matrix A with distinct eigenvalues $\lambda_1, \lambda_2, \ldots, \lambda_m$ $m \le n$ is diagonalizable if and only if for all $i = 1, 2, \ldots, m$ $\gamma_i = \alpha_i$.

Proof:
$(\Rightarrow)$:

Suppose A is diagonalizable, that is,

$$P^{-1}AP = D \tag{C.45}$$

where

$$D = \operatorname{diag}(r_1, r_2, \ldots, r_n) \tag{C.46}$$

and P is an orthogonal matrix. Because A and D are similar, they have the same eigenvalue spectrum, and we can write

$$\prod_{i=1}^{n}(\lambda - r_i) \tag{C.47}$$

as the characteristic polynomial where the r_i are not necessarily distinct. Because P is invertible, we can write

$$AP = PD$$

and it is clear that the j-th column of P belongs to E_{r_j}, the eigenspace associated with eigenvalue r_j. If r_j has algebraic multiplicity α_j, that is to say, it appears α_j times in the product

$$\prod_{i=1}^{n}(\lambda - r_i) \tag{C.48}$$

then there will be α_j of these vectors belonging to E_{r_j}. Furthermore, because P is invertible and these vectors are columns of P, they must form a linearly independent set. This means that the dimensionality of the associated eigenspace (the geometric multiplicity) is at least as large as α_j, that is,

$$\gamma_j \ge \alpha_j. \tag{C.49}$$

However, we already know from Theorem 7 that $\gamma_j \leq \alpha_j$, so we must conclude that

$$\gamma_j = \alpha_j. \tag{C.50}$$

($\Leftarrow$):

Suppose the $\gamma_j = \alpha_j$ for all $j = 1, 2, \ldots, m$. The characteristic polynomial for A is

$$\prod_{i=1}^{m} (\lambda - r_i)^{\alpha_i} \tag{C.51}$$

with

$$\sum_{i=1}^{m} \alpha_i = n. \tag{C.52}$$

The equation $\gamma_j = \alpha_j$ tells us that

$$\dim(E_{r_i}) = \alpha_i \quad \text{for all} \quad j = 1, 2, \ldots, m. \tag{C.53}$$

Let B_i with γ_i vectors be a basis for the eigenspace

$$E_{r_i} \quad i = 1, 2, \ldots, m. \tag{C.54}$$

Then Theorem 6 tells us that the union of these spaces forms a linearly independent set that spans $\mathbb{R}^n$ because

$$\sum_{i=1}^{m} \alpha_i = n. \tag{C.55}$$

Theorem 3 says that these n linearly independent eigenvectors are sufficient to give us the diagonalizability of A.

Theorem 9: Symmetry if Orthogonal Diagonalizability

Matrix A is symmetric if and only if A is orthogonally diagonalizable.

Proof:
($\Leftarrow$) (doing the easy implication first):

If $Q^T A Q = D$ with Q orthogonal, then

$$A = QDQ^T \tag{C.56}$$

and

$$A^T = (QDQ^T)^T = QD^TQ^T = QDQ^T \tag{C.57}$$

because $D^T = D$, and so we have $A = A^T$, that is, A is symmetric.

($\Rightarrow$):

First, we show that if A is an $n \times n$ symmetric matrix with real entries, then the eigenvalues of A are real, and a real eigenvector is associated with each real eigenvalue. Suppose an eigenvalue of A is written as the complex number

$$\lambda = a + ib \tag{C.58}$$

with associated eigenvector

$$x = u + iv \tag{C.59}$$

where $i^2 = -1$. From the definition of eigenvector we have

$$A(u + iv) = (a + ib)(u + iv) \tag{C.60}$$

and left multiplication by $(u - iv)^T$ we see that

$$(u - iv)^T A(u + iv) = (a + ib)(u - iv)^T(u + iv) \tag{C.61}$$

which simplifies to

$$u^T Au + v^T Av = (a + ib)(u^Tu + v^Tv) \tag{C.62}$$

and noting that $A = A^T$ we see that

$$u^T Av = v^T Au. \tag{C.63}$$

Because

$$u + iv \neq 0 \Rightarrow u^Tu + v^Tv > 0 \tag{C.64}$$

we must have $b = 0$ as there is no term on the left side involving i. Going back to our previous eigenvector equation with $b = 0$, we get

$$A(u + iv) = a(u + iv) \tag{C.65}$$

and so $u+iv$ is an eigenvector if $Au=au$ and $Av=av$ with either $u\neq 0$ or $v\neq 0$. In summary, any eigenvalue is real, and we can find a real eigenvector as its companion.

Next we need to show that, for any eigenvalue λ, its geometric multiplicity γ is equal to its algebraic multiplicity α. As a start, we recall that, if the dimension of E_λ (the eigenspace for λ) is γ, then we know that α (the algebraic multiplicity of λ) is at least γ, that is, $\gamma \leq \alpha$. We follow an approach that is reminiscent of Theorem 7. Let $v^{(1)}, v^{(2)}, \ldots, v^{(\gamma)}$ be an orthonormal basis for E_λ, and let us extend this to a full orthonormal basis

$$B=(v^{(1)}, v^{(2)}, \ldots, v^{(\gamma)}, v^{(\gamma+1)}, \ldots, v^{(n)}) \tag{C.66}$$

that spans $\mathbb{R}^n$. As in Theorem 7, we can write

$$B^{\mathrm{T}}AB = \begin{bmatrix} \lambda I & V^{\mathrm{T}}AW \\ 0 & W^{\mathrm{T}}AW \end{bmatrix} \tag{C.67}$$

where

$$B=[V \quad W], \tag{C.68}$$

$$V=[v^{(1)} \quad v^{(2)} \quad \ldots \quad v^{(\gamma)}], \tag{C.69}$$

and

$$W=[v^{(\gamma+1)} \quad v^{(\gamma+2)} \quad \ldots \quad v^{(n)}]. \tag{C.70}$$

However, we can utilize the symmetry of A as follows: $B^{\mathrm{T}}AB$ is symmetric because it is orthogonally similar to the symmetric matrix A, implying

$$V^{\mathrm{T}}AW = 0 \tag{C.71}$$

and hence

$$B^{\mathrm{T}}AB = \begin{bmatrix} \lambda I & 0 \\ 0 & W^{\mathrm{T}}AW \end{bmatrix}. \tag{C.72}$$

We know that

$$\gamma = \dim(E_\lambda) = \text{nullity}(\lambda I - A) \tag{C.73}$$

and because rank + nullity for a square $n \times n$ matrix equals n, we see that

$$rank(\lambda I - A) = n - \gamma. \tag{C.74}$$

Recalling a previous observation that

$$rank(XY) = rank(Y) \tag{C.75}$$

when X is invertible, we see that

$$rank(B^T(\lambda I - A)B) = rank(\lambda I - A) = n - \gamma. \tag{C.76}$$

However,

$$B^T(\lambda I - A)B = \lambda B^T B - B^T AB$$

$$= \lambda I - \begin{bmatrix} \lambda I_\gamma & 0 \\ 0 & W^T AW \end{bmatrix} \tag{C.77}$$

$$= \begin{bmatrix} 0_{\gamma \times \gamma} & 0 \\ 0 & \lambda I - W^T AW \end{bmatrix}.$$

We are using γ subscripts to denote numbers of rows and columns in these partitions.

Noting that

$$rank\left(\begin{bmatrix} 0_{\gamma \times \gamma} & 0 \\ 0 & \lambda I - W^T AW \end{bmatrix}\right) = rank(\lambda I - W^T AW) \tag{C.78}$$

$$= n - \gamma$$

we see that $\lambda I - W^T AW$ has full rank, and so λ cannot be a root of its characteristic polynomial. This means that $\gamma = \alpha$ for the λ eigenvalue. As the foregoing argument can be applied to all the different eigenvalues, we have algebraic multiplicity equal to geometric multiplicity for all eigenvalues. We can now invoke Theorem 8, which tells us that A is diagonalizable because $\gamma_i = \alpha_i$ for all $i = 1, 2, \ldots, m$.

It remains only to prove that A is orthogonally diagonalizable. Note that application of Gram–Schmidt (as done in Theorem 7) gives us an orthonormal base with preservation of eigenvector u inside E_λ but with no guarantee that the basis vectors will be eigenvectors for the other eigenspaces. Our approach is to apply the strategy of Theorem 7 to each and every different eigenspace. Each application produces a set of orthonormal eigenvectors acting as a basis for each eigenspace, and then Theorem 6 guarantees that the union of these eigenvectors gives a linearly independent set for all $\mathbb{R}^n$. There is one last detail, showing that we have orthogonality for two basis vectors belonging to different eigenspaces. This is also facilitated by the symmetry of A. For $u \in E_\lambda$ and $v \in E_\lambda$ with $\lambda_u \neq \lambda_v$, we have

$$\lambda_u \langle u, v \rangle = \lambda_u u^\mathrm{T} v$$

$$= (Au)^\mathrm{T} v$$

$$= u^\mathrm{T} A^\mathrm{T} v$$

$$= u^\mathrm{T} A v \qquad \text{(C.79)}$$

$$= u^\mathrm{T} v \lambda_v$$

$$= \lambda_v \langle u, v \rangle.$$

As $\lambda_u \neq \lambda_v$, we can only have this happen if

$$\langle u, v \rangle = 0, \qquad \text{(C.80)}$$

that is, if u is orthogonal to v. The theorem is finally proven.

Theorem 10: Spectral Decomposition

Consider $\lambda_1, \lambda_2, \ldots, \lambda_k$ to be the k distinct eigenvalues of $n \times n$ symmetric matrix A, with $v_1, v_2, \ldots, v_k$ representing the respective multiplicities of these eigenvalues. Assume the existence of Q and D such that

$$A = QDQ^\mathrm{T} \qquad \text{(C.81)}$$

and so

$$A = \sum_{i=1}^{n} d_i q^{(i)} q^{(i)\mathrm{T}}. \qquad \text{(C.82)}$$

Define

$$S_j = \{i \mid d_i = \lambda_j\} \qquad \text{(C.83)}$$

as the set of indices i for which

$$d_i = \lambda_j.$$ (C.84)

Then A has the unique spectral decomposition given by

$$A = \sum_{j=1}^{k} \lambda_j E_j$$ (C.85)

where

$$E_j = \sum_{i \in S_j} q^{(i)} q^{(i)\mathrm{T}} \quad \text{for} \quad j = 1, 2, \ldots, k.$$ (C.86)

Proof:
The spectral decomposition is a simple extension of the previous

$$A = \sum_{i=1}^{n} d_i q^{(i)} q^{(i)\mathrm{T}}$$ (C.87)

decomposition in that all $q^{(i)} q^{(i)\mathrm{T}}$ terms with coefficients $d_i = \lambda_j$ are gathered together and summed to yield the E_j term. The important issue here is that this produces a unique decomposition. To prove this, we start with an alternate $\bar{Q}$ and $\bar{D}$ such that

$$A = \bar{Q} \bar{D} \bar{Q}^{\mathrm{T}}$$ (C.88)

and then we proceed as we did earlier.
Define

$$\bar{S}_j = \{ i \mid \bar{d}_i = \lambda_j \}$$ (C.89)

and let

$$\bar{E}_j = \sum_{i \in \bar{S}_j} \bar{q}^{(i)} \bar{q}^{(i)\mathrm{T}}$$ (C.90)

where $\bar{q}^{(i)}$ $i = 1, 2, \ldots, n$ are the n columns of the matrix $\bar{Q}$. Then the new formulation for A is

$$A = \sum_{j=1}^{k} \lambda_j \bar{E}_j$$ (C.91)

which can be seen as the same expression as C.85 if we can show that

$$\bar{E}_j = E_j \tag{C.92}$$

for all $j = 1, 2, \ldots, k$. If we focus our attention on any particular E_j, then

$$E_j = \sum_{i \in S_j} q^{(i)} q^{(i)\mathrm{T}}$$

$$= Q_j Q_j^\mathrm{T} \tag{C.93}$$

where Q_j is a matrix made up of those columns from Q with indices belonging to the S_j set. Similarly,

$$\bar{E}_j = \sum_{i \in S_j} \bar{q}^{(i)} \bar{q}^{(i)\mathrm{T}}$$

$$= \bar{Q}_j \bar{Q}_j^\mathrm{T}. \tag{C.94}$$

Now, because both Q and $\bar{Q}$ are orthogonal matrices, we have

$$Q_j^\mathrm{T} Q_j = I \tag{C.95}$$

and

$$\bar{Q}_j^\mathrm{T} \bar{Q}_j = I. \tag{C.96}$$

Both Q_j and $\bar{Q}_j$ span the eigenspace associated with λ_j. In fact, we can consider the $\bar{Q}_j$ column vectors to provide a basis for column vectors in Q_j, implying that there exists a matrix H_j such that

$$Q_j = \bar{Q}_j H_j. \tag{C.97}$$

Note that

$$H_j^\mathrm{T} H_j = H_j^\mathrm{T} \bar{Q}_j^\mathrm{T} \bar{Q}_j H_j$$

$$= (\bar{Q}_j H_j)^\mathrm{T} \bar{Q}_j H_j \tag{C.98}$$

$$= Q_j^\mathrm{T} Q_j$$

$$= I.$$

So, H_j is an orthogonal matrix, and hence,

$$\begin{aligned}
E_j &= Q_j Q_j^{\mathrm{T}} \\
&= \bar{Q}_j H_j (\bar{Q}_j H_j)^{\mathrm{T}} \\
&= \bar{Q}_j H_j H_j^{\mathrm{T}} \bar{Q}_j^{\mathrm{T}} \\
&= \bar{Q}_j \bar{Q}_j^{\mathrm{T}} \\
&= \bar{E}_j
\end{aligned} \tag{C.99}$$

as required.

Theorem 11: $rank(M^{\mathrm{T}}M) = rank(M)$

For any $m \times n$ matrix M, the symmetric matrix $M^{\mathrm{T}}M$ has

$$rank(M^{\mathrm{T}}M) = rank(M). \tag{C.100}$$

Proof:

Because $M^{\mathrm{T}}M$ has dimensions $n \times n$, we can write

$$rank(M) + nullity(M) = n = rank(M^{\mathrm{T}}M) + nullity(M^{\mathrm{T}}M) \tag{C.101}$$

We observe that

$$NULLSPACE(M) \subseteq NULLSPACE(M^{\mathrm{T}}M) \tag{C.102}$$

because

$$Mx = 0 \implies M^{\mathrm{T}}Mx = 0. \tag{C.103}$$

Now consider all x such that

$$M^{\mathrm{T}}Mx = 0. \tag{C.104}$$

Note that

$$\langle Mx, Mx \rangle = (Mx)^{\mathrm{T}}(Mx) = x^{\mathrm{T}}(M^{\mathrm{T}}Mx) = x^{\mathrm{T}}0 = 0. \tag{C.105}$$

Hence $Mx = 0$, and so

$$x \in NULLSPACE(M). \tag{C.106}$$

Consequently,

$$NULLSPACE(M^T M) \subseteq NULLSPACE(M)$$

$$\Rightarrow NULLSPACE(M^T M) = NULLSPACE(M)$$

$$\Rightarrow nullity(M) = nullity(M^T M)$$

$$\Rightarrow rank(M) = rank(M^T M)$$

(C.107)

as required.

If we substitute C^T for M and note that

$$rank(M^T) = rank(M)$$

(C.108)

we can derive the corollary.

Corollary

For any $m \times n$ matrix C, the symmetric matrix CC^T has

$$rank(CC^T) = rank(C).$$

(C.109)

Theorem 12: Positive Definiteness Is Invariant Under a Similarity Transformation

Suppose A is positive definite. Let

$$B = P^T AP$$

(C.110)

with P invertible. Then B is also positive definite.

Proof:

Because A is symmetric, we have

$$B^T = P^T A^T P = P^T AP = B$$

(C.111)

and so B is also symmetric. Furthermore, for any $x \neq 0$, we have

$$x^T Bx = x^T P^T APx = (Px)^T A(Px) > 0$$

(C.112)

as A is positive definite and $Px \neq 0$. So, B is also positive definite.

Theorem 13: Positive Definite if All Eigenvalues Positive

For symmetric matrix A, the following statements are equivalent:

a. A is positive definite.

b. All eigenvalues of A are positive.

Proof:
$(a \Rightarrow b)$:

Suppose λ is an eigenvalue of A, and $x \neq 0$ is its associated eigenvector. Then

$$x^T A x > 0 \quad \Rightarrow \quad x^T A x = x^T \lambda x = \lambda \| x \|^2 > 0 \qquad \text{(C.113)}$$

and for $x \neq 0$ this can only be true if $\lambda > 0$.

$(b \Rightarrow a)$:

Because A is real and symmetric, we can find orthogonal invertible Q and a diagonal matrix of eigenvalues D such that $Q^T A Q = D$, and so

$$A = Q D Q^T. \qquad \text{(C.114)}$$

For any $x \neq 0$, let $y = Q^T x$. Then,

$$x^T A x = x^T Q D Q^T x = y^T D y = \sum_{i=1}^{n} y_i^2 \lambda_i \qquad \text{(C.115)}$$

and so $x^T A x > 0$ if all $\lambda_i > 0$.

Theorem 14: Possession of "Square Root" if Positive Definite

An $n \times n$ symmetric matrix A is positive definite if and only if there exists an invertible $n \times n$ matrix B such that

$$A = B^T B. \qquad \text{(C.116)}$$

Proof:
$(\Leftarrow)$:

$B^T B$ is symmetric, and for any $x \neq 0$, we have

$$x^T A x = \| B x \|^2 > 0 \qquad \text{(C.117)}$$

if B is invertible.

$(\Rightarrow)$:

A real and symmetric implies the existence of orthogonal invertible Q and a diagonal matrix of positive eigenvalues D such that

$$A = Q D Q^T. \qquad \text{(C.118)}$$

Because each eigenvalue is positive, we can define $\sqrt{D}$ as the diagonal matrix with diagonal entries

$$\left(\sqrt{\lambda_1}, \sqrt{\lambda_2}, \ldots, \sqrt{\lambda_n}\right). \tag{C.119}$$

Then let $B = Q\sqrt{D}Q^{\mathrm{T}}$, noting that

$$B^{\mathrm{T}}B = \left(Q\sqrt{D}Q^{\mathrm{T}}\right)^{\mathrm{T}}\left(Q\sqrt{D}Q^{\mathrm{T}}\right)$$

$$= Q\sqrt{D}^{\mathrm{T}}Q^{\mathrm{T}}Q\sqrt{D}Q^{\mathrm{T}}$$

$$= Q\sqrt{D}^{\mathrm{T}}I\sqrt{D}Q^{\mathrm{T}} \tag{C.120}$$

$$= QDQ^{\mathrm{T}}$$

$$= A$$

as required.

Theorem 15: $A^{\mathrm{T}}A$ and AA^{T} Share Positive Eigenvalues

Suppose A is an $m \times n$ matrix with $rank(A) = r$. Then the positive eigenvalues of $A^{\mathrm{T}}A$ are equal to the positive eigenvalues of AA^{T}.

Proof:
Suppose $A^{\mathrm{T}}A$ has λ as an eigenvalue with multiplicity γ. Then the $n \times n$ matrix $A^{\mathrm{T}}A$ is symmetric, and we can find an $n \times \gamma$ matrix V with orthonormal columns such that

$$A^{\mathrm{T}}AV = \lambda V. \tag{C.121}$$

Now, let $U = AV$ and consider the following:

$$AA^{\mathrm{T}}U = AA^{\mathrm{T}}AV$$

$$= A\lambda V$$

$$= \lambda AV \tag{C.122}$$

$$= \lambda U.$$

So, λ is also an eigenvalue of AA^{T} (although the associated eigenvector may be different from the eigenvector associated with λ for $A^{\mathrm{T}}A$); in fact, the number of components are different in U and V if $m \neq n$).

The multiplicity of λ is also γ for AA^T because

$$rank(U) = rank(AV) = rank((AV)^T AV)$$

$$= rank(V^T A^T AV) = rank(\lambda V^T V) \qquad \text{(C.123)}$$

$$= rank(\lambda I_\gamma) = \gamma.$$

Theorem 16: The Singular Value Decomposition

If A is an $m \times n$ matrix with

$$rank(A) = r, \qquad \text{(C.124)}$$

then there exists orthogonal $m \times m$ matrix U and orthogonal $n \times n$ matrix V such that

$$A = USV^T \qquad \text{(C.125)}$$

where S is a diagonal matrix with nonnegative entries on its diagonal and:

  (i) S is a full rank $r \times r$ diagonal matrix if $r = m = n$.

  (ii) S is $r \times n$ and has extra zero-filled columns if $r = m < n$.

  (iii) S is $m \times r$ and has extra zero-filled rows if $r = n < m$.

  (iv) S is $m \times n$ and has extra zero-filled rows and columns if $r < m$ and $r < n$.

Proof:
The most difficult case is the last one. We go through this and consider the other cases to be small modifications of this case. Considering Theorems 13, 14, and 15 and the rank of A, we note that both $A^T A$ and AA^T have r positive eigenvalues. Let S be the $r \times r$ diagonal matrix with main diagonal entries that are the square roots of these eigenvalues. We can then observe that there exists an $n \times n$ orthogonal matrix

$$V = [v^{(1)} \quad v^{(2)} \quad \cdots \quad v^{(n)}] \qquad \text{(C.126)}$$

such that

$$V^T A^T AV = \begin{bmatrix} S^2 & 0 \\ 0 & 0 \end{bmatrix}. \qquad \text{(C.127)}$$

We can partition V so that

$$V = [V_1 \quad V_2] \tag{C.128}$$

with $n \times r$ matrix V_1 containing the first r columns $v^{(1)}, v^{(2)}, \ldots, v^{(r)}$, and $n \times (n-r)$ matrix V_2 having the remaining $n-r$ columns $v^{(r+1)}, v^{(r+2)}, \ldots,$ $v^{(n)}$. Considering the previous equation, we have

$$V_1^T A^T A V_1 = S^2 \tag{C.129}$$

and

$$V_2^T A^T A V_2 = 0. \tag{C.130}$$

Note that V_2 has columns in the null space of A, that is,

$$A v^{(k)} = 0 \quad \text{for} \quad r+1 \leq k \leq n. \tag{C.131}$$

To see this, observe that $A v^{(k)} \neq 0$ implies that

$$\|A v^{(k)}\| > 0 \tag{C.132}$$

and so,

$$v^{(k)T} A^T A v^{(k)} > 0. \tag{C.133}$$

This would put a nonzero entry on the diagonal of $V_2^T A^T A V_2$, contradicting

$$V_2^T A^T A V_2 = 0. \tag{C.134}$$

Thus,

$$A V_2 = 0 \tag{C.135}$$

a fact that we will need later.
Define

$$U = [U_1 \quad U_2] \tag{C.136}$$

to be an $m \times m$ orthogonal matrix with the $m \times r$ matrix

$$U_1 = A V_1 S^{-1} \tag{C.137}$$

and the $m\times(m-r)$ matrix U_2 containing a set of orthonormal columns that makes U orthogonal. This will ensure that

$$U_2^{\mathrm{T}}U_1 = 0 \qquad\qquad (C.138)$$

yielding

$$U_2^{\mathrm{T}}AV_1 = 0 \qquad\qquad (C.139)$$

because

$$U_2^{\mathrm{T}}AV_1S^{-1} = 0 . \qquad\qquad (C.140)$$

With U, V, and S defined, we can derive $U^{\mathrm{T}}AV$ to be

$$
\begin{aligned}
U^{\mathrm{T}}AV &= \begin{bmatrix} U_1^{\mathrm{T}}AV_1 & U_1^{\mathrm{T}}AV_2 \\ U_2^{\mathrm{T}}AV_1 & U_2^{\mathrm{T}}AV_2 \end{bmatrix} \\
&= \begin{bmatrix} S^{-1}V_1^{\mathrm{T}}A^{\mathrm{T}}AV_1 & S^{-1}V_1^{\mathrm{T}}A^{\mathrm{T}}AV_2 \\ U_2^{\mathrm{T}}AV_1 & U_2^{\mathrm{T}}AV_2 \end{bmatrix} \qquad (C.141) \\
&= \begin{bmatrix} S^{-1}S^2 & S^{-1}V_1^{\mathrm{T}}A^{\mathrm{T}}(0) \\ 0 & U_2^{\mathrm{T}}(0) \end{bmatrix} \\
&= \begin{bmatrix} S & 0 \\ 0 & 0 \end{bmatrix}
\end{aligned}
$$

as required.

APPENDIX D: DIFFERENTIATION WITH RESPECT TO A VECTOR

Let us suppose that we have a function $f(w)$ that is defined on a domain of vectors (points in a Euclidean space) and the function range is the real line. That is, $f(.): \mathbb{R}^n \to \mathbb{R}$.

We start with the definition:

$$
\frac{\partial f(w)}{\partial w} \equiv \begin{pmatrix} \dfrac{\partial f(w)}{\partial w_1} \\ \vdots \\ \dfrac{\partial f(w)}{\partial w_n} \end{pmatrix} . \qquad\qquad (D.1)
$$

Start with the simple case where $f(w) = w^T v$. So, $f(w)$ is just an inner product, a function mapping w onto the real line. Now,

$$f(w) = w^T v = \sum_{i=1}^{n} w_i v_i \qquad (D.2)$$

and substitution of this sum into the definition gives

$$\frac{\partial f(w)}{\partial w_i} = v_i \qquad (D.3)$$

for the typical component of the differentiation, which has a final value of

$$\frac{\partial f(w)}{\partial w} \equiv \begin{pmatrix} v_1 \\ \vdots \\ v_n \end{pmatrix} = v. \qquad (D.4)$$

As the inner product is a symmetric function, we can write

$$\frac{\partial w^T v}{\partial w} = \frac{\partial v^T w}{\partial w} = v. \qquad (D.5)$$

The same sort of expansion and subsequent differentiation can be applied to larger products, and it is easy to demonstrate that

$$\frac{\partial w^T A}{\partial w} = A. \qquad (D.6)$$

However, in this case, we do not have symmetry in the product, so a separate expansion gives

$$\frac{\partial A^T w}{\partial w} = A^T. \qquad (D.7)$$

A more involved (and tedious) expansion of $w^T A w$ can show that the derivative of a quadratic form is given by

$$\frac{\partial w^T A w}{\partial w} = (A + A^T) w. \qquad (D.8)$$

Note that, if A is symmetric, then this reduces to a result that looks like the familiar differentiation of a simple quadratic in w:

$$\frac{\partial w^T A w}{\partial w} = 2Aw \tag{D.9}$$

and, of course, if A is the identity, then

$$\frac{\partial w^T w}{\partial w} = 2w. \tag{D.10}$$

Index

12-6 potential, 19
21st century, 242
3-D regular polytope, 239
3-D rotation matrix, **228–231**

A

absorption, 310
absorption permeability, 310
ACD/ChemSketch, **104**
acceptor, 72, 94, 245
acceptor stem, **74**
accuracy, 286
ACD/Labs, **104**
actin-binding, 107
action at a distance, 26
active site, 287
ad hoc classification, **310**
adenine, 72, 78
affinity, 295
air density, 15
air drag, 14
Airbus A330–200, 31
algebraic multiplicity, **351–352**, 365, 367, 370–371
algorithm
    classification, 106
    comparing geometric relationships, 247
    divide-and-conquer, 112
    geometric, 84
    learning, 290
    Needleman and Wunsch, **180–186**
    pattern analysis, 84
    regression, 106
    Smith–Waterman, **190–193**
    superposition, 246
alignment
    gap, 171, 180
    global, 171

penalty, 180
score, 171
alignment for structure comparison, **250**
alimentary tract, 95
alkaloid, 108
allosteric movement, 245
alpha carbon, 38, 43, 201, 218, 247, 249, 251, 259, 261–262, 269, 279
alpha helix, **44**, 63, 92, 208
    left-handed, 208
    right-handed, 208
alpha-hairpin, 87
AMBER, 21
ambient temperature, 134
amine group, 42
amino acid, 38, 52
    cysteine, 64, 65
    exact match, 173
    glutamic acid, 62
    mutation, 62
    relative frequency of occurrence, 173
    valine, 62
amino acid categories, **39**
    nonpolar hydrophobic, **39**
    polar uncharged, **39**
    polar-charged negative, **39**
    polar-charged positive, **39**
An Inconvenient Truth, 112
analogs, 244
Anemonia sulcata, 52
angel hypothesis, 27
angels, 26
angiogenin, 63
aniridia, 108
anomalous dihedral angles, 208
antibodies, 37
anticodon, 74, 126

9781584886839